THE CHOSEN WARRIORS

BY

JOSE L. FIGUEROA AND MICHAEL J. ORTIZ

ISBN: 978-1-968894-38-2

Acknowledgment

I would like to thank God for granting me the talent to write this story. Next, my family for loving me and pushing me to never losing hope and to always follow my dreams, I couldn't do this without you guys. Special thanks goes to my family watching over me in heaven, I miss you and think of you guys every day, this is for you. Words can't begin to describe how much I love you all.

Humbly grateful,

Michael J. Ortiz.

DEDICATION

First and foremost, I would like to thank the Lord above for granting me the opportunity to share this book with the world. ***To my parents****—thank you for always pushing me to excel in everything I do. Mom, you will forever be in my heart. Rest in peace. I love you.* ***To my little sister****, one of my toughest critics—thank you for making sure I didn't end up with a typical story. Your honesty helped shape this book into something unique.* ***To my beautiful wife****—thank you for encouraging my ideas and editing this book to ensure it was ready for publication. You've been my backbone when I doubted myself. Thank you, my love.* ***To my daughters and son****—you've been my inspiration to never give up. If I can do it, so can you. Anything is possible if you work hard to achieve it.* ***To my co-writer****—thank you for helping this novel become better than I could have ever imagined.* ***To the publishers****—your professionalism and years of experience are evident in the quality of the final product. Thank you for helping this book see the light of day.* ***And to anyone I may have missed****—I apologize. Thank you all for having an impact on my life, whether you know it or not.*

Very appreciative,

Jose Luis Figueroa

"The journey of a thousand miles begins with a single step."

— Lao Tzu

DISCLAIMER

This is a work of fiction born from imagination and purpose. Though real places may be mentioned, they serve only as backdrops to a story that is not bound to reality. Any resemblance to actual people or events is purely coincidental and unintended. Some artwork in this story was generated using AI tools and is original to the author.

Table of Contents

CHAPTER 1.
LIFE UNEXPECTED

It was written that….

For eons, angels and demons have walked the Earth, their eternal struggle concealed within the fabric of humanity's existence.

Chosen warriors have always stood between the realms, unseen protectors maintaining balance. Few know the truth; fewer still dare to challenge it. This is the story of how the current warriors came to be... and of the forces that threaten to unravel everything.

A busy city street filled with people walking around and shopping. A couple of business-looking gentlemen stopped to look across the street, noticing a smoky grey aura emanating from two men dressed in black hoodies and blue jeans. Those people look back toward them as their faces change to their true form, a demon, for a split second, then instantly transform back.

The two businessmen staring begin emitting an angelic glow to let the beasts know they're keeping an eye on them, so they don't try to harm any humans. The four beings continue to move toward their initial destination.

The story begins on Mar. 13, 1998; it was a cold and semi-windy day, and the Heart family was at home enjoying a nice, quiet day. A short woman with a light complexion and long brown hair in her mid-30s named Aggy is glancing out the window as she washes the dishes and sings the song that's playing on the radio. The woman is listening to a popular 80s song on the station. Suddenly, the phone rings.

Aggy: "Hello?" *She asks in a cheerful singing tone.*

A suave, relaxed voice is heard on the other end with a sense of urgency.

The voice: *chuckling* "Hi Aggy, this isn't a social call. You guys need to get ready, I've received news that you're about to be attacked."

Aggy: *her cheerful expression changes to seriousness* "What or who and why?"

The voice: “It is unknown at the moment, just gear up and get ready. Whatever or whoever they are, they’re coming with everything they’ve got. Remember, protect the objective at all costs.”

Aggy: *feeling a bit offended and upset* “You make it seem as if we wouldn’t. He’s our flesh and blood, how...”

The voice: *quickly cutting her off* “I wasn’t insinuating that you two wouldn’t, just reminding you that he has a prophecy to fulfill and our enemies will try to convert him over to their side or take him out. You have under half an hour to prepare. I want a full report after this is handled. By the way, I called Wolf. He’ll be here as soon as possible to help out.”

Aggy: *a sense of urgency in her voice* “Albert, shit’s about to hit the fan. Let’s get our gear; hopefully, we can finish them off before Louie comes downstairs. We have under 30 minutes.”

Albert: *is Aggy’s husband, a man in his mid-40s with a caramel complexion and curly black hair who looks physically toned, speaks to Aggy in a concerned tone while he is getting the weapons from hidden compartments in the living room, as Aggy searches in the* *kitchen's secret compartments to help Albert* "Aggy take Louie to the Wolf Queen, so she could take care of him, just in case...."

Aggy: *she interrupts* "That's too far. I won't be back in enough time to help you out."

Albert: *clearly emphasizing* "Just go, don't worry about me."

Aggy: "Don't be stubborn, Albert. We'll handle this together."

Albert: *hearing sounds coming from all around* "Fuck! It's too late. They're already here."

Aggy: "I'll take him after we're done."

Albert: "Don't forget to get the ammo bag."

Albert and Aggy grab their ammo, get their weapons ready, and then pray as they prepare to protect their home from the intruders.

Aggy and Albert: *they begin to pray in unison* "Heavenly Father, protect us from the evil we fight in your name. Please give us the strength to do your bidding and protect our son, in your name, we pray, Amen."

The couple finishes praying and watches as the floor bursts open; the smell of sulfur fills the home; flames arise from the opening. Beasts, spirits, and other supernatural beings appear, a hand with long nails is seen rising from the crater, followed by the other hand, and then the demon climbs out. As a loud commotion was heard, Louie, a 12-year-old boy with curly hair on top and faded sides, a light complexion, green eyes, and a small frame, ran down the stairs with fear in his eyes. He is the only child of Aggy and Albert.

Louie: “Did you hear that bang, and what’s that foul odor?” *he asked with concern.*

Albert: “Go back up! It’s too dangerous for you to be here!” *he yelled to Louie as Albert continued to battle the demons.*

Louie: “Why? I can help you!” *he questioned with sadness in his voice.*

Aggy: “Go to your room until we tell you otherwise!” *she scolded.*

Albert: *takes a quick glance at his surroundings* “Damn, they’re coming from everywhere!” *he stated in a worried tone.*

Aggy: “Let’s show them that kind hearts aren’t to be trifled with.” *She tries to hype her husband with the battle cry.*

Shots are fired, and the smoke of the guns mixes with the smoke from the fire. Beasts continue to charge at the couple. At the top of the stairs, Louie watches in horror as his parents fight what seems to be an impossible fight. He tries once again to ask.

Louie: “Mom, Dad! How can I help?” *he screams out, hoping they would allow him to assist.*

Albert: *looks over at Aggy* “Grab Louie and get him out!” *he exclaims.*

Taking the opportunity that he's distracted, a demon attacks him, scratching his face and sending him further away from Aggy.

Aggy: "I'm not gonna leave you!" *She replies.*

Creatures begin to head toward the teenager. Albert sees what they are trying to do and covers his wife's blind side so she'll have enough time to get to Louie. Aggy is caught by surprise when she's suddenly hit from behind; luckily, it pushes her closer to the stairs. Frantically, she pulls Louie, and then they run to the front door.

Aggy: "Go!" *She tells her son with urgency.*

Louie: "But, I want to stay here and help...?" *in a disappointed tone.*

Aggy: "No, just go, we'll be there shortly." *She tells him, trying to reassure him, even though she knows deep in her heart that it might not be possible.*

Louie: "You promise." *with tears in his eyes.*

Aggy: "Yes baby, we'll try." *once again trying to reassure him, before she goes back inside, she kisses her son on the cheek and gives him some comforting words, a tear slides down her face and before Louie can see her tear she quickly wipes it away* "Remember Louie, we love you and we'll always be with you."

She smiles, then goes back in. Just as she slams the door closed, crackling and crashing are heard. The house is slowly caving in as Louie watches in dismay. His home is suddenly engulfed in flames, with thick black and gray smoke. The smoke was so dense that it began choking him outside.

Louie: *as he coughs* "Mom, Dad!!! Noooo!!" *he yelled as loud as he could.*

He tried to head toward the door, but the flames were too intense for him. Reluctantly, Louie tries again. Then, an explosion throws him back several feet, slamming him up against a parked car, which bursts all the windows. After what seemed like an eternity, the fire department and the paramedics arrived at the site. The paramedics try to keep Louie calm and attend to his wounds. The teen is on his knees, crying and praying, even though there is no way his parents could've survived the explosion. The firemen are still fighting the flames.

Louie: *hoping that his parents somehow survived* "My parents are inside still, please save them!" *he pleads to the firefighters.*

Fireman: "We'll do our best, kid." *The firefighter replied, trying to get rid of him so he could get back to work.*

The firefighters fought profusely until the flames were under control, then they quickly searched the entire house. They looked around, seeing various bodies, both human and unknown. Sadly, they found no survivors and no idea where the fire had started. Inside the house, they call for the coroner. Moments later, the coroner arrives along with the police. The officers walk over to the teen, asking him what exactly occurred, who was in there, and what happened, and doing everything possible to distract him as the coroner does what has to be done. The coroners used two stretchers from the ambulances to take them inside the house. Minutes later, they come out with closed body bags with still steaming, and the smell of burnt flesh fills the air. The smell is so strong that it makes a couple of the paramedics gag. They carry the corpses past Louie, the boy cries as he runs to it, reaching out to them.

Louie: "Noooo!!" *he yelled as a flood of tears engulfed his face.*

Twelve years later...

At a cemetery, the air was filled with the scent of rain, and the wind was beginning to pick up as a twenty-four-year-old man stood over a tombstone in the shape of a heart with two names on it. He holds his tears in as his hands reach the tombstone. It had the following engraving:

Albert Heart, Jan. 10, 1943- May 12, 1998. Abigail Heart, July 12, 1951- May 12, 1998. Loving husband, father, and friend. Loving mother, wife, and friend. Forever in our hearts.

The man: "Don't worry, Mom and Dad. I shall continue your work." *A tear slides down his cheek* "All evil will pay for what they've done to you. From this day forward, Louie is dead. Now, evil will fear the name... Blak Hart!"

Several years in the past...

In another state, it is 6 am. A tanned, slender gentleman with yellow and brown eyes, who has an incredible muscle tone, is gearing up for work. He has a peppered mustache and beard, wears thin reading glasses, and has thinning hair in his mid-40s. His name is Will. His wife, Nora, in her mid-30s, medium build, with long dark brown hair and a light complexion, gets up from the sofa to approach him.

Nora: *playfully asking* "Where are you going?"

Will: "Honey, I gotta go. The boss called and said I'm needed. He said Hunt and Huntress need assistance."

Nora: "Ok, be careful. I love you. Say goodbye to the kids."

Will: "Kids, I gotta go."

Their twin children, Michael Cross (a short seven-year-old boy with scruffy brown hair, thin reading glasses like his father, and a very slight tan skin) and Rayne Cross (a slightly tall and slim girl with long brown hair and a pale complexion), run to him.

Cross and Rayne: "Why? Watch a cartoon with us."

Will: "I would love to, but I can't. My job called, and I have to go in."

Cross and Rayne: *sadly replies* "Ok, fine."

The kids give Will a bear hug. They are holding him extra tight.

Will: "Ok, ok, I have to go. I don't want to be late. I love you. Behave."

Cross and Rayne: "We love you too. We'll behave."

On the highway, Will drives on a road that has little to no traffic. The sun hasn't risen yet, so it is very dark. Will drives as fast as he can to meet with Hunt (his coworker). He tries to call him, but there is no response.

Will: "Something seems off. He usually picks up on the fifth ring. I hope he's ok."

He continues to drive until his headlights illuminate something or someone that jumps into his path. He swerves and gets the shotgun from his bag. He spins the car around to get the perfect shot. Then he realizes nothing is there. Just as he was going to proceed, deer surrounded the car. Their eyes are glowing red; they begin to morph into beings that almost look like centaurs. In actuality, it is a wendigo. (a beast that has a hunger that never seems to be satisfied)

Will: "What the fuck? What are they doing around this area? They usually stay in the woods; they are rarely found on highways."

The kids are getting ready for school.

Cross: *he just remembered something that he noticed a day or two ago, calling his Nora over* "Mom, I saw that man again."

Nora: *looking at him with a concerned expression* "Where did you see him this time?"

Cross: "In front of the school, watching us. He looked creepy."

Nora: "Next time you see him, stay as far away from him as possible. Once you're far enough, keep an eye on him. Tell me what he was doing."

Cross: "Ok, ma. But why does he keep showing up?"

Nora: "I don't have a good feeling about it." *thinking, lately there have been an excessive amount of paranormal activity, he might be one of the ones that have materialized.*

As the kids begin to walk toward the bus stop, Cross notices a man dressed in a black funeral suit with pale skin and dark eyes. He realizes it's the man who was in front of the school. Cross grabs Rayne's hand. Rayne looks at him with a confused stare. The man looks in his direction and begins to walk toward them. The boy notices and then begins to panic, but he tries to remain calm for his sister and tells her.

Cross: "Let's walk to school today."

Rayne: "Why? The bus is about to get here."

Cross: "It's just a nice day to walk. I'll buy you candy from the store before we get to school." *he glances toward the stranger, making sure they aren't going to be followed*

Rayne: "A walk sounds nice." *excited about the candy.*

Back on the highway, two of the wendigos ram into the car. Will switches weapons and goes for his assault rifle. He shoots them as they dodge to the side, giving him enough room to open the car door. He stood outside the car, shooting as many as he could, but was outnumbered. They continue to attack him from all angles, and blood starts pouring from his wounds. Just when it seems that all hope is lost, he finds a way. Will shoots the gas tank and causes a leak. He runs several feet away, then shoots the ground next to the gas, making a spark. The spark creates a fire that goes directly to the gas tank, causing the car to explode. That gave him enough time to escape. He shoots the wendigos that survived, then heads into the woods to find a nearby town several miles away.

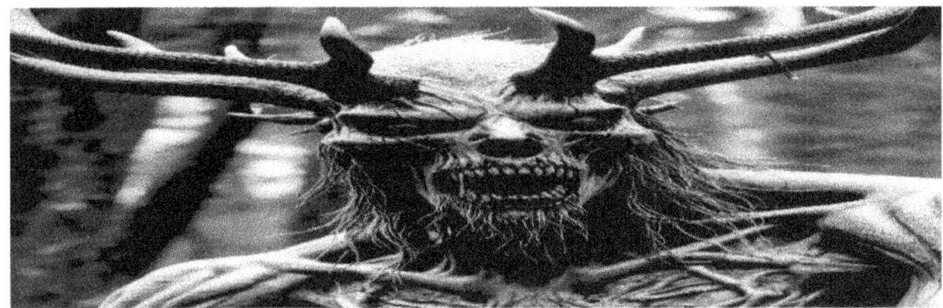

He still hears more of them chasing after him. He breaks into the first car he sees. He hotwires the vehicle, then tries to drive home. Will finally arrived home bruised from the explosion, cuts from the battle, and battered, while covered in blood. As he opens the door, he stumbles onto the floor of the house. Will calls out to his wife.

Will: *starts off to yell in a low tone, then gets louder and louder* "Nora, Nora, Noorra help!"

Nora: "Oh my God! What happened?" *She gasps in shock.*

She runs to him, cleans, and bandages his wounds. Shortly after that, Nora gives him coffee as he begins to explain what happened calmly. Hours later, Rayne and Cross arrive from school.

Cross: "Dad," *he says in a scared manner*, "something is outside. It looks like it's searching for something."

Will looks out the window, recognizing that it's one of the wendigos that has been chasing him.

Will: "Nora, take the kids to the basement! Now!" *he demanded.*

Nora: "Ok, but what about you?" *She asked, worried about her husband.*

A wendigo breaks through the wall, the pieces of the wall, and pressure severely injures Nora, throwing her across the room, and landing hard on the floor. *She screams as her bones break in several different places.* Not too far from her, pieces of the wall flew in multiple directions, and several of them hit Cross and cut both sides of his mouth. *appearing as a Chelsea smile.*

Will: *seeing the incidents unfolding, then seeing his wife on the ground* "Nora! Nora!" *he screamed, trying to get some energy to fight even though his body had taken a toll from his previous encounter.*

Will, enraged, begins to change. His eyes started glowing yellow, and his fangs started to appear. His nails are now changed to sharp claws. Nora didn't have enough time to get the kids to the basement. The wendigo stood over the kids. Will, seeing this, charges at the wendigo. He grabs it by the neck with his claws, crushing the windpipe. Several more wendigos storm into the house. As he fights, Cross runs to his parents' room and goes into the weapon closet. He gets his father's chrome revolver.

Cross: "Rayne, go to the basement!" *he commands.*

Cross begins to shoot the wendigos with precision. He eliminates all the beasts on his father. By the time he empties the revolver, it is too late. The wendigos ripped apart Will. One of the Wendigos looks toward Will.

A Wendigo: *in a deep raspy voice* "You're done!" *looks toward Cross* "You'll get what you deserve soon enough." *he sprints out through the hole in the wall.*

Cross rushes to his father's side, crying. He holds Will's body, yelling.

Cross: "Dad! Dad!"

Nora grabs Cross's hand. Cross looks around and realizes that they are left alone. Rayne opens the basement as far as she can. She pushes the debris, then crawls over to her mom and brother. All that can be seen is the carnage left with the bodies of the wendigos, and blood all over the walls.

Some time has passed, and Cross and Rayne have continued to have supernatural experiences. Things that they weren't able to see, they can now see without a problem. Slowly, their third eye begins to open.

Twelve years have passed; the kids are now high school seniors. The twins, Cross and Rayne, have manifested some more gifts. Both their eyes now glow the same color as Will's, Cross can sense and see the supernatural. While Rayne can see only silhouette figures and hear the supernatural.

Several towns away...

A couple argues over how to raise their child. Gil is a husky man with a scruffy beard, about six feet three inches tall. His skin looks dirty and rough. Gail, the wife of Gil, looks like a young 30-year-old when in actuality she's in her late 40s, with a soft tan complexion, physically fit, and stands at about five feet seven inches tall.

Gil: "We have to let him know."

Gail: "Not yet, he's not old enough." *She says, raising her voice.*

Gil: "I started when I was his age." *pointing at his chest.*

Gail: "I don't care, he's not you."

Gil: “It’s in his genes, you can’t deny him his birthright. All the men in my family were taught at this age.”

Gail: “Just because our decedents were who they were doesn’t mean he has to follow in their steps.” *replying in a stern voice, as she walks away, she turns to him* “Remember what I just said, don’t try to start.”

Gail leaves to run errands, while Gil plots to corrupt his son. He had to hurry up before Gail got back.

Gil: “Darren, come to the basement.” *he beckons.*

Darren: “On my way.” *he said excitedly.*

While Gil waited, he poked his finger, causing it to bleed. Then he began to draw some markings and symbols on the floor.

Darren: “Yeah, Dad, what happened?”

Gil: “Nothing, just wanted to teach you something.”

Gil teaches his son a few chants. Afterwards, they try to see the effects as he lights the red candles. Gil starts the chants, then lets Darren continue to chant, not noticing that the flames on the candles are getting higher and higher. All the lights in the basement are flickering, causing Darren to stop.

Darren: *a bit frightened* “Why are the lights doing that?”

Gil: “It is just showing how powerful we are.”

Gil continues to chant, and suddenly his wife comes down the stairs.

Gail: “I thought we discussed this. He was not supposed to learn this!”

Gil: "No, you spoke about it, I disagreed, so here we are."

Gail: "Darren, go upstairs, you're not in trouble. Your father is."

Gil: "No, stay here!" *demanding.*

Darren: *doesn't know what to do, then decides to listen to his mom* "Ok, mom."

Darren runs to the top of the stairs and then stops, now to find out what they are arguing about. He stays there listening in on the conversation. He hears footsteps heading toward the stairs. Gil finishes the chant; a black smoky figure appears from the symbol. It speaks and pushes Gail against the wall. She begins to chant and curses this entity, sending it flying toward Gil and knocking him down. The entity disappeared into the mist. Gail yells at Gil.

Gil: "You see, you've babied him so much that you made him into a coward!"

Gail: "This is it; I'm done with your antics. It's over. I'm taking Darren."

Gil: "If you value your life, I suggest you rethink your actions before you regret them."

Gail: *Gil grabs her* "Let me go, we're leaving, and there's nothing you can do about it."

Gil: "Really, I know a lot more than you think I do. Don't make me do this, Gail. I don't want to kill you. Please don't make me do it."

Gail: "Do your best. I know more than you do. I am more powerful than you, so do what you must, but remember I won't be holding back this time!" *forcefully chops at his hand with a martial art move, then roundhouse kicks him, slamming hard to the wall, creating an echoing boom.*

Getting up and brushing off that attack, Gil takes a machete that was hanging on the wall, then chants. Darren hears a deep voice chanting along with his father as Gail laughs in a creepy tone. Darren begins to tremble in fear. He tries to go, but his fear has him frozen. Gil charges at Gail. Without hesitation, Gail dodges and chants.

Her hand begins to emit an electric charge. She runs to Gil and hits him with an electric blast, then starts punching him. Gil retaliates by cutting her; he used a dark black blade, which is now burning red with sparks of fire and smoke coming off of it.

Gil: "You give me no choice. While our son is at the top of the stairs listening to us, he will now know who we really are." *evil laughter* "Ha, ha, ha. Once I finish you off, no one will be able to tell me how to raise my son."

Gail: "You are not leaving here alive, that I can guarantee. As for my son, he will be with me. He'll have a normal life, the one thing our parents never gave us—the choice to decide."

Gil: "Over my lifeless carcass." *His voice begins to deepen* "I won't stop until I taste your blood, bitch. You are no match for us."

Gail charges up as Gil slices and stabs her. She finally grabs him and begins to shock him at full capacity. He shrieks in pain. The shock sends him toward the stairs. Darren sees a black figure get thrown first, then his father. Gil looks at him as Gail runs over to his beaten body.

Gail: "Darren, go upstairs now!"

Gail's eyes are white, and her hands are emitting lightning sparks. She looks back at Gil and delivers a final blow. The electrical charge blew out all the lights and fried Gil. The basement has a stench of burned flesh that reaches the boy. Darren begins to cry as his bloody and bruised mom walks up the stairs, *holding her ribs*, she explains to him why it had to happen and tells him.

Gail: "Pack up all the clothes you can; we're leaving." *She rushes him.*

Twelve years later, Darren is in his senior year of high school. He's a loner. Due to his past, he finds it hard to trust anyone, fearing betrayal, but having a quiet darkness brewing within. He was secretly practicing the few things his father taught him the whole time.

In the present...

Several states away, Blak Hart begins his hunt for demons in Louisiana. There have been reports that shadow figures have attacked people in the state, and people have seen glowing red eyes watching from the woods. As he hunts, he begins to hear the demons talking amongst their selves.

First demon: "In New York, there are three teens who are filled with untapped power. One of the hoards is going over there before anyone else realizes and takes their power."

Second demon: "I don't think they should, even though they don't know how to use their powers. One of our bosses is already out there setting everything up so they can join us willingly."

Out of the trees, Blak Hart jumps down. He cuts one of the demons' heads off as the second one attacks. Out of reaction, he grabs the second beast, disintegrating it, not knowing how he was able to do that. All that was left in his hand were ashes.

Blak Hart: "How the hell was I able to do that? It got the job done, but what's going on with me? It doesn't matter right now; I have to get to those teens before anyone does. All missions can be done later; this one is now a top priority. Since I'll be in NY anyway, I can take care of one of the missions over there. I can ask my boss, Ark, who these kids are and what they look like. Hopefully, he has a clue what is happening to me." *he thought.*

Meanwhile, in New York, Cross is engaged in a fight. The fight all started when Rayne had an argument with someone; the guy insulted her a bit more. Since it was affecting her, he tried to hit her. Before his fist got anywhere near Rayne, Cross came out of nowhere and started beating him. The hallway was crowded with students watching this event. Rayne pulled her brother off the teen. As they were walking, she was accidentally bumped into one of the onlookers.

Rayne: "Remember what mom said when you feel like that, walk away."

Cross: "Like what?" *trying to seem calm when clearly he wasn't* "I was fine, he was going to hurt you. He also said something about my face." *still upset.*

Rayne: "I know. He was talking about your eyes, not your scars. Your eyes were yellow. I don't want you getting in trouble for nearly killing someone else." *She states, worried about him.*

Cross: "Oh, either way; don't worry. I'll try not to let things escalate to that point again unless it's completely necessary."

Rayne: "You keep forgetting, I'm not as defenseless as I used to be. I can fight my own battles now." *getting a little annoyed.*

Cross: "It's a brother's job to protect his sister, no matter how good you could fight. I will always want to protect you." *he reminded.*

Rayne walks Cross to his next class; Rayne sees the kid she bumped and laughs. Cross tells her to stop and that she should apologize later on. She shrugs her shoulders and goes to her class. Cross sits down and feels someone is staring at him about two seats away. He looks over and sees the kid that Rayne bumped. Class finishes, and the student who has been watching Cross the whole time walks by him with his head down. Cross goes after the kid and stops him. The kid begins to tremble.

Cross: "Relax, I'm not going to hurt you. Just want to apologize for my sister's carelessness." *assuring the boy.*

The kid: "It's ok, I'm used to getting treated that way."

Cross: "No one should be treated that way."

The kid: "Well, what about the kid you hit? Did he deserve it?"

Cross: “That’s a whole different story, and yes, he deserved it. Anyone who raises their hand to hit a female deserves what they get. This is the main principle that I stand by.”

The kid: “You haven’t changed a bit. So he was about to hit Rayne?”

Cross: *getting defensive and curious* “How do you know me or my sister?”

The kid: “She’s in my lunch period. Damn, you don’t recognize me?” *feeling embarrassed and hurt.*

Cross: “Oh, ok. Yes, he was going to hit her, and by the way, she’s my sister. I’m...” *the kid interrupts him.*

The kid: “I know who you are, you’ve made a name for yourself through your fights and your heroism. You really don’t remember?” *trying to see if he begins to remember* “It’s me, Darren.”

Cross: “Oh shit, Darren! You changed your appearance from the last time I saw you, with peach fuzz and all.” *chuckling.*

Darren: “I’ve been in most of your classes this year.”

Cross: “I’m sorry, it’s not like you to stay in the corner. I’m the one who usually keeps to myself, just as you used to.”

Rayne: “Do you feel better now?”

Rayne looks at Darren and laughs.

Cross: “Stop acting like a bitch. Rayne, can you believe this is Darren? The kid you bumped and still didn’t apologize to.”

Rayne: "That's why I bumped him. He's been back for a couple of weeks now and told me not to tell you until the time was right. Well, sorry, Darren, I guess this is the time." *She said in a sarcastic tone.*

Darren: "It's ok. So, what have you two been up to since the last time?"

Rayne: "The same shit we were doing when you decided to disappear on us again."

Cross: "Relax, Rayne, hopefully he has a good reason for it. Right?" *he questioned.*

Rayne: "Like? What excuse could he possibly have?"

Darren: "I really felt like shit for my vanishing act, but I just had to. I'll explain later on, away from school." *defending his actions.*

Blak Hart arrives in New York, takes care of his mission in Manhattan, and then calls his boss. Ark tells him that the teens live in Long Island and sends the pictures of them along with their bio.

Ark: "While you're over there searching for them, I have a couple of things you can take care of." *he looks down at a few files in his hands.*

Blak Hart: "Where exactly are these missions?"

Ark: "Montauk, Southampton, Riverhead, Blue Point, and finally in Amityville. Do them exactly in that order."

Blak Hart: "You said a couple, this is a little more than a couple. I'll be more sightseeing than working." *he teased his employer.*

Ark: "What are you complaining about? You could possibly do two or three of them in one day. Not to mention, the pay on these has a good amount of zeros in it." *annoyed at his employee.*

Blak Hart: "Fine, I'll do them, just keep in mind. If I find these kids in between these missions, the other missions will be postponed. Is that clear?" *trying to sound like he has some authority when he knows that he doesn't.*

Ark: "As long as the missions get done in that order, it won't matter much. When are you going to start?"

Blak Hart: "Tonight, I'll start up on the missions after I verify that the address on the bio is still correct."

In Cross's house, the twins have been seeing more spirits than normal on the property. Rayne heard several of them telling her that someone was looking for them and that she should stay away from him, while others told her not to listen to them. The one looking for her and her brother is going to be there to protect and help out. Rayne runs to Cross and tells him what she heard. He asked her to take him to the voices.

Cross: "I don't see anyone."

Rayne: *sarcastically* "Duh, they're ghosts, not living people, and they all left. What did you do?"

Cross: "Nothing, I guess I scared them off. By the way, what kind of ghosts get scared of a teenager?" *laughing at the entities.*

Outside the house, a figure can be seen standing close to the fence. Blak Hart looks around the property, casting away spirits and chasing low-class demons out of the yard. Rayne has a strange feeling and looks out the window. She gets a glance at a stranger in the backyard. She runs to Cross's room.

Rayne: "Someone or something is outside in the backyard." *worried.*

At that very moment, Cross gets a flashback of the moment he told his father almost the same thing.

Rayne: "Be careful."

Cross: "I will if they're anything like your ghosts, I'll be fine." *trying to make light of the situation.* "Stay inside."

Rayne: "This isn't funny."

Cross goes outside, slowly creeping around the property, and doesn't see anyone out there, but he can sense an energy other than his out there. He searches and searches, but nothing is out there. As he walks back in, Blak Hart has a corpse of a demon in one hand, while hiding in a tree, holding onto a thick branch with his other hand. He jumps down, burns up the body, and gets ready to do one of the missions.

The next day, Blak Hart calls Ark. He reports that all his bounties are hiding out in the area of the high school. The boss lets him know not to get caught and take care of the mission. The reason it was supposed to be in a certain order was that these were the ones sent to destroy the kids. So now the mission has changed: protect the teens at all costs. The only dilemma was how Blak Hart was going to do the side missions along with the main mission without scaring away the teens.

For several weeks, the teens have gotten to reconnect. They have bonded so much; it gave Cross the sense that Darren is trustworthy, like the way it was when they were kids. The three sat on the stoop outside and began to reminisce.

Cross: Do you remember when we were at the playground running around with our toy swords and guns, laughing while having fun? Then we climbed to the top of the monkey bars, silently enjoying the view.
They start remembering that moment like a flashback.

Younger Cross: "Hey... do you think we'll ever change?" *curious.*

Younger Rayne: "Where's this coming from?"

Younger Darren: "Come on, like anything's going to change how we are. We're still friends, right?" *Cross and Darren smile.* "Our friendship is too strong; nothing could get in the way of it."

Cross: "Well, duh."

Rayne: "Hey, what about me?"

Darren: "How can I forget about Cross's cute little sister?" *slightly blushing.*

Rayne quickly turns to hide her blushing face.

Cross: *joking* "It's ok, sometimes I forget about her."

Rayne: "I'm telling mom!!!"

Cross: "No, don't do it. Come on, I was kidding. I'll buy you a treat before we go home."

Rayne: "A treat sounds nice; this isn't going to work every time."

The three continue to play when Darren's mother pulls up.

Rayne and Cross: “Hi, Mrs. Gier.”

Mrs. Gier: “Hi, kids. Darren, let’s go. I still have errands to run.”

Darren: “Ok, Mom.”

Cross: “We’ll see you again, right?”

Darren: “Of course, why wouldn’t you?”

Cross: “I’ve been feeling weird lately.”

Darren: “You’re just being weird.” *smiles at Cross.* “See you next time.”

Darren runs to the car as the twins wave goodbye. Their flashback stops, and they all look at each other smiling.

Meanwhile, Blak Hart heads to the school in search of the demons. He finally gets close to the school and turns the corner. He sees the aura of a hellhound, then he climbs up a nearby store for a better vantage point. Blak Hart walks around the roof of the store, calculating the proper angle to attack. Just then, the hound hears the board of the roof creak and then looks in his direction as it begins its transformation.

Blak Hart: *amazed at the beast* “That’s a new one, I gotta ask Ark about this later.”

The beast charges at Blak Hart as he jumps up in the air. It snarls and claws at him with great strength. Blak Hart draws his sword and tries to slice the hellhound, but the creature dodges and scratches the hero’s chest, but he wasn’t fazed. He smirks, then hacks and slashes the animal. It screeches in pain as the beast bites down on Blak Hart’s shoulder.

Blak Hart: *laughing at the creature* "Is that the best you got?" *trying to provoke* "You're gonna have to try harder than that!"

Blak Hart cuts its head off with one swing of his sword, but its jaw stood clenched deep in his shoulder. He shrugs it off and rips off the jaw still attached to his shoulder as he feels something watching him. He turns around and sees another beast running towards the school. Blak Hart runs after the creature. Subsequently, as all this is going on, the two boys are leaving the school.

Cross: "Hey Darren, what class do you have next?"

Darren: "Lunch, what about you?"

Cross: "Nothing too important, study hall, mind if I join you for lunch?"

Darren: "Are you sure you should skip class? I don't think Rayne would like that."

Cross: "Let me worry about my sister."

Darren: "Alright then, remember you said that."

The two boys walk towards the café when Cross realizes the kid he fought a couple of days ago is just waiting for him. By the looks of it, he was planning round two. Cross pushes Darren to the side.

Darren: "What the hell was that for!" *shocked.*

Cross: "Ssshhhh! How about we take a stroll outside?"

Darren: "What? Why?"

Cross: "It's a nice day, why not?"

Darren: "Alright then, I've never left school for lunch."

Cross: “Hehe, you’re in for a treat.” *smiles.*

The boys walk out of the school, and out of nowhere, they suddenly feel a huge gust of wind. Cross sees an aura of a huge creature running from something as if its life depended on it. Darren was thrown to the ground by the force of the breeze.

Cross: “Did you see that!?”

Darren: “No, I just felt it. What was that?”

Cross helps Darren up as a man with curly hair and ripped clothes covered in blood runs past them with a sword on his back, having some residue of blood still dripping from it. Cross feels like he knows this man, but he can’t figure out where. Cross starts to run after the man when he hears a voice screaming out his name.

Rayne: “Cross, where do you think you’re going!!!!!!?” *angry.*

Cross: “Rayne, my wonderful sister. Hehe” *trying to figure out what to tell her.*

Rayne: “HOW DARE YOU TRY TO LEAVE SCHOOL WITHOUT ME!!!”

Darren: “I told him not to.” *hoping he would get some brownie points.*

Rayne: “You hush; you’re getting on my nerves!”

Darren: *embarrassed and begins to look down* “I’m sorry I... just wanted to.... help you.”

Rayne: *very defensive* “If I want your help, I’ll ask for it, understand!”

Cross: "Look, Rayne, I gotta check something out. You can yell at me later, ok?" *still focusing on the stranger that ran by them.*

Cross took off after the man as fast as he could; he was thinking to himself.

Cross: "I have to know. Why do I feel like I know him?"

Rayne and Darren are left behind. Rayne gets a bad feeling about her brother, so she starts to head in the same direction Cross went.

Darren: "Rayne, wait, let me go with you."

Rayne: "Hurry up, let's go, my brother can't be left on his own for long." *preparing herself for whatever might happen.*

Darren: "Alright, why not?" *wondering if this was a smart choice.*

They start running after Cross, while next to Rayne, Darren accidentally grazes her hand and then begins struggling not to blush as he runs next to Rayne. They find Cross standing in a field containing long green grass and a plethora of various kinds of flowers all over the place. Darren runs up to Cross, and while still out of breath, Rayne and Darren ask Cross.

Darren: *trying to catch his breath* "What's wrong?"

Rayne: "Are you ok? What are you looking at, Cross?"

Darren: "Why are you just standing here?" *Cross replied in a whispering tone.*

Cross: "Something is watching us. It might be ready to hunt us."

Meanwhile, Blak Hart catches up to the creature and starts slashing it. The beast uses its claws to attack and then bites him. It gives him devastating gashes, causing him to bleed profusely. Since Blak Hart was very close to it, he plunged his sword into the creature's chest. The animal screams in agony, throwing Blak Hart into the forest. It pulls the sword from its chest and runs through a meadow, where it spots the teens standing and staring at it.

Rayne: "What are you feeling?"

Cross: *worried* "I'm not sure, but whatever it is, it's angry."

Darren: *beginning to regret following them* "I can feel it too. He seems pissed."

Cross: *warning them with intensity* "It's coming this way, run!"

The creature circles the teens, sniffing and licking its lips. The kids try to run, but it is right behind them. Cross realized Rayne was falling behind. The beast was about to bite down on her. Cross turns around and rushes toward it as the animal bites down, Rayne and Darren look back and see Cross in the creature's jaws, trying to hold it back.

Cross: *Coughing blood* "Get out of here!"

Rayne: "No, I'm not leaving you!!!!"

Cross: "I wasn't asking, get out of here now!!!!" *Coughs up more blood*

Darren: "Cross, why did you do this?!!!"

Cross: "Shut up and get my sister to safety!!!!!"

Rayne: "Nooo, I'm not going anywhere!"

Darren grabs Rayne while she's hitting him and starts running, respecting Cross's wishes. Rayne begins to cry and pleads for her brother's life. Darren runs and doesn't look back, now realizing that he's already in front of the school.

Darren: "How did we get here so fast?"

Rayne: "It doesn't matter; my brother needs help." *crying*

Darren: "Stay here, please, I don't want.... something to.... happen to you." *pleading and blushing*

Rayne: *angry and yelling* "I CAN TAKE CARE OF MYSELF!"

Darren: "YOUR BROTHER TRUSTS ME TO PROTECT YOU SO FOR THE LOVE OF GOD, STAY HERE AND SHUT UP!!" *demanding*

Rayne: "Darren...." *shocked that he raised his voice at her and she secretly liked him taking charge*

Darren: *lowers his tone* "Stay here, I will be right back."

Darren runs back to Cross, getting there a lot faster than before, just to see a horrific sight. He was being thrown around like a rag doll, the sound of his bones breaking with every throw. Darren tried to grab him, but the beast kept blocking him. The boy is frozen in fear and horrified when he realizes his friend is still alive and awake as he's being thrown and mauled. The animal turns around, spits out Cross, and charges at Darren. He tries to run, but it is already on him. The creature is about to bite down on Darren's head when a voice is heard.

Cross: "HEY, I'M NOT DONE WITH YOU!!!" *yelling while spitting up blood*

The creature turns to Cross, then quickly rushes at him, throwing him into some trees nearby. Cross screams in agony as more of his bones break. Darren sprints toward him. As he stands above the mangled body of Cross, he is so badly beaten that Darren doesn't recognize him. He falls to his knees and punches the ground.

Darren: "WHY DID YOU SAVE ME!?" *crying and angry at his friend for putting his life on the line for him*

Cross: *Coughs* "I know you could take care of my sister."

Shocked that Cross was still conscious even through all that pain and suffering, Darren tries to keep Cross still. His nerves kicked in; now he worries about Rayne and how she'll handle this scenario, and how he will find Cross the help that he so desperately needs.

Darren: "Don't speak. You need to save your strength."

The creature spots Rayne and speeds in her direction. Darren calls out to Rayne, but she can't hear him. He tells Cross he'll be back, then runs to defend Rayne. Just as he's about to punch the beast, seconds before it reaches Rayne, out of the clear blue, Blak Hart jumps out and throws the animal onto a grey vehicle parked several feet away and pounds the creature's head to a bloody pulp.

Blak Hart: "What the fuck are you kids still doing around here?!! Get out of here!"

Darren: "That thing attacked us."

Rayne: "My brother is critically hurt; he needs help."

Blak Hart walks over to Cross, looks at his wounds, and gives him a small vial.

Blak Hart: "You must drink it all; it won't fully heal you, but it'll slow down your bleeding greatly."

Darren: "What are you doing?"

Rayne: "What did you just give him?"

Blak Hart explains to the teens in more detail the effects the liquid in the vial will have on him.

Blak Hart: "The vial will slow down the bleeding, and it will help his body to try and heal itself as much as it can until he gets the help he needs."

He stopped a driver and pulled him out of the car. He put the teen in the back seat, then Rayne sat right next to her brother. Rayne told Darren to tell her mom to call the hospital.

Darren: *thinking* "My car, he totaled it."

Immediately after, Darren headed to Cross's house to notify their mother. Blak Hart starts speeding through the streets, going through red traffic lights.

Blak Hart: "Sorry about the bumps. If we don't rush him to the hospital, well... You know what could happen. Hold on, kid. We're almost there."

They hop onto the highway, being followed by several squad cars. They finally get to the hospital. Blak Hart drops them off at the emergency room, then tells them what to say.

Blak Hart: "Tell them a wild animal attacked him. I'll give you my cell number, in case anything else happens, just call. I'll pass by around two pm tomorrow, make sure Darren is here."

Rayne: "How did you know our names? None of us said our names."

Blak Hart: "Don't worry about that right now, save your brother. Tomorrow, I'll answer your questions and more."

Blak Hart speeds away, being closely chased by the police. They chased him for several miles, then the officers began shooting at him. Shot after shot, he dodges until one of the bullets hits the gas tank, making it leak; they pop out two tires, sending him crashing into the side of the bridge. The spark of a lit cigarette instantly exploded the car. The police get out of their vehicles with guns drawn in case someone decides to attack. In the background, the fire engines and ambulances can be heard arriving on the scene.

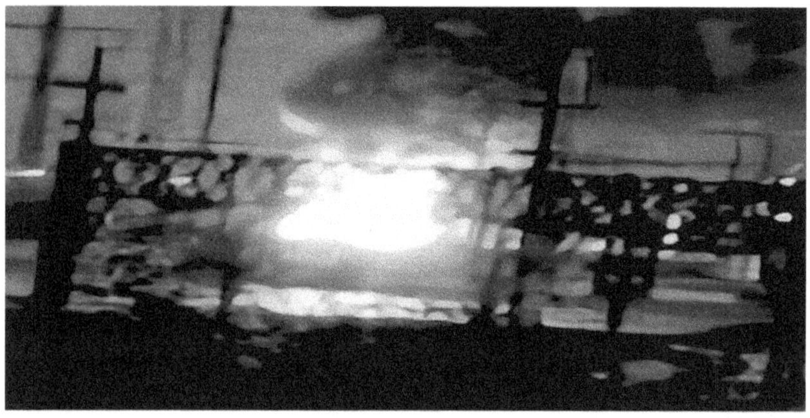

Fifteen minutes later, the fire is put out. The firefighters and police look in the vehicle for any remains of the driver, but, to their surprise...

No one is found! They call for a helicopter to search for anyone who might appear to have come out of an accident.

Two blocks away, there is a man with tattered clothes walking away from the drama next to the railing of the bridge, still steaming from the explosion and his hair all burnt off. He stops to watch the action, then laughs at them.

Man with tattered clothes: "They look like a bunch of chickens with their heads cut off."

He takes out his pack and he walks away smiling as he lights the cigarette.

CHAPTER 2.
THINGS HAPPEN FOR A REASON

Back at the hospital, the doctors go over Cross's X-rays. They find it very unusual that, for the severity of Cross's injuries, he was still very aware of his surroundings. They also found it odd that his body was healing faster than humanly possible. Rayne asks them how severe his injuries are. They stated that he had a horrible concussion, his ribs were broken, yet somehow, in minutes, his body healed itself back in place.

Rayne: "Why are Cross's strands of hair white?" *her voice trembled slightly*

Doctor: "That can be attributed to a traumatic experience. His concussion was severe, and the broken bones realigned nicely. He's going to be here for a few days, so we can monitor him and make sure that he is fine. Is your mother coming anytime soon?"

Rayne: "She's waiting for someone to give her a lift over here."

Meanwhile, Ark calls to notify Blak Hart that he's sending a special doctor to Cross.

Blak Hart: "What's his name?"

Ark: "His name is Vin. Let's say, he has experience in cases that are in our field. His cover is working in a therapist's office as an acupuncture specialist."

Blak Hart: *chuckles* "What's an acupuncture specialist gonna do in a situation like this?"

Ark: "He has the gift of healing. Any other questions, wiseass?" *witty banter*

Blak Hart looks at his cell and flips off Ark.

Ark: "That's very mature." *knowing what Blak Hart did and smirks*

Blak Hart: *chuckles* "You should know me by now. Tomorrow, I'm going to visit Cross at the hospital to explain what is happening to him."

Ark: "Don't explain too much, I'm still trying to figure out what gifts he will be getting."

Later that night, Cross is fast asleep and starts to dream of four warriors fighting off a huge hoard of what seemed like demons. The fire is surrounding them as one of the warriors jumps up and starts shooting his twin revolvers, the other seems to have a flamethrower as he burns the demons to a crisp. The next warrior rushes at the demons, then, rips them apart with what looked like a whip. As the last warrior begins to attack, Cross is pulled forward. He can now see through the warrior's eyes as the warrior slashes through the hoard with a weapon that looks like a scythe, it looks like it was made of bone, and a cold mist of smoke is emitted from it. Hours later, he awakens from his dream just as Darren and Rayne walk in.

Darren: *slowly and quietly walks into the room in case Cross would be sleeping* "How are you feeling?"

Cross: "Still a little weak and in pain, but I'll be fine."

Rayne: "Seriously, how are you really feeling?"

Cross: *flexing, trying to act like he's fine* "Like I've been hit by a train over and over, other than that I feel better than yesterday. You guys wouldn't believe the dream I had last night."

Cross tells them what the dream was about. Darren looks at Cross with confusion. Darren doesn't know how to process the fact that Cross's dream sounds too familiar.

Darren: "I had a similar dream, most of it was the same, the only changes were that I was shooting next to a warrior with a scythe, and the warrior was using a whip-like weapon. Then, when I woke up, I felt like someone was on my property looking for something or someone, and I didn't know what to do."

Cross: "All you would have to do is relax, and most likely it'll go away soon enough. Did you see anyone there?"

Darren: *worried and scared* "No, just felt like it."

Rayne: "Wow, that's some good meds. It's so good, it's affecting Darren." *thinking that the whole thing is ridiculous*

Shortly after, Dr. Vin walks in, greets the teens, and then states who he is and the reason why he's there. He then checks Cross and leaves. Blak Hart walks into the room, and everyone stops talking. Rayne looks at him with a puzzled look, while Cross thanks him for his help.

Cross: *wondering* "Could this guy be one of the people in the dream? I never had one of these dreams until he came around."

Blak Hart: *trying to be as professional as he can* "Well, let me introduce myself. My code name is Blak Hart for reasons that you all will see, eventually. As for your question, Rayne, I was sent here to protect the three of you from supernatural forces. I know all your names. Mister brittle bones on the bed is Michael Cross," *teasing Cross*, "you're Rayne, his twin sister, and last but not least, you're Darren Gier, their friend."

Cross: "Well, that ruins our introductions." *chuckles while grabbing his side from pain* "So, why exactly are you here? We don't need your help; whatever comes our way, we'll be able to handle it."

Blak Hart: "Yeah, cause you guys were doing such a bang-up job without me, and you're in here cause you like hospital food. Rrrright... Anyhow, the sudden changes you three have been experiencing are part of your destiny. If you allow me, I can help the transition become a little smoother through training and eventually in battle. This isn't about punching shadows or waving crystals. You'll be facing things that feed on fear, twist reality, and don't die easily. Training's not optional- it's survival."

Darren: "Battling what? How can you train us on things that you know little to nothing about?" *he scoffed*

Blak Hart: "Well, genius, I don't know everything, but I know enough to help you and get us ready to battle whatever comes our way. Once the three of you are ready, the only way we can defeat these things is together."

Rayne: "What do you mean we?"

Blak Hart: "So far, I've been doing this fight alone. Cross, when you heal up, then call me so our training can begin. By we, Rayne, I mean the four of us together, contrary to what many may think, you, Rayne Cross, will be one of the great warriors by my side if you allow me to train you."

Cross: "We don't have your number. Why should we trust you?"

Rayne: *excited that she was finally considered as a potential warrior* “Me, a warrior? I’m not exactly combat-ready. And he gave me his number yesterday.”

Darren: *a little bit jealous* “Why did he give you his number?”

Blak Hart: “Rayne, you’ve got fire, instincts, and guts. That’s more than half the battle. Alright, Romeo, calm down. First off, I gave her my number just in case any of you got attacked again. If I couldn’t be trusted, you would be able to sense that I shouldn’t be. Rayne would’ve had a problem with me being here. I’ll be leaving now. If you would like my help, you can call me. When you’re ready to stop surviving and start fighting.”

Cross: “Don’t be too hasty, when I get released, I’ll call you.”

Blak Hart: “Alright, no problem. See you then. And Darren, don’t do anything stupid.”

That night, Cross and Darren had another dream. This time, it was more vivid; they were able to see a person and hear him say...

Stranger: “Ha, Ha, Ha... Now it’s time I finish this little game.”

Now they were able to see a bald man who looked just like Blak Hart, one with a broken mask and his arm cut off, another with a mask covering everything from the nose down, and a female who had an uncanny resemblance to Rayne. Then, the warriors speak...

Bald man: *Growls* “Try it, you bastard!”

The masked warrior: “Calm down, let him make the first move.”

Warrior who looks like Rayne: “I can’t heal him.”

The masked warrior: “Don’t worry about me, let’s take care of this son of a bitch!”

Bald man: “Ha, Ha! Now, something we can finally agree on.”

Warrior who looks like Rayne: “I’m gonna tear him apart.”

Bald man: “Get in line. This is gonna be fun. Hehe.”

Warrior with a broken mask: *he says while in pain* “Guys look sharp, they’re coming!”

Stranger: “Kill them!”

Both forces charge at each other, the warriors giving everything they’ve got. Then, a huge explosion goes off. The air was thick with ash and fury. Sparks danced like fireflies as steel met bone. Cross could feel the ground tremble beneath him – or was that his heartbeat?

Just at that moment, both Cross and Darren suddenly awaken. Both teens were wondering if the other had the same dream. Darren calls Cross on his cell.

Darren: “Hey, Cross, are you awake?”

Cross: “I am now,” *annoyed* “what do you want? It’s 4:23 in the morning.”

Darren: *feeling horrible for waking his injured friend up* “Sorry, I just had another weird dream, did you?”

Cross: “Yeah, I did; can we talk about it in the morning? These meds are kickin’ my ass.”

Several days later, Cross was home, still recovering and trying to make sense of what had happened so far. Meanwhile, coming out of school, Darren and Rayne are walking home together. Their friendship is slowly evolving into becoming more than just friends. They continued to talk and walk when they heard rustling in the woods. Out of the woods jumps a fiery beast with large red eyes, a body the size of a horse, and the appearance of a hound. It tackles Darren to the ground, swinging its claws, scratching him while trying to chomp down. Rayne, without a second thought, picks up a thick branch and attacks the beast. She swings with all her might; her eyes begin to change from their regular color to yellow, and a surge of strength begins to rise. She continues attacking, hit after hit, the weird-looking hound is injured, and small drops of blood are seen falling from it while it is still over Darren. It appears that Rayne has the beast stunned. Nevertheless, the young warrior keeps hitting until the beast's wounds grow bigger, bleeding more than before. Finally, it runs off, leaving Darren with minor cuts, bruises, and a hurt ego. She helps him up, checks his wounds, and then they proceed to walk home. Further down, they approach another beast. This one looks a little more muscular and somewhat larger. Rayne decides to attack it before it makes the first move toward one of them. As she prepares to battle the beast, it leaps into the air and pounces on her. It's hitting Rayne over and over, pounding her body with its massive paws, then scratches, leaving bloody wounds. Once Darren saw the blood, he went into a rage that Rayne has never seen. He runs, then attacks the beast with anything he could find. The hound lunges at him, but Darren rolled, grabbed a rock, and swung. Blood splattered, Rayne was still. He had no time to think, only move, so he sprints to her aid, picks her up, and takes her to a temporarily safe location. Darren gives her a kiss on the cheek before charging toward the beast. Using every ounce of strength, he punches with all his might. To his surprise, he knocks the demon off balance. Running around the beast without hesitation, he continues until the beast was disoriented. It stumbled off, leaving a blood trail. Darren walked toward Rayne with pride, but sadly enough, she passed out from her concussion. Darren felt disappointed that she wasn't able to see his shining moment of heroism, but now, with his newfound confidence, Darren knows he can defend Rayne if she would let him. Since she's not conscious, he has to carry her home, replaying his awesome display of courage in his mind over and over. A huge smile is on his face.

They finally arrive at the house, and Cross darts over to them. Darren explains what happened to them, then Cross grabs his phone and calls Blak Hart.

Cross: "Hello, Blak Hart, we had an incident."

Blak Hart: "I'll be there in ten minutes; I will need, in full detail, a description of what occurred."

Cross: "I'll make sure Darren and Rayne give you all they can remember."

Blak Hart: "Ok, thanks. See you in a few." *finishes off the beast in front of him, then darts off*

Ten minutes later, the front door creaked open, and Blak Hart stepped in. His silhouette was framed by the fading light. His shirt hung in tatters, blood streaked across his chest, but his skin was flawless. Not one scratch. They also noticed his double-barreled pistols holstered one on each leg and an unusual sword on his back pulsed faintly, its engravings glowing like embers, cradled in its' sheath.

Blak Hart: "Alright, give me your reports."

Rayne and Darren explained the moment in detail, everything from Rayne's heroic moment to the description of the beast Darren fought. Blak Hart pulled out a tiny tablet and contacted Ark. With his help, they would be able to find out what these beasts are.

Somewhere in heaven, in a white room, rests a huge table, a crest carved in front of the seat where the beings are supposed to be on it, and an orb in the middle watching over the teens. The Angels talk about the three teens and Blak Hart.

Head Angel: "Their powers are beginning to awaken."

The second Angel: "There has to be a reason for their awakening."

An unknown Angel: *a bit worried* "We should wipe their memories, it's not their time yet."

Head Angel: *angered* "WHAT WOULD THE KING SAY IF HE HEARD YOU SAY THAT?"

Angel trainee: *trying to defuse the situation* "Calm down, the King woke them up for a reason, let's just keep watching."

An unknown angel: *raising his voice* "They're human and weak. They'll just end up hurting themselves. I don't know why the Ki..."

Head Angel: *now standing from his seat, slamming his fist onto the table* "SHUT YOUR MOUTH BEFORE I DO IT FOR YOU. HOW DARE YOU QUESTION THE KING!"

The third Angel: "This isn't the time for petty squabbles; leave them alone for now. There must be a reason for this event, so let's deal with more serious matters and come back to this on a later date."

Back at the house, Ark contacts Blak Hart. He sends him pictures of different variations of the beasts. These are low-level demons' hound dogs. They attack and bring their prize back to their owners.

Ark: "Under no circumstances are you to leave the teens unless it has to do with their scholastics."

Blak Hart: "No problem, I already figured that's the way it would be."

Ark: "Alright, good, and if it gets bad, give me a call." *shocked that he didn't give him attitude or a wise ass remark*

Blak Hart: "Ok will do...."

Ark: "Blak Hart, I'm serious, no more lone wolf crap. Is that understood?"

Blak Hart: "Yeah, yeah, damn Ark! Give me some fucking credit here. I'm doing everything by the fucking book so far, and you're still acting this way, fuck it, whatever. Yeah, I got it."

Ark: "I'm trusting you, this is very important."

Blak Hart looks at the phone and once again flips him off.

Ark: "Not too many people can pull off the bald look, but you wear it well."

Blak Hart looks at his phone with anger, his eyes begin to glow red. When Ark was about to speak, Blak Hart crushes it, almost disintegrating it and disconnecting the call. He was a little offended by the comment. Since the car crash incident, his hair wasn't growing back in the burnt areas, making him slightly self-conscious.

Cross: "That escalated suddenly. How do you guys work together if you fight like that?"

Blak Hart: "It's a love/ hate relationship, I love the work, he hates me, but I get the job done. So, it's a win-win." *smirks and chuckles* "Hehe."

Rayne: "I think I can get used to these guys." *she says, liking the fact that they were unapologetically themselves*

Cross takes Blak Hart for a tour of the house as Rayne talks to Darren.

Darren: *He looks out the window, he sees his mother arriving* "My mom's outside, I'll see you tomorrow."

Cross: "Alright, Darren, see you tomorrow."

Rayne walks up to Darren, kisses him on the cheek, and tells him to be safe and call her later. He blushes and nods, yes. Then, he walks out to his mom's car.

Rayne begins to read a poetry book called *An Emotional Journey*©, and suddenly the phone rings. Nora answers the phone.

Nora: *calls her daughter over* "Rayne, it's for you, it's Luna." *Luna is Rayne's best friend. She is a beautiful young lady with soft white skin, physically toned and sky-blue eye color, and her hair is white as snow with strands of blond peering at certain points*

Rayne: "Be right there." *rushing over to her mother to get the phone*

Rayne talks to her friend Luna about what's been happening. Her friend is interested in the paranormal and the unexplainable. She always loves to hear about Rayne's adventures.

Luna: *eager to listen to her friend's new adventure and more* "So are you a superhero or something?"

Rayne: "Really, Luna, a superhero? This is just my life, there's nothing super about it."

Luna: "I wish it were my life, I deal with boring stuff all day. Your life sounds like so much fun." *sounding a little down*

Rayne: "Believe me, Luna, you don't want this life."

Luna: "If you say so. By the way, what's going on between you and that Darren guy? You two look close."

Rayne: "We are just friends," *blushing* "I think."

Luna: "Yeah, right, I see how you two act around each other, you have been bitten by the love bug." *giggling*

Rayne: *blushing* "Hehe yeah right, you're being weird, Luna."

Luna: "How is that different from any other time, plus you gotta see how he looks at you. That boy is in loooove."

Rayne: "You think so." *her cheeks continue to stay red while curiously wondering if she might have noticed what Luna was saying*

Luna: "Of course. By the way, how's Cross? I haven't seen him in a while." *blushing*

Rayne: "He's ok, why? Let me guess, you have a crush on him." *teasing*

Luna: "No, I don't. I was just wondering." *her cheeks brightened a rosier color*

Rayne: "Then why are you blushing?" *taunting her* "I wouldn't mind having a sister-in-law like you. Then we can truly be family."

Luna: "Shut up. I'm not blushing, I don't like him, and, in my eyes, we're already family." *blushing and wondering how she knew*

Cross: *calling his sister in a loud tone* "Rayne, let's go, food's ready!"

Rayne: "Alright, I'm coming! Luna, I will call you later."

Luna: "Alright, talk to you later." *still blushing just thinking about how it would feel like him kissing her*

Cross served up pork chops and mashed potatoes to everyone, shortly after they said grace. There was an awkward silence for quite some time. Nora kept staring at Blak Hart as he ate, and Rayne kept telling her to stop.

Cross: *thinking* "He eats like a soldier, fast, efficient, always scanning the room. Who the hell is this guy, really?"

Rayne: "Mom, what are you doing?"

Nora: *still staring while thinking* "He just looks familiar; I can't put my finger on it."

Cross: "So staring at him is doing what exactly?"

Nora: "Stop being a smart ass, Michael! Blak Hart, I apologize, it's just that until I figure out why you look so familiar, it's going to bug me all night."

Blak Hart: *trying not to laugh* "Hehe, I like her." *looking at Nora* "It's fine, I get like that too when I have to figure something out."

Cross: “So you think that was funny, huh?”

Blak Hart: *laughing* “Actually, yeah. There’s nothing wrong with speaking her mind.”

Blak Hart asks to be excused and heads to the bathroom. On his way, he sees a picture with a familiar face. His eyes locked onto the photo, his breath caught. The edges of memory curled around him like smoke. He grabs the picture and rushes back to the dining room. Then he says, “No way, who are these men in the picture?”

It’s silent for a bit as the teens put their heads down, and Nora stands up and tells Blak Hart.

Nora: “That’s my late husband and his hunting buddy Hunt.”

Blak Hart: “The guy next to your husband, you said, is Hunt? It must be a code name, cause I know his real name. How long did you know Hunt?”

Nora: *wondering why he was asking so many questions* “For years, we grew up together. Hunt is Cross’s godfather.”

Blak Hart: *finally he put certain things in his mind together* “Wow, so that means that your husband’s code name was Wolf. Hunt was my father, and Wolf was my godfather. So, you’re the Wolf Queen, my godmother.” *due to the realization of the information, he begins tearing up* “Sorry, I didn’t recognize you; I’ve been through a lot. I’m sorry about Wolf. I miss him dearly. I was hoping to find him,” *sighs* “I’m sorry for bringing up hurtful memories.”

Nora: “It’s alright, I knew you looked familiar.”

Cross: “What?! What just happened? Who were you to my father and Hunt? Why didn’t you say anything?”

Blak Hart: "If I would've known she was alive, I would have arrived many years ago."

Rayne: "So, who are you to us then, and what do you want?"

Nora: "Rayne, Cross, relax. He's family. I thought you died with your parents."

Blak Hart: "At times, I wish I did." *A tear slides down his cheek* "Most of me did that day. I was supposed to have gone with Wolf that day... Anyway, Wolf was like a second father, and you were/are like my second mother."

A tear slides down Blak Hart's face, he turns away, and wipes his face. Nora realizes and has Cross and Rayne look for a box that Wolf had set aside for Blak Hart decades ago. Nora continues to talk to Blak Hart...

Nora: "I remember holding you in my arms and babysitting you when your parents went hunting. Louie, everything may not be as it once was, but we are family. You are always welcome here. By the way, thank you for saving my kids."

Blak Hart: "I only remember bits and pieces, but you and Wolf were one of the greatest memories, along with the time you brought the kids over. When I would stay over, the kids would always want to play dominoes or a board game that took all day to finish."

Cross: "Found it!"

Cross brought down a huge box, which appeared worn down and beat up, with a note attached to it with the name *Luis Heart* on it. Cross put it on the table and slid it to Blak Hart.

Cross: "Here you go, bro."

Blak Hart: “Th-thank you....” *a little confused*

Nora: “We will leave you to it, take your time.” *She walks away*

Rayne: “We will be in the next room if you need us.”

Nora, Cross, and Rayne head to the living room as Blak Hart, the paper was yellowed, edges curled. The ink had bled slightly, but the words were clear, deliberate, loving, and final. He opens the letter as another tear runs down his cheek, as he begins to read the letter:

Dear Louie,

If you’re reading this, that means you’re over eighteen years old, and we have evolved into something greater to fight evil. You have made us very proud. By now, Nora and Will would have trained you and have taught you the things we weren’t able to. In this box, you’ll find files we left for you. These files are for you to double-check and make sure that these cases were done; if not, they need to be completed. Sorry to leave you our unfinished work, but you’re the last one left. Make sure you get together with Wolf and think of a game plan. Also, the location of some very powerful tools for you to equip in your arsenal. Remember, we love you and miss you greatly.

With all our love,

Mom & Dad

Agatha & Albert Heart AKA Huntress & Hunt

Louie lowers his head, then looks up at Nora as tears run down his face. He picks up a picture of his parents holding him as a baby. He pushes the box to the side and looks at the picture. Once he was done, he put it in his wallet. He tilts his hat down so no one can notice him crying. Once he has composed himself, he walks into the living room without saying a word. Nora stands up and hugs Blak Hart. She holds on to him as Rayne joins in the hug.

Rayne: "Cross, stop being a jerk and hug him."

Cross: "Alright, alright, I'm coming." *feeling kind of awkward since he was unsure what to do at the moment, considering that just a few hours ago, they didn't know who he was*

Cross walks up to Blak Hart and gives him a huge hug and tells him, he's glad he found them. Then he thanks him for saving them. Blak Hart looks at him and smiles. Blak Hart begins to have an emotion he hasn't felt in quite some time... The love of family. The only worry he has in his mind; the kids have to learn quickly. The minute his enemies find out he still has family members left, they will come for them next. He excused himself and pulled Cross to the side. He tells him that there are a lot of things that Cross doesn't know, and if he doesn't mind, then he could begin to explain it. Cross has a concerned look on his face, then braces himself to hear what Louie has to say...

Blak Hart: "I've had to kill many different things, even things that look like people. We all have to train on fight moves, strategic moves, and defensive positions. My fear is that when any supernatural being finds out that Wolf's family and Hunt's kid are still alive and are working together, shit's about to hit the fan. I want to make sure that when it happens, we're already including your mom. I'm not losing any family members again. This I vow."

Cross: *somewhat emotional from Louie's speech* "I guess it's about time I show you this, come follow me."

Cross brings Blak Hart to the basement, and he starts looking around. He notices that there are tons of boxes with Wolf's name on them.

Blak Hart: "There's a lot of your dad's stuff down here. Have you gone through any of it?"

Cross: "No, it's kind of hard to, all the memories just come rushing back." *putting his head down as flashes of his dad and him go out hunting for deer and other great memories*

Blak Hart: "So is this what you wanted to show me, or is it something else?"

Cross: "No, it's over here, come on."

Cross takes Blak Hart to a hidden room with a strange wooden door, which has symbols all over it, and he called it the Wolf's Den. Cross took out a key from a hidden compartment built into the wall. As Cross unlocked the door, a cold breeze rushes out, sending chills up their spine. When he turned the lights on, they saw two crests carved onto the ground, an elaborate room filled with office equipment, weapons, and items that seemed to be locked in several glass cases, emphasizing importance or danger in nature. They venture further into the room and notice pictures of their fathers together in what resembles a military wardrobe. Hunt and Wolf had plenty of medals on their uniform signifying they were high-ranking officers.

Blak Hart: "I didn't know they served in the military. This was never in the bio. Why was this kept from me?"

Cross: "Neither did I. I wonder what branch of the military they were in."

Blak Hart: "Judging from the weaponry and uniform, it appears to be Special Operations. The gear seems to be a mix of all military styles, and the medals look like they were awarded to them from every branch of the United States military possible, those known and unknown."

Cross: "Have you ever seen those bands, around their arm before?"

Cross points to a band around their right arms. Wolf's was blue and black, while Hunt's was black and red with a symbol or crest on it. The picture's old, so it was hard to tell.

Blak Hart: "I'm not sure, it seems familiar, but I just can't remember it, though."

Cross walks over to the desk and grabs a huge book; he brings it to Blak Hart. The book was tattered and looked like it had blood on it at some point; it had claw marks on the cover.

Blak Hart: "What's that?"

Cross: "It's a book with everything our fathers fought and saw."

Blak Hart starts looking through the book; it seems to be some sort of journal. Cross turns the computer on and opens a file where a video of their fathers commences to play.

Cross: *he hovers over the play button* "You ready?"

Blak Hart: *nods slowly* "No, but hit it anyway."

The video has Hunt starting, followed by Wolf:

Hunt and Wolf: *at the same time* "Well, boys, if you're watching this, that means we're gone."

Wolf: “Cross, I’m glad you didn’t give up. Just trust Louie, if he’s anything like his father, you’re in for a wild ride.”

Hunt: “Learn to trust, my son. No matter how I died, remember you are not alone. Even though they’re not blood, they are still technically family. They will always be there for you; sorry I couldn’t be. Cross, please help my boy. He may do stupid things, but remember he will be the one who’ll always have your back.”

Wolf: “Ok, kids, in my den, check the books in the shelves next to the ‘Claw Sharpener’ if you need more info.”

Hunt: “Boys, remember, we may not be with you physically, but spiritually we’re always with you. Also, there are beings out there looking out for you without you knowing. Keep that in mind.”

The video ends, both Cross and Blak Hart have tears running down their cheek. They wipe it off, then continue to look at files on the desk. Cross found several files in the bin labeled *Bounties at Large*; these are the files with bios of Samuel Wright, Rampage, and Gilbert Gier. Cross recognizes one of the names. It was Darren’s last name, Gier.

Cross: “Could this be Darren’s father?”

Blak Hart: “I don’t know. Read it and find out.”

Cross reads the file; it states that Gilbert Gier was in the Navy Special Operations squad. He was part of an elite group that wanted nothing more than to start the apocalypse. He was wanted for conspiring with an unknown evil spirit. Information on the evil spirit was still being researched at that time. He worked with a powerful warlock, who assisted him whenever he needed it.

Blak Hart: “Well, what does it say?”

Cross: "That Gilbert was working with an evil spirit to get the end of the world started, he had a strong military background."

Blak Hart: "I don't think we should tell Darren about his father, but I should pay his mom a visit. Gail most likely knew of his extracurricular activities."

Cross: "I really don't think she would know anything, she's the sweetest, kindest person I know."

Blak Hart: "Lesson one, just because a person has shown you one appearance, it may not be who that person truly is, always research. The bio I have on her would surprise you."

Cross: "Can I see it?"

Blak Hart: "When we go to the dining room, I'll show you the file. Make sure Darren doesn't see it."

Cross: "Why? He needs to know."

Blak Hart: "Now isn't the right time. I have to call Ark so he could fill in the gaps. While I go to Gail's house, ask your mom if she has anything that could help us out."

Later that day, Blak Hart pays Gail a visit. She automatically recognizes him. She invites him in, and they sit in the living room.

Gail: "Are you here to see my son or speak to me about my ex-husband?"

Blak Hart: "The second one, I just wanted to know how much you knew of his activities and what was your involvement, if any. Also, I wanted to ask, how did you know my parents?"

Gail was shocked that Blak Hart knew of her contact with his parents. She tells him how, she used to work with his parents together on certain missions, while her husband hated them. She always wondered why he hated them, when they were always polite to him. The reason Hunt didn't get him was as a favor to Gail. She used to keep tabs on him and let them know what he was up to. Slowly, but surely, Blak Hart was feeling horrible for misjudging her. She tells him that her specialty is dabbling with white and black magic, depending on what the situation calls for.

Blak Hart: "In other words, you were their informant. If you don't mind me asking, do you have any of his memoirs, files, or any paperwork still? And if you do, may I go through them?"

Gail: "I kept them in my office just in case. If you like, I can make copies of all his paperwork. Whatever you don't want, you can chuck it."

Blak Hart: "That would be great, thanks."

Gail: "Is there anything else I can help you with?"

Blak Hart: "Actually, yes. I was wondering if you would be able to create a protection potion for me. I have several missions I have to do, and I would like a little extra protection from injuries."

Gail: "Any way I can help the last living member of the Heart family, I will do happily. Your parents have helped me out so much that I am eternally in their debt. How soon do you need it, and how many?"

Blak Hart: "By tomorrow night, if possible, and eight vials."

Gail: "Wow, yeah, you are your father's son. You can pick it up by noon tomorrow."

Blak Hart: *smiles* "Thank you. I was wondering if you could keep this between us, until I feel Darren's ready."

Gail: "You're welcome. You're right, he's nowhere near ready for all this. Now that you're friends with Darren, please promise me that you'll protect him. By the way, I heard rumors that in the Pine Barrens in Suffolk County and in the New Jersey Pine Barrens, there were weird sounds, people being attacked by some huge deer-like creatures. Personally, I think it could be wendigos, but I could be wrong."

Blak Hart: "I give you my word. Once again, thank you for all that you've done for my parents and for everything you're doing now."

Gail finishes the copies and hands them over to Blak Hart. He looks at her and smiles, thanking her once again, then heads off to the Cross household. He walks in and asks Nora if he could work in the Wolf's Den. She gave him permission to go, but reminded him.

Nora: "If you need something and can't find it, call me, I'll get it for you. Whenever you need an office, feel free to work in the den. You don't have to ask me every time, just let me know you're there so I won't get startled."

Blak Hart: "Thank you."

Blak Hart goes into the Wolf's Den and sits down at the desk where his father used to sit at. He starts looking through the paperwork, and while reading, he finds something interesting. He found out that Gilbert was working on an experiment that had people locked up in several cells and deprived them of food. These guinea pigs were used in rituals, then injected with something and caged again. They would keep giving them certain commands daily. These commands were told to them as if they were dogs, repeating them daily so when it was time, they would carry out their orders. He and a coconspirator would kidnap people and keep doing this over and over again. At a certain point, when they felt the subjects were ready, they would abduct the subjects' family members. Shortly after, they would trap them in the cell and watch as the unwilling participants would maul their family member without hesitation, only leaving minor bones and skulls. Some of the subjects escaped, never to be found. Some of the ones that stood died because the ritual was too strong, as the rest of the subjects that survived got stronger with every ritual done. Blak Hart was feeling disgusted and angry. He thought, how could anyone be so heartless?

Shortly after, the teens arrive from school, they head toward the den after Nora tells them that Blak Hart was in there working.

Cross: "What are you up to?"

Blak Hart: "I'm working on some information that I got from one of my informants."

Rayne: "Are there any missions for us?"

Blak Hart: "There might be, I'll just have to talk to Ark and find out. Then he can give us missions so we can all get paid."

Darren: "Is there something I can do to help out?"

Blak Hart: "Actually, there is. You can alphabetize all the files in the bin labeled Missions Completed."

Darren: "Ok, I'll get started right now."

Blak Hart: "As for you two, may I have a word with you?"

Cross & Rayne: "Sure, where?" *together answered, wondering what he wanted to say*

Blak Hart: "Outside."

They walk outside, Blak Hart tells them what was said in his visit to Gail's house. Rayne is in disbelief, Cross, on the other hand, was reacting mildly shocked, but was trying not to show it. He also told them that Gail agreed not to let Darren know until they felt he was ready.

Rayne: "He should know about his parents; he probably has an idea anyway."

Cross: "Rayne, they're right. He's not ready; something like this would blow his mind."

Rayne: "Fine, but when it's time, I want to tell him."

Blak Hart: "It's not our place, that job is for his mother to do. Before things get real, we need to train for what's to come."

Cross: "When can we start?"

Blak Hart: "The training will start after Darren is done filing."

They return to the den just to find Darren looking at one of the files. He sees them and abruptly stops. He shows them the file.

Blak Hart: "What did you find?" *worried it might be about his father*

Darren: "One of the files had a picture that looked very familiar."

Blak Hart, Cross, and Rayne all looked at each other, wondering if it would have anything to do with his family.

Blak Hart: "Is it a family member?"

Cross: "Could it be a friend of the family?"

Rayne: "Is it someone from school?"

Darren: "I can't put my finger on it, he just looks familiar."

Cross: "Can I see it?"

Darren hands the file to Cross. Cross agrees with him, and Rayne recognizes the picture.

Rayne: "It's the substitute teacher, Professor Bryan Astor. He was in my art class last week."

Cross hands the file over to Blak Hart to see what he makes of it. Blak Hart started reading through it and let the teens know what it says about Bryan Astor.

Blak Hart: "It says here that our family was hunting him because of the following: human sacrifices of all ages, satanic rituals, and he has the ability to control minor-level demons. He was one of the people who caused several areas in the United States to have group suicides, serial murders, and entities that don't belong in this realm to manifest. He is a powerful warlock. He also has a rap sheet a mile long, and he's wanted in every state in the United States. He's on the most wanted list, starting from the FBI, CIA, and all other government agencies. He's been off the grid for quite some time; the manhunt has been in effect for the past thirty years. His real name is Collin Dirksen; his whereabouts were unknown until now."

Darren: "How could a school hire someone without doing a background check?"

Cross: "That just means that whoever hired him, covered it up, and is somehow working with him in his extracurricular activities."

Blak Hart: "The most important question is who hired him, and what's their involvement in the whole thing."

Rayne: "Before you start researching this guy, let's start the training."

Cross hits a button on the bottom of his father's desk, and the floor of the Wolf's Den opens up. Cross jumps down as the stairs appear. Cross starts turning on the lights as Blak Hart and the others come down the stairs. Rayne starts looking around, and Blak Hart notices that there are claw marks and huge holes in the wall.

Rayne: "I know I haven't been down here in a while, but these holes, scorch marks, and claw scratches weren't here before."

Cross: "Yeah.... let's get the training started."

Blak Hart took off his baseball cap to prepare to spar. Darren realizes that Blak Hart is hairless, compared to the last time when he took the teens to the hospital.

Darren: "Dude, you're bald! What happened to your hair?!"

Cross looks at Darren using hand gestures to stop, but at that very moment, Blak Hart's eyes turned a glowing blood red. It scared Darren, then, in a panicked state, Cross yells to him.

Cross: "Run, don't just stand there, run!"

Blak Hart pulls Cross to the side, not realizing the change in his eyes, and he lets Cross know that he wasn't going to hurt him.

Cross: "Your eyes show a whole different story."

Blak Hart: "What do you mean?"

Cross: "Can't you tell when your eyes change? Like when you get angry or annoyed."

Blak Hart: "Damn, that's something new. I just knew my body was powering, and I felt when my body was ready for battle. Thanks for telling me. How did you know I was annoyed or upset?"

Cross: "Your eyes did the same thing when you were arguing with Ark."

Blak Hart: "Good observation. No wonder our parents were great friends, if their minds were anything like ours." *smiling*

He calms down, then lets Darren know what happened to his hair and told him that it still bothers him a little when it is spoken about. He was now relaxed and ready to spar with Cross. They begin sparring, Cross punches, and kicks the way his father showed him. Blak Hart dodges and blocks him, while blocking, Cross hit him so hard that it moved him back two feet. Next was Rayne and Darren. Rayne was giving it her all. Darren just kept dodging; he didn't retaliate for fear that he would hurt her. Blak Hart realized that with all the dodging Darren was doing, he has a speed that was unnatural. It looks like Darren has a new ability that's awakening. Louie decided to stop them, and switch sparring partners. He sparred with Darren; he warned him to dodge as many punches as he can. Blak Hart told him not to worry; if he isn't able to dodge that, he would be able to stop instantly. The sparring begins, Blak Hart swings and swings. Every time he was about to make contact with Darren's face, he always stopped an inch away, letting him know Darren needs to block and dodge faster. Now it was Darren's turn to try to hit him. Darren gives it all he's got, all except for one, actually, hit Blak Hart.

Blak Hart: "Everyone had a good sparring session. We all need to work harder on certain things to get to where we need to be. Rayne, your power hits were good, but the fighting form needs improvement. Whenever you're ready, I'll be willing to teach you that and deadlier moves. Cross, you're up to par with the style you were taught. Would you like to learn another style? If you do, I can teach you all the ones I learned to make you a deadlier adversary. Last but not least, Darren. You have just awoken a new ability; you need to learn to control your speed. Once you master it, you would be a force to be reckoned with. If you would like my help, I would be more than happy to show you."

Cross, Rayne, & Darren: "How soon can you start training us?"

Blak Hart: "Tomorrow, I'll bring someone to help us out."

Darren: *looking at Cross and Rayne, then he calls them over* "He was about to beat my ass not too long ago."

Rayne: "You do know that if he really wanted to hurt you, none of us could hold him back, and he would've done it."

Cross: "But he didn't."

Rayne: *trying to change the conversation, she looks at Louie* "What about today?"

Blak Hart: "I have to gather some info. If you three really want to start, Cross can show you his form. Darren, have Nora throw plastic cups at you. Your training would be from dodging all the way to catching them."

Cross: "So is there any investigating I can do?"

Blak Hart took out his phone and began texting Cross. *When you get a chance, ask your mom for the password and look on your father's old computer, search for people who are known to be in Collin Dirksen's circle.*

Cross's cell beeps, he reads it, and looks up at him.

Cross: "Really, why?"

Blak Hart pulls him to the side, and he proceeds to tell him why he can't say it in front of everyone.

Blak Hart: *in a low tone* "This is the first mission I am giving, as the next of kin in the wolf blood line, we were destined to work together. Just getting you ready, you are the most prepared to go on a mission with me. When they're ready, we'll do several minor assignments before I can assure Ark that each of you is ready for your own missions. You're the only one I can confidently vouch for."

Cross: "That makes sense. If they ask, what do I tell them?"

Blak Hart: "Let your mom know, she'll have your back."

Cross: "Why do you think she would help?"

Blak Hart: "Who do you think used to help your father out? She didn't get the nickname Wolf Queen just because he was a Wolf. She earned it due to battles and all the people she helped save."

Cross: "Good point."

Blak Hart: "In case of anything, just call me."

Blak Hart calls Ark and asks to meet. Ark is hesitant, but Blak Hart keeps insisting. Ark tells him that, he's in a meeting. He'll meet him in a diner in the next town over; he'll get there shortly once it's done.

Meanwhile in heaven, the Angels have another meeting concerning the teens and Blak Hart.

Head Angel: "They're learning fast."

The second Angel: "Do you think they should trust Darren?"

The third Angel: "The boy is not like his father; he can be trusted."

The Head Angel: "I trust your judgment." *putting his hand on the third angel's shoulder*

The second Angel: "We can never be too careful, brothers."

The Angel trainee: "Relax, brothers, we are here just to observe, nothing more for now."

An unknown Angel: "We should have erased their memories; this is only going to get worse!"

The first Angel: "Our orders came from the king himself! Do you dare question his word!?" *First angel walks up to the unknown* *angel and grabs him* "Shall I remind you what happened to the last person who went against the king's word?"

The second Angel: "Brothers enough," *separates the two* "we don't have time for this."

The third Angel: "I agree, what's the point in fighting each other."

The Angel trainee: "Why must this keep happening." *covers his face*

The second Angel: "Brothers, are you done arguing? If this meeting is done, I have to go. There are assignments I must complete."

An unknown Angel: "What could be more important than the situation at hand?"

The second Angel: "Anything is more important than listening to arguments that were already set into action by our king. Anyway, if you must know, the prince sent me to do something."

The unknown Angel: "What did he send you out to do? He could've sent a more qualified Angel to do it instead."

All the Angels watch as the second Angel walks away from the unknown Angel. The head Angel knows the reason why he left, to acknowledge the second Angel for doing the right thing, he walks up to him.

The Head Angel: *patting him on the back* "I'm proud of you for walking away rather than entertaining our siblings' foolish actions."

The second Angel looks at him, smiles, then thanks him as he walks off. The unknown Angel continues to argue as the rest of the Angels ignore him and all part ways.

The Unknown Angel: "You all will listen to me sooner or later, you'll see; you're all making a big mistake. I will laugh at all of you when I prove to you all that I've been right the whole time and the king..." *he begins yelling* "WILL APOLOGIZE TO ME, DID YOU HEAR ME. HE WILL BE APOLOGIZING TO ME FOR BEING THE ONLY ONE WHO KNEW SOMETHING HE DIDN'T." *laughing menacingly.*

CHAPTER 3.
WHAT ARE THEY LOOKING FOR?

At the diner, Louie is waiting to meet up with Ark. This is the first time that Blak Hart meets Ark face to face. These two have always done business over the phone through video chat, and for some reason, the picture was never clear enough to get a good look at him. Ark has been watching the Hart family for several decades, without the knowledge of Blak Hart. For Blak Hart, he's always been the boss, assisting him in finding jobs to get rid of evil, getting Blak Hart that much closer to finding out who sent the order to kill his parents.

Blak Hart was eating lasagna when a well-dressed man walked in. He had dirty blonde hair that draped over most of his face, sky-blue eyes that showed when his hair shifted, and he stood about 6'3". He was well-spoken and carried a very relaxed demeanor. He sat down by Louie. Blak Hart looked at him.

Blak Hart: "That seat is reserved; I suggest you go elsewhere. I am not liable for what I might do if you stay."

Ark: "Learn to ask who the person is and what their purpose is. Don't automatically jump to conclusions. You never know if the person is an ally or an adversary, Blak Hart."

Blak Hart: "How do you, oh." *smiles* "Heh, Ark. My bad, I mean, I apologize. I was just making sure your seat was reserved."

Ark: "I greatly appreciate it, but don't always assume that everyone is against you. Expect it, but act on it only when you're truly sure."

Blak Hart: "Understood. Are you gonna order something, or just watch me eat as we have our first meeting?" *looking at Ark*

Ark: "I don't need food to survive, but I only eat at certain restaurants. The food has to be exactly to my liking."

Blak Hart: "Order something small. If it's to your liking, order an actual meal. If you don't, then we leave."

Ark: "You don't have to stop eating on my account."

Blak Hart: "If the food here isn't up to your standards, then I would like to eat at those selected restaurants that have your approval."

Ark orders a Danish cheesecake with strawberry and blueberry filling. The waitress brings him the pastry and asks if he would like coffee with it. He says yes and proceeds to have the conversation with Blak Hart.

Ark: "The food is good; I'll order more later on. What was so important that you wanted to meet in person?"

Blak Hart: "What do you know about Collin Dirksen? We found his file, and I want to know the info you purposely left out of the file."

Ark: "I always put all the information I have into the files." *clears his throat*

Blak Hart: "Sure, sort of like the file of the demon that was able to change into whatever it touches and can also mimic its opponent's power? Yeah, that was very informative. Come on, Ark, we've been working together since I was thirteen years old. I thought I had already earned enough of your respect and trust to get the complete file. I believe I at least deserve that." *a bit annoyed.*

Ark: "You have." *he slides a file to Blak Hart*

Blak Hart: "Thanks." *pats Ark on the back*

Ark: "Blak Hart, just remember you're not alone in this anymore; those kids are all you have now. Please be careful. All your actions will affect these teens in some way, shape, or form."

Blak Hart: "Well, you've always had my back. Or has that changed? If you still do, then I'm good." *smirks at Ark*

Ark: "I'll always 'have your back,' as you call it, but I hope you do realize that I won't be able to do it as much as I used to. That's why I sent someone to continue to train you from where your parents left off."

Blak Hart: "I know, I know. It's greatly appreciated, but when I'm outnumbered…"

Ark: "Those teens will be your team. Train them well, trust in them, for they do trust you. Would you lay your life for them, well, would you?"

Blak Hart: "I am training them, and yes, I am ready to put my life on the line to protect them. They're family, the only ones I have."

Ark: "Alright then."

Blak Hart: "You know my birth name; in fact, you know almost everything about me. While I know very little about you, if you don't mind me asking, who are you?"

Ark: "I feel you're not ready yet for the answer."

Blak Hart: "After all the bullshit I've been through, I believe I've earned the right to know who has been helping me and why.

You said you knew my parents, so if they trusted you and you trusted them with your personal info, then why don't you trust me?"

Ark: "Are you sure you want the answer?"

Blak Hart: "Hell yeah, I want the answer, I'm tired of being left in the dark. How else can I fight evil and protect, if all you're giving me is partial info?"

Ark: "You've been doing pretty good so far."

Blak Hart: "Stop trying to avoid the question."

Ark: "You have to trust me. When the time is right, I will tell you. All I can tell you is that I was always around your family. Your parents and I have worked together in the past. I knew you even before you were born. Thanks for bringing me here. The food was good. See you next time. You're closer to the truth than you think. Just don't rush it; some answers burn more than they heal. Your father once asked me the same question. I didn't answer him either. But he understood, some truths are earned, not given."

Blak Hart: "Wait, that's it? I need more than that."

As mysteriously as Ark arrived, he vanished. Blak Hart was left with more questions about Ark's involvement with his parents. Louie looked at the file lying on the table.

Blak Hart: "I'll deal with this info and decide what to do. After this, I'll talk to Nora to see if I can take Cross with me on this mission."

Blak Hart walked back to the house, noticing he was being followed by a car with dark-tinted windows. He took a shortcut through the woods to try to lose them. When he came out, the car was waiting for him. The window rolled down, and a man started to call out to him. The stranger's voice was very soothing, almost hypnotic. It started to rain. The rain hissed against the car's roof, but inside, it was silent. Too silent.

Strange Man: "Hey, you're Blak Hart, right?"

Blak Hart: *defensive* "Depends on who's asking, so who the fuck are you?"

Strange Man: "I'm just someone who wants to help."

Blak Hart: "I'm sure I can take care of myself, so you have to try better than that."

Strange Man: "Do you trust that Darren boy? You know what his father has done? The apple doesn't fall far from the tree, you know."

Blak Hart: "How do you know about Darren or his father?!"

Strange Man: "Here." *throws a file at Blak Hart.* "My card is in there if you ever want to talk."

Blak Hart: "Who the fuck are you...?"

When Blak Hart looked up, the car and the man were gone. He ran to the house, confused and drenched from the rain. Cross let him in the door. He greeted everyone, then told them he'd be right with them after he spoke to Nora. He asked her if she would allow Cross to accompany him to his parents' property and look for the basement or room they had used for missions. With a concerned look, she began to tell him:

Nora: "Louie, that's a long trip. I know he would help you keep your emotions in check. Please be careful and keep him safe. We don't know if the house or land is still being watched."

Blak Hart: "No problem, you know I'll protect him as if he were my own flesh and blood. Thank you."

Nora: "You're welcome. Please call me when you guys arrive and are getting ready to head back."

Blak Hart: "You can count on it."

Blak Hart headed to the Wolf's Den to tell Cross about the meeting.

Cross: "About time. What took so long?"

Blak Hart: "There's a lot of strange shit going on."

Cross: "Really, like what?"

Blak Hart told Cross about the strange man who approached him and showed him the file and the business card. He asked if it looked familiar. Cross studied the card as if he'd seen it before.

Cross: "I can't figure it out, but I've seen it before."

Blak Hart: "We'll worry about this later. We have more important things to take care of."

Cross: "Oh yeah? What did Ark say?"

Blak Hart told Cross how the meeting with Ark went and showed him the files Ark gave him. They started heading to the Wolf's Den, where Rayne and Darren were waiting.

Rayne: "So what's going on?"

Darren: "Hey Louie, what's up?"

Blak Hart: "Hey, guys, give me a sec to set everything up."

Cross took the files about Darren and put them into his father's desk, then locked it up as Blak Hart spread everything on the wall so they could get a better look.

Blak Hart: "Alright, let's see what we got here."

Blak Hart stepped back, showing all the mugshots and locations that Collin Dirksen had been. They were marked on a map along with his full bio. Blak Hart led the meeting.

Blak Hart: "All the information we need to know about Collin Dirksen is on the wall. If you take a look, there have been more children and teenage disappearances since 2006. That's the year he started working in the high school. Not to mention, sightings of strange creatures in the woods next to the school went up. This is gonna be your first mission. Darren and Rayne, I need you two to go and find out his habits, who he talks to, preferably people with authority in the school. Don't do anything, just observe and note everything down. I want a report of everything seen and heard."

Rayne: "What about Cross, what's he gonna be doing?"

Blak Hart: "He's going to be doing another mission. He will be assisting me on a recon mission."

Darren: *feeling left out and jealous,* "Why can't we go with you and help out as well?"

Blak Hart: "We can get more work done if we split up. I need his keen sense of smell in the mission. You two are perfect for the mission in the high school for your stealth and speed."

Cross pulled Blak Hart aside.

Cross: *thinking Louie thought he wasn't suitable for that mission,* "So you think I'm not as stealthy or as quick as them."

Blak Hart: "That's the only way I can explain it to them without hurting their feelings. I can't let them know we've been doing missions together already."

Cross: "Oh, ok. I thought you didn't think I was capable of doing that mission."

Blak Hart: "I need you on the important mission with me, because this involves going back to my parents' run-down house to try and find their hideout. This is going to be the first time I've been there since their death, so you see, this is an important one for me. It's personal. I don't need any childlike mistakes or rebellious actions since it's pretty far away."

Cross: *feeling honored that he was trusted with something so important,* "You trust me with this… but what if I mess up?"

Blak Hart: "Then we learn. But I trust you because you're ready, even if you don't feel it yet."

Cross: "No problem. I know it's gonna be hard to deal with, emotion-wise. I understand. Whatever I can do to help you out, you've got it."

Blak Hart: *smiles, acknowledging he's trying to be supportive not only in the mission but as a family member's emotional rock.* "Thank you."

Blak Hart focused back on Darren and Rayne to continue the rundown. He planned out how they would be able to complete the mission in the most efficient way possible. They trained and trained until the scenario simulations were nearly flawless. He told Darren to trust her instincts, while Rayne must respect his new ability of premonition. If they learned to trust each other's abilities, everything would go smoothly.

Blak Hart: "We'll be leaving now, but just remember the training and trust yourselves along with each other."

Darren and Rayne: *together* "No problem. You can count on us."

Blak Hart: "Cross, let your mom know that we're leaving."

Cross walked over to Nora and kissed her on the cheek. She cried and told him to be careful.

Nora: "Listen to Louie, and remember, you are his backup. So when things go south, you're all he has."

Cross: "I know, ma, I know. Remember whose son I am. I am Michael Cross, son of Wolf and the Wolf Queen." *chuckling* "I'll call you when we arrive."

Nora: *with a sense of pride at what her son said, a tear slid down her face, knowing these two would be trouble for anyone who got in their way.* "I love you." *Kisses Cross on the forehead*

Cross: "Love you too, ma." *yells in Louie's direction, quickly drying his tears,* "You ready yet?"

Blak Hart: "Be there in a sec."

Blak Hart went into the bathroom to freshen up before the trip. He washed his face, and as he looked up at the mirror, the bathroom lights flickered. The air grew heavy, cold. His breath fogged the mirror as the reflection twisted into something monstrous. The reflection had small horns on the top of its head, a face covered in black veins, and dark red eyes. Its ears were pointy; the reflection spoke to Blak Hart in a raspy voice.

The Reflection: "Why keep up with this fight? You saw what they did to your parents. If you were to join them, they could revive your parents. That Darren kid and his friends can help you rule, and if they don't, they can perish. Their choice. We'll meet again real soon," *smiles at Blak Hart, showing sharp razor-like teeth*

Blak Hart: *eyes glowing red* "Never. I'll finish you off along with all your kind. When I find you, I promise your death will be slow and painful. Those teens are my family. Fuck with my family, and you fuck with me. Be prepared to die by my hand!"

Blak Hart walked out of the bathroom. Cross and Nora were standing outside of it. Hearing him arguing with someone, Nora looked at him and pulled him to the side.

Nora: "Who were you arguing with?" *notices his eyes*

He explained to Nora what happened and what was said. She told him to wait for a moment, then went to her room. She found a file with the name *Luis Heart* on it, brought it to him, and told him the contents.

Nora: "In the file, it has the possible prophecy of you being the greatest evil ever known."

Blak Hart: *horrified,* "You know that's not me. How could such a prophecy be set for me? How could I turn into the very thing that I despise, the thing that killed my parents? What if they're right? What if every time my eyes glow red… I'm one step closer to becoming that thing? I can't let myself think like this; that's what they want, and I'm not letting them. No chance in hell."

Nora: *trying to calm him down and give consoling positivity,* "I'm possible. I know you're not like that, but this is just to show you why they are constantly after you. You see now why I keep telling you to protect my son. The path for the two of you is several; he can be the partner in saving everyone or in destroying it. These were made centuries ago, but you are a Heart after all, which means if things were etched in stone, you are the only one who can break from it and make your own path. If you allow me, I can create a spell to strengthen your senses. It will allow you to see a person's true self."

Blak Hart: *now calming down, knowing that Nora will do her best to avoid the horrible outcome because it is tied to her son as well.* "No problem. Any help against evil is greatly appreciated."

She chanted and began to go into a spiritual trance. Nora whispered in a forgotten tongue, her fingers tracing runes in the air. A soft glow enveloped her, pulsing with each word. After she finished, she gave him a gem. Blak Hart took the gem, and once it touched his hand, the gem dissolved, disappearing into his skin. He asked why it went into his skin, and she told him it was supposed to in order to bind with him. Before he could thank her, he felt a surge of energy flowing through his entire body. Quickly after, he called Cross so they could leave.

Cross: "What did you and my mom talk about?"

Blak Hart: "I'll explain it to you on the ride."

Blak Hart and Cross were on their way to the Heart property. Louie told Michael what happened with the mirror and what Nora spoke about.

Cross: "Do you believe that an alternate prophecy is for us to be the ultimate evil?"

Blak Hart: "I wouldn't put it past destiny at all. Our destiny is whatever side we decide. There are two destinies that could happen depending on which side we choose. If we become evil, I will take over hell with you by my side and destroy heaven. Since we've been doing good, we are destined to become part of an elite group to keep the balance between good and evil."

Cross: *sarcastically,* "That isn't that stressful at all, having all those possibilities depending on what side you're on."

Blak Hart: "It can be, but all I can do is try to make my own path."

Cross: "I'll do the same. How's that going for you?"

Blak Hart: *smirks* "I'll tell you when I figure it out."

Cross: *smiling,* "My father used to tell me that I was made to do great things. I still don't understand what he meant, but hopefully, working with you, I can figure it out."

Blak Hart pulled up to a cemetery and got out of the car. Cross followed as he walked in. Louie headed all the way to the back of the cemetery, where a heart-shaped headstone stood alone, dead flowers all around it except for one black-and-red rose. Blak Hart cleaned off the dead flowers.

Blak Hart: "Hey, Mom, hey Dad, sorry it's been a while since I last came to visit you, but look who I found, it's Wolf's son."

Cross: "Hey, I'm sorry I didn't get to meet you when I was older, but from what your son and my mom have told me, you two were kick-ass parents. You gave me a kick-ass God brother, so thank you so much."

Blak Hart: *hiding his tears behind his shades,* "We have to head to our old house; I promise to visit when we are done." *Blak Hart kissed his hand and placed it on the headstone, then headed back to the car.*

Cross followed behind him, and they headed to the house. When they finally arrived at the property, the rubble hadn't been cleaned off. The property was still barricaded with faded yellow tape reading: *Danger, foundation unstable.*

Blak Hart broke the yellow tape and walked onto the land. He saw that all the doors and windows were boarded up. Cross went around back and spotted a way in.

Cross: "Blak Hart, come over here, I see a way in."

Blak Hart saw the opening but couldn't get in. Cross started pulling at the wood covering the opening. Blak Hart then kicked the wood to the side, almost hitting Cross in the head.

Cross: "Thanks for the warning."

Blak Hart: "My bad, are you ok?"

Cross: "Yeah, lucky I moved out of the way."

They both noticed that they still couldn’t fit, so Blak Hart looked around and realized that one tree was still standing. It wasn’t burnt. He walked up to it and knocked on it.

Cross: "Does it sound hollow?"

Blak Hart: "Yeah. Wait a minute, I remember my father had installed a secret entrance involving one of the trees. Let's see if this is it."

Blak Hart punched the tree and nothing happened. Upon hitting it, he got cut. There was blood on the tree, and surprisingly, the trunk opened up.

Blak Hart: "What the fuck? I hit the damn thing and nothing. What could've opened it?"

Cross: "Show me again how you hit it."

Blak Hart hit the tree again, reopening the wound; several drops of blood landed into what appeared to be a small funnel. The tree opened up wider.

Cross: "Maybe that's what opens it up, blood."

Blak Hart: "Maybe. You try it."

Cross: "Ok, let's see."

Cross hit the tree and got cut, but nothing happened. He tried again, still nothing.

Cross: "It could be just your blood that opens it up."

Blak Hart: "Hey, Mom, hey Dad, sorry it's been a while since I last came to visit you, but look who I found, it’s Wolf's son."

Cross: "Hey, I'm sorry I didn't get to meet you when I was older, but from what your son and my mom have told me, you two were kick-ass parents. You gave me a kick-ass God brother, so thank you so much."

Blak Hart: *hiding his tears behind his shades,* "We have to head to our old house; I promise to visit when we are done." *Blak Hart kissed his hand and placed it on the headstone, then headed back to the car.*

Cross followed behind him, and they headed to the house. When they finally arrived at the property, the rubble hadn't been cleaned off. The property was still barricaded with faded yellow tape reading: *Danger, foundation unstable.*

Blak Hart broke the yellow tape and walked onto the land. He saw that all the doors and windows were boarded up. Cross went around back and spotted a way in.

Cross: "Blak Hart, come over here, I see a way in."

Blak Hart saw the opening but couldn't get in. Cross started pulling at the wood covering the opening. Blak Hart then kicked the wood to the side, almost hitting Cross in the head.

Cross: "Thanks for the warning."

Blak Hart: "My bad, are you ok?"

Cross: "Yeah, lucky I moved out of the way."

They both noticed that they still couldn't fit, so Blak Hart looked around and realized that one tree was still standing. It wasn't burnt. He walked up to it and knocked on it.

Cross: "Does it sound hollow?"

Blak Hart: "Yeah. Wait a minute, I remember my father had installed a secret entrance involving one of the trees. Let's see if this is it."

Blak Hart punched the tree and nothing happened. Upon hitting it, he got cut. There was blood on the tree, and surprisingly, the trunk opened up.

Blak Hart: "What the fuck? I hit the damn thing and nothing. What could've opened it?"

Cross: "Show me again how you hit it."

Blak Hart hit the tree again, reopening the wound; several drops of blood landed into what appeared to be a small funnel. The tree opened up wider.

Cross: "Maybe that's what opens it up, blood."

Blak Hart: "Maybe. You try it."

Cross: "Ok, let's see."

Cross hit the tree and got cut, but nothing happened. He tried again, but still nothing.

Cross: "It could be just your blood that opens it up."

Blak Hart drops more blood into the funnel, and the ground opens up, revealing a set of stairs heading down. A little concerned about what might be down there, Blak Hart decides to go in first.

Blak Hart: "I'll go in first to make sure it's clear, then I'll call you to come down."

Cross: "Alright, but hurry up, I've got a bad feeling."

Blak Hart: "Me too, I'll make it quick."

He headed downstairs, pulling out his gun. He got to the bottom, took several steps as a few of the lights turned on. The air was thick with mildew and memory. Each step echoed like a whisper. The lights buzzed, flickering as if resisting the presence of the living. It was a long hallway, and Blak Hart slowly started walking. A row of lights kept turning on, leading to a door. He got close to it and noticed there were symbols on the door, but no doorknob. Just as he approached the door, Cross called out to him.

Cross: "Hey, you ok down there!? Hello!?"

Blak Hart: "Yeah, how is it up there!?"

Cross: "It's quiet, too quiet, and the weather is starting to get really ugly out here!"

Blak Hart: "I'll go in first to make sure it's clear, then I'll call you to come down."

Cross: "Alright, but hurry up, I've got a bad feeling."

Blak Hart: "Me too, I'll make it quick."

Blak Hart: "I'll go in first to make sure it's clear, then I'll call you to come down."

Cross: "Alright, but hurry up, I've got a bad feeling."

Blak Hart: "Me too, I'll make it quick."

He headed downstairs, pulling out his gun. He got to the bottom, took several steps as a few of the lights turned on. The air was thick with mildew and memory. Each step echoed like a whisper. The lights buzzed, flickering as if resisting the presence of the living. It was a long hallway, and Blak Hart slowly started walking. A row of lights kept turning on, leading to a door. He got close to it and noticed there were symbols on the door, but no doorknob. Just as he approached the door, Cross called out to him.

Cross: "Hey, you ok down there!? Hello!?"

Blak Hart: "Yeah, how is it up there!?"

Cross: "It's quiet, too quiet, and the weather is starting to get really ugly out here!"

The sky grew dark as a greyish cloud covered the house. Suddenly, a huge storm began, pouring rain, and then hail started coming down. The sound of the thunder was like explosions, and the wind picked up, growing faster and faster.

Meanwhile, back at the high school, Darren and Rayne were about to start their mission. They walked in and spotted Collin

Dirksen, aka Professor Bryan Astor. They tried to be as stealthy as possible while watching his movements and everyone he spoke to. At the ninth period, he headed to the lunchroom. Darren and Rayne trailed him, but once inside the cafeteria, they lost him.

Rayne: "Where did he go?"

Darren: "He was right in front of us. Could he have noticed us following him?"

Rayne: "Can't be, we were careful."

Collin looked from one of the exit doors and spotted them searching for him. He found it odd that these teens had been after him all day but hadn't approached him like students would a teacher.

Collin: "Why would they come after me? My cover has been flawless for the past five years. They couldn't have figured anything out, they're too stupid to put it together. Just for safety precautions, I'll send someone after them to get them distracted."

Collin walked to the storage room, drew several symbols on the concrete with his blood, and conjured up several ghouls. He told them to chase the teens out and do whatever they could to scare them or make them join. The ghouls hid in the restrooms, waiting for the teens to pass by.

Rayne and Darren surveyed the lunchroom. Since they didn't see Collin, they left, figuring that he had left. They tried to find him, looking past every room. As they passed the bathroom, the ghouls ran out and began chasing the teens all the way out of the school. The teens continued running, the ghouls cornering them between the two school buildings. The ghouls started attacking relentlessly. Just out of instinct, Rayne retaliated with a gust of air, pushing them back.

When Darren looked at her, Rayne's eyes were glowing yellow. Her nails were razor-sharp, and she looked at Darren, letting him know:

Rayne: "I know you saved me from whatever that was last time, now I need you to do the same thing. Only this time, don't try to save me, just help me. Together we can do this, just like the way we did in training."

Darren: "I don't want you to get hurt."

Rayne: "I won't get hurt as long as you help me and stop trying to be my hero."

Darren: "I can't promise that I won't try to save you, but I will help you get out of this jam."

They prepared as the ghouls charged at them. They scratched, bit, and hit the teens with all their might. Darren started punching and kicking, knocking some of the ghouls back as Rayne's claws tore through flesh, black blood sprayed like ink. A ghoul lunged. She ducked, spun, and slashed. Another scream. Darren's fists cracked bone. Then, teeth sank into his shoulder. He screamed in pain as it bit him again with all its might.

Rayne, seeing Darren in pain, started hacking and slashing at the creatures blocking her from Darren. One of the ghouls grabbed her and dug its claws into her back. Darren heard her screaming in pain, and he started yelling out:

Darren: "Leave her alone, I'll kill you for hurting her!!!"

Darren let out a huge scream. His eye turned white with a streak of yellow, and electricity shot out of his body, shocking the ghouls.

It fried the one that was biting him, then instantaneously caused its head to explode while destroying the one on Rayne's back.

Rayne glanced at Darren.

Rayne: *irritated,* "What the hell was that? I didn't need rescuing!"

Darren: "I... I don't know. I didn't hurt you, did I? I wasn't saving you, just helping you like you wanted."

Rayne: "Oh, ok. Thank you. Let's get the other ones, they're getting away."

Darren: "Remember, Louie said to stay out of trouble."

Rayne: *being devious.* "Well, we had to defend ourselves, sooooooo..."

Darren disagreed and pulled her away, telling her that the best place to be right now was home. A little annoyed, Rayne agreed and told him that her mom would help them with their wounds.

Back at the Heart property, Cross called out to Blak Hart again.

Cross: *in a nervous tone,* "Blak Hart, Cross wants to go inside the secret underground base now!"

Blak Hart: "Come on down, fuckin' door has no knob. How the fuck am I supposed to get in?"

The boom of thunder sounded again, followed by lightning. Cross zoomed downstairs with a nervous look on his face. Just as he got next to Blak Hart, the huge metal door slammed shut behind them.

Cross: "Do you know how to open the door?"

Blak Hart: "Don't you think if I could, I would've done it by now? And forget the door, are you scared of thunder? How are you supposed to fight scary, evil shit when thunder is your main fear?" *laughing*

Cross: *changing the subject* "What are those symbols on the door?"

Blak Hart: "Those symbols tell me how to open the door without fear of the thunder. Naaaaa, just fucking with you. I have no clue." *still laughing*

Cross: "That's not funny. So how do we get in? Just for your information, I can fight evil, scary things, but thunder I can't hit, so ha." *retorts back.*

Blak Hart: "Ok, I'll stop bothering you about the thunder. Did you hear that?"

Cross: "Hear what?"

Blak Hart pointed to the door. As Cross looked at it, Blak Hart hit the wall, creating a boom, making Cross scream and jump in fear.

Cross: *grabbing his chest from fear,* "Not funny!"

Blak Hart: "That's it, now I'm gonna stop. I couldn't resist." *laughing hysterically*

Cross: "Instead of trying to scare me, why don't you try to open the door?" *defensive*

Blak Hart: "What do you want me to do, hit this door that looks thicker than the outside one?"

Cross: "No, try to see if it opens with blood too."

Blak Hart was prepared to hit the door, then used the cut on his hand to squeeze so the blood would drop onto the floor. Cross noticed that there weren't any funnels or drains on the floor, so he told Blak Hart to smear the blood on the door.

Blak Hart: "Why the fuck would I put blood on the door?"

Cross: "I only thought of it because, to get in, you had to use your blood; the door smells like someone had used their blood on it. So why not give it a shot?"

Blak Hart tried the suggestion. He smeared the blood where the symbols were. The cyphers glowed a bright red color, then clanking and the sound of metal gears grinding echoed. Finally, the door opened, and the sound of stone rubbing against stone could be heard. A strong gust of thick, dusty air rushed from within. The musty smell filled the area as they slowly walked in. More lights turned on as if there were motion sensors.

Blak Hart: "I guess the electricity is still working. It appears to be similar to your father's office. I guess this was our fathers' other workstation. There's another door, if I remember correctly, that leads to the basement."

Cross: "Should we check here first..." *looks at Blak Hart and notices his eyes were tearing up.* "Are you ok?"

Blak Hart: "Yeah, I'm just getting flashbacks."

Cross: "Let's just look around down here and slowly work our way up."

Blak Hart: "Sounds good."

Blak Hart and Cross started looking around the room. They spotted a few files that looked familiar. They found a folder that completed some they already had, then they saw the file with the name on it, *Rampage*. It had been set aside along with Collin Dirksen's file. They continued to look around and found a lever with stains on it. Blak Hart opened his cut, then smeared the blood on it and pulled the lever. Nothing happened, so he moved to the next one. He did the same thing, and it opened a panel, revealing another hidden room with multiple types of weapons.

Cross cut his hand and pulled the lever that Blak Hart couldn't open, and it opened up with all of Wolf's weapons. As they looked through the weapons and gear, they heard someone else's footsteps walking around in the house. It sounded like someone was looking for something, throwing things. Blak Hart and Cross looked at each other, wondering who or what could possibly be in the house. The mystery was that no sound was made when the person entered, and they didn't hear anyone breaking any of the boards that blocked the windows and doors.

Blak Hart: "Wait here. I'll check. If you hear bangs, then come up. You'll be my element of surprise."

Cross: "I'm here with you to help out no matter what."

Blak Hart: "I know, but how about I check first? If you hear bangs, then you come up."

Cross: *chuckles* "That sounds better."

Blak Hart went upstairs through the basement and still heard that person rummaging through the rubble. He crept up the stairs, then ended up in the kitchen. He peeked around the corner of the wall and saw a muscular man with a mask that resembled an iron skull with sharp teeth. Just when he was going to turn around and notify Cross of what he saw, the man with the iron skull turned around and noticed him. He began to charge toward Blak Hart.

Blak Hart dodged the attack, then began to ask him questions as he retaliated.

Blak Hart: "Who are you and what the fuck are you doing in my parents' house?"

Cross ran upstairs just to see Blak Hart and the person in mid-battle.

The Man Fighting Blak Hart: "I am Rampage. Your bloodline owes me everything! I will take what is owed to me by any means necessary!"

Blak Hart: *his eyes turning red,* "Over my dead body!"

Rampage: *with confidence,* "That can be arranged."

Cross: "Hello, I'm Cross. Fuck with him, you fuck with me."

Rampage: "Two for the price of one, oh what fun."

Cross charged at Rampage as Blak Hart punched him in the stomach, sending him several feet away. Cross tackled Rampage to the ground and started punching him in the head, making his fists bleed. The man grabbed Cross by the neck with his tail and started to strangle him. Blak Hart rushed to help Cross, but Rampage let out a loud, soul-shattering roar that caused flames to engulf his head. His eyes glowed a fiery red, and he spoke in a deep, roaring voice.

Rampage: "RAAAAAAAR! YOU TWO ARE PISSING ME OFF!"

Blak Hart: "Put him down! Or I will rip you to shreds!"

Rampage: "HAHAHA! YOU'RE IN NO POSITION TO MAKE THREATS, I CAN SNAP HIS NECK BEFORE YOU CAN THROW A PUNCH!!!"

Blak Hart: "The real reason you won't let him go is because you know that once he's not your shield anymore, I'll whoop your ass."

Rampage accepted his challenge and threw Cross like a sack of potatoes. Blak Hart looked at Cross and saw that he wasn't moving. Right before Rampage rushed at him, Blak Hart thought that Cross was dead. His eyes began to glow a crimson red with streaks of orange, his body bulking up. Around his head was a bright light, fire surrounding it, as he clenched his fists. The hands were engulfed in flames. Blak Hart growled and headed toward Rampage. The loud growl helped Cross awaken from the previous impact.

Rampage: "What kind of demon are you? No matter, just something new for me to destroy."

Blak Hart: *spoken in a dark, raspy voice,* "Demon, I am not. What I am... is your imminent death."

Blak Hart received a hit from Rampage, pushing him back a foot or two. He countered with a hit of his own, the force so strong that it cracked the metal mask Rampage had on. The mask cracked, revealing half of Rampage's face. He looked young, and his eye turned blue while the other eye stayed glowing red. He charged at Blak Hart, throwing a barrage of punches and kicks. Blak Hart dodged left and right, most of them, then at the right time, he threw a burst of his own scorching punches and kicks that hit in quick succession. Rampage was sent through the wall, causing the house to start collapsing.

Blak Hart tried to go to Cross as he lay there with his eyes open but motionless. Blak Hart called out to Cross.

Blak Hart: "CROSS, ARE YOU OK?! COME ON, TALK TO ME!"

As Blak Hart stepped forward, Rampage grabbed his legs and threw him through another wall, then stood over him, punching Blak Hart into the ground. With each punch, you could hear his bones breaking. Blak Hart screamed in pain as every punch hit him, blood squirting out of many of his wounds.

Rampage: "RAAAAAAAAAAAR! I'LL MAKE SURE THERE'S NOTHING OF YOU LEFT TO BURY!!!!!"

Suddenly, a loud yell came from Cross's direction. Rampage looked toward him and realized that he wasn't there. Surprisingly, he felt a sharp pain in his stomach. When he looked down, it was Cross's arm through Rampage's stomach. Cross had fur all over his body, glowing yellow eyes, sharp teeth, and claws. Cross threw Rampage to the floor, then commenced to slash and claw, taking huge chunks of flesh. Cross bit down on Rampage's neck, and the beast let out a blood-curdling scream, waking Blak Hart from his knocked-out state.

Blak Hart tried to stand but fell to his knees. He leaned on what was left of the wall to hold himself up and called out to Cross.

Blak Hart: "CROSS! IS THAT YOU!?"

He didn't get a response. Cross kept ripping and tearing into Rampage. Blak Hart made his way to Cross, limping and holding his sides. Blak Hart realized that Rampage seemed lifeless and was gutted like a pig. Blak Hart put his hand on Cross's shoulder, making Cross jump back.

Blak Hart: "Easy, it's me, Blak Hart."

Cross: *in a deep, growly voice,* "You're hurt."

Blak Hart: "Eh, it's ok, I heal pretty fast. Plus, it could have been worse. I could look like him." *points to Rampage*

Cross: "I did that?" *looks down at his blood-soaked claws* "No, no, not again."

Blak Hart: "Have you transformed like this before?"

Cross: *puts his hand on his head.* "Yes, but I have a hard time controlling it." *He stared at his blood-soaked claws. The memories came rushing back: the student's screams, the shattered mirror, the night he woke up in the woods with no recollection. He wasn't just scared of the beast inside. He was scared if it liked the taste of violence.*

Blak Hart: "Well, we can help each other learn to control our new abilities. I think I might have to buy you a flea collar for the next time you transform like this." *smiles.* "Have you had your rabies shot yet?" *teasing*

Cross: "Sounds good, and as for the last comment... ha, ha, very funny. At least you don't have to worry about being bald anymore, just turn your head on fire and there you go." *cries from laughing so much*

They looked back at Rampage as he got up. All his injuries were healing at a rapid pace. Blak Hart and Cross tried to charge at him, but Rampage looked at them angrily and faded away.

Blak Hart: "This is bullshit, we needed to question him, damn it!" *As he processed the entire incident and the newfound abilities, he thought,* "Every time I burn, I hear the voices of my ancestors.

Not words, just rage. I think this power remembers them. It remembers their struggle, their reason for fighting."

Cross: "Another time, bro. Another time."

At the Cross household, Darren and Rayne were being attended by Nora. She treated what was left of their wounds, noticing that the cuts and bruises were slowly healing on their own.

Nora: "What ability did you get? Is it the nails and teeth growing, eyes glowing, any fur or something completely different?"

Rayne: "A little bit of all of them."

Darren: "My speed and strength got more powerful, along with some lightning abilities."

Nora: "That's interesting. Have you called Louie to let him know?"

Rayne: "No, I don't want to bother them."

Nora healed Darren to the best of her ability and sent him home. The phone rang, it was Cross. He was just letting Nora know that he was ok and had a lot to tell her about the trip. Nora let out a sigh of relief and told him that she loved him and to be very careful. Once she hung up, Rayne called for her mom, crying.

Nora: "What happened?"

Rayne: "I'm trying to take a shower; the water is hovering in mid-air, shimmering like glass. My tears joined and started swirling around me, suspended in motion, and I think I was controlling it. Mom, what's happening to me?"

Nora: "The ability that has awoken for you is called hydro kinesis. That's the ability to manipulate anything containing water.

As for Darren, he has electro kinesis. It's going to be interesting to see what the other two have."

Rayne: "Do you or did dad have this?"

Nora: "The wolf tendencies he had, as for the hydro, I don't know." *thinking* "I have to research the Cross bloodline to see what additional abilities they might have had."

At the Heart property, Louie and Cross checked each other's wounds. Blak Hart was almost completely healed, while Cross wasn't too far behind.

Cross: "So, what are we gonna do now?"

Blak Hart: "Make copies of all the files, take all the gear and weapons, and head back."

Cross: "How do you think the others made out?"

Blak Hart: "I have a feeling that their job didn't go smoothly, but I have faith in them."

Cross: "What's with you turning into flames all of a sudden?"

Blak Hart: "This is the first time that it got like that. I've had it before, but never like that. Oh sure, let's not forget the fact that you were a werewolf for a short period of time."

Cross: "Well, my father was nicknamed Wolf for a reason, now you know why. But your father wasn't called fireball, so what's the deal?"

Blak Hart: "Ark most likely knows the reason; I think it's time you two meet. I'll arrange it for the next meeting."

Cross: "You think he'll mind?"

Blak Hart: "After what just happened, he doesn't have a choice. While I call Ark, you need to call your mom and let her know that we're gonna be late. Tell her we're going to meet with Ark for information on the files we found and the confrontation we just had."

Cross: "I hope she doesn't get upset."

Blak Hart: "Your mom is cool as hell; she understands what happens in this line of work."

Cross got a call from Rayne. She just wanted to know how he was and told him everything that transpired. After the call, he looked at Blak Hart and then told him what was said. Louie told Cross that he knew and informed him that Ark wanted them to call the minute they got to the restaurant.

Blak Hart: "Tomorrow, we must all start training so we can learn how to use our newfound abilities properly."

Cross: "It's not going to be easy, but I think it'll be interesting to see how we all can use it in battle together."

Blak Hart: "The hardest part is using it without hurting anyone on our team."

Cross: "For now, all we have to worry about is getting back to New York without any incidents."

They got into the car and prepared for the three-hour trip. The weapons were ready to go, just in case. As they got onto the highway, lightning danced across the horizon. Blak Hart looked at Cross as they stared at it. A chill went up their spines, knowing this wasn't the weather, it was a warning. To get his nerves in check, Cross read more on Rampage and compared it to Wolf's and Hunt's notes. So far, it was very informative, but the notes weren't matching up with the file.

Cross: "In the notes, it says that Rampage doesn't attack unless he's provoked."

Blak Hart: "Hard to tell from what we saw. So that means he's probably working for someone."

Cross: "Who? Also, what is he looking for?"

Blak Hart: "Both are good questions, hopefully, Ark has some answers for us."

They continued to drive on. Out of nowhere, several hell hounds rammed the car, almost causing an accident. Blak Hart swerved and told Cross to get his guns out and start shooting. Cross then commenced to shoot the hell hounds, aiming with precision, killing them. Afterward, they proceeded to the restaurant.

Blak Hart: "I guess whatever we're stumbling on has to be pretty big, so big that they're trying to take us out of the equation."

Cross: "Why are we targets?"

Blak Hart: "We are probably the only ones that can stop them and whatever they're planning. This is why I keep meeting Ark; with his help, we'll have an idea of what we're up against."

Cross: "Do you have a game plan for this?"

Blak Hart: "First, find out what is after us, then plan out an attack strategy. And before all that... eat, of course."

Cross: "Who taught you how to plan and fight?"

Blak Hart: "My father was beginning to teach me before he died. Shortly after his death, I moved around a lot. I had to go from town to town, trying to survive on my own to stay away from being placed in a foster home. I found odd jobs in dojos, several police agencies, fire departments, and some military bases by cleaning up, warehouse chores. In turn, they would pay me and show me defensive tactics along the way. When I thought I had learned enough from an area, I moved on. On one of my journeys, I was walking into a government facility to ask for employment when I first met up with Ark. He took me under his wing and believed me when I told him how my parents died. The rest I learned from him. So, this odd relationship we have is more like father and son. Just don't tell him I said that. We have each other's back, but we get on each other's nerves occasionally."

Cross: "Wow, that explains a lot. So, are you taking us under your wing?"

Blak Hart: "It seems like this is the way it was supposed to be. I'll lead you guys, and we all work as a team to rid the world of the demon scum of the Earth, or just keep the balance. Some shit like that."

Cross: "Do you really think of us as a possible team?"

Blak Hart: "The way shit happened back there, I have no doubt in my mind that we can pull it off."

They finally arrived at the diner. Blak Hart called Ark and told him that they just got there and would be waiting in the same booth as last time.

Ark: "I want you to try out one of my spots. Head over to Riverhead next to the big mall and wait for me there. Unless you know the diner that's next to an auto body repair shop. If you do, then go there. If I'm not there, just wait for me at the mall entrance. It might take me an hour, just be patient."

Blak Hart: "Alright, no problem. We will meet you at the diner. While we're waiting, we'll just look at the files we got from my parents' property."

Ark: "Well, just fill me in on what you two have found."

Blak Hart: "You got it. See you in a few. Later."

Cross: "What did he say?"

Blak Hart: "Change of plans, he wants to meet at a new location. Knowing him, the food there must be beyond a four-star quality. He's very picky about where he eats and the quality of food."

Cross: "Damn, so we'll be eating like royalty tonight."

They got to the diner and found a booth. They sat down while waiting for the waitress.

Cross: "I'll be back; I have to use the restroom."

He went to the bathroom, and as he walked in, a man in a suit came from behind him and put a gun to his back.

Strange Man: "Do as you're told and you won't get hurt. You yell, you die."

Cross: "You do know that I didn't come here alone."

Strange Man: "I know you're not. Don't worry, you're not the one I came here for. You are just my leverage for a mini meeting with Blak Hart. Nowadays, it's so easy to get to someone by threatening or attacking a family member or a friend of the person you are targeting."

The strange man came out with Cross in front of him. The man pushed the gun to Cross's back, reminding him.

Strange Man: "Remember, say one word and you're dead."

Cross nodded his head as they continued to walk toward Blak Hart. Louie looked at Cross and gave him a warning. Blak Hart charged at the strange man, and then a shot was heard. Cross pushed the gun to the side and yelled, warning him. The stranger pulled the trigger, missing Cross but hitting Louie in the shoulder. Cross rushed to Blak Hart as Blak Hart continued charging at the stranger. The stranger, with a wave of his hand, sent Louie flying to the other side of the room. Cross tackled the stranger, only to be thrown several feet away. Right as he was about to hit Cross, Blak Hart got up. His head was on fire, his hands once again engulfed in flames, and every drop of his blood that fell traveled directly to his hands, creating an iron gauntlet. Before Blak Hart could get to the stranger, Ark appeared from thin air and uppercut the stranger, sending him crashing through the diner entrance door.

The stranger returned and commented on Ark’s hit.

Strange Man: "Nice hit, who are you? No mortal has that kind of power. Unless you're..."

Ark ran toward the stranger and continued hitting him, demanding answers.

Ark: "I am Ark, and what is your purpose here? Aren't you an angel?"

The stranger looked at Ark and told him they would meet again real soon before disappearing in a burst of light.

Blak Hart and Cross looked at Ark.

Blak Hart: *clapping* "Way to go, Ark. What do you mean that guy is an angel?"

Cross: "What? Are we fighting angels too? I thought they were on our side."

Ark: "Angels ARE on our side. I just had to step in. You two aren't ready yet to fight someone with such power."

Blak Hart: "We weren't doing that bad." *feeling a bit underestimated* "Oh, yeah, by the way, Cross, this is Ark."

Ark: "Louie, Michael, try the cheesecake. The taste is heavenly."

Cross: "Is the owner going to be upset that the diner got damaged?"

Ark: "I know the owner personally. By the way, the stranger you met is Amon. He is a very powerful angel; he has defeated over forty legions of demons. He is usually hanging around an angel named Chax. These two were about to be cast out but somehow proved their loyalty to the king. This is the first time that I know of Amon being on Earth. Why would he be here? Why is he coming after you?"

Blak Hart: "This is shit you should know. Why is an angel attacking Cross to get to me? Unless..." *he began to think of a theory* "Maybe he is secretly working with the enemy and part of his job is to take us out or take us to someone who can force us to switch sides."

Cross: "How's your shoulder?" *he stares at the diner's broken door* "I wasn't scared of the gun, I was scared of being the reason you got hurt."

Blak Hart: "I heal fast, remember. I'm doing good."

The waitress came by, took their order, then left. Blak Hart began to ask Ark for information on the files, Rampage, and what his real name was.

Ark: *before they started talking about Rampage, he wanted to address Louie's theory* "That's an interesting hypothesis. I'll look into it. At this moment, it's not important." *the conversation shifted directly to Rampage* "Rampage is usually a calm individual; he doesn't usually fight unless he is provoked. If he attacked you, then that means he is under someone's control. Get Collin, squeeze him for information, then we can go from there. Remember, he has abilities of his own, don't get reckless. Train hard, then, after you gather up all the intel, make your move. This mission is top priority."

Blak Hart: "All this talk of work is getting me hungrier, what else do you suggest?"

Ark: "Everything here is good. Whatever you can think of, they will make. Cross, as it seems, you will be Blak Hart's right-hand man, so prepare yourself. Just like Albert, Louie is a handful."

Cross: "I know, but I know that no matter what happens, he will help us."

Ark told them that he had to go but left several files on the table. He notified them that these files would inform them about several bounties that were to come. They needed to train harder than ever because, from now on, they were targets for anything evil.

Blak Hart: "Don't think that you're gonna get away without answering my question. I want to know what your real name is, not your code name either."

Ark: "What's the rush? I will reveal my true self to you in time. Would knowing who or what I truly am change how you think of me?"

Blak Hart: "No, but I just want to know. After all we've been through, why is it so difficult for you to tell me?"

Ark: "It isn't the right time for me to let you know. When I do, it will be when I battle by your side."

Blak Hart: "When is that?"

Ark: "Just eat your food and shut up." *smiles at Blak Hart*

Ark waved farewell and walked out the door. Blak Hart, feeling a bit annoyed that Ark never answered his question, began to eat.

Blak Hart: "Damn, the food is incredible. Ark wasn't lying when he said the food would be heavenly."

Cross: "I don't know what you two were disagreeing on, I was too busy enjoying the food. Who's paying for all of this?"

Blak Hart: "Damn it, AAAArrkkk! I guess I have to."

Blak Hart went to pay for the food. The cashier told him that it was on the house. Cross told Blak Hart to go before they changed their mind. The cashier explained that for hunters of their caliber, the only way they could show their appreciation for Cross and Blak Hart's work was by giving them the best service and not charging them. Blak Hart and Cross both thanked the cashier and proceeded to head back to the Cross household.

They finally arrived at Cross's house at three in the morning. Cross noticed a dark figure moving around in the backyard. He told Blak Hart to check in on his mother and sister as Cross went to check it out. While Louie went to check on Cross's family, Cross sneaked around back and saw a tall smoke figure looking into the house. He ran toward it and realized that the apparition was a diversion. He yelled to Blak Hart.

Cross: "Beware, someone or something is about to attack you!"

Blak Hart: "Thanks for the heads up."

While Blak Hart searched the house for the supernatural, Cross rushed inside. More dark figures manifested themselves, this time they were solid and much bigger than the ones previously seen. They tried to detain Cross by attacking him with a series of hits, throwing branches, rocks, and anything they could find. Blak Hart heard Nora chanting and repelling the entities. Blak Hart called to her.

Blak Hart: "Nora! We're back, tell me where you are so we can help you!"

Nora: "Tell Cross to go and protect Rayne, I'm in the laundry room. There are quite a few here, please be careful when attacking them. They seem to absorb energy. Don't use too much energy on the attacks. If you overuse your energy, they'll drain you."

Blak Hart: "Don't worry, I'll handle it."

Blak Hart called Cross but got no response. He called out again, but nothing. He sprinted to the back and saw Cross covered in fur, clawing his way out of the hoard of shadow beasts. Right as Blak Hart darted over, Cross finished off the last one. Blak Hart informed Cross of the situation.

Blak Hart: "We have to hurry up, the house is infested with specters. Your mom has been trying to protect Rayne from them, but she's getting drained. She needs our help. Go to Rayne and help her while I go help your mom." *he informed Cross on what Nora told him*

Cross: "After I help Rayne, we'll come back and assist you and mom."

Blak Hart: "Sounds like a plan."

Cross ran up the stairs to help Rayne while Blak Hart helped Nora out of the house. Nora tried to put a barrier around them, but she was too drained. Blak Hart tried to make his own barrier, but it didn't work, so he started running as fast as he could, kicking any specters in his way. Unexpectedly, there was a loud crash coming from the back of the house. Blak Hart ran back inside to find out what the commotion was all about. He looked out the window and saw Cross and Rayne, covered in blood, getting up from the ground.

Cross: "Rayne, are you ok?"

Rayne: "I'm fine, how about you? You're bloodier than I am."

Cross: "I killed off a little more than you did. I'm healing pretty quick."

Blak Hart: "How many are left upstairs?"

Cross: "Not many. If you want, we can help you."

Blak Hart: "Nah, I'm good. Just go and protect your mom and let her know you two are fine."

Blak Hart went upstairs to look for the remaining entities. From the looks of it, Nora's chants handled the rest.

Blak Hart: "The house is clear; you can return inside."

Blak Hart: "Aawww, did I mess up your plans? Well, suck it up, let's end it now!"

The greenish-gray being tackled Blak Hart to the ground and told him:

The Greenish Gray Being: "This is the last time you interfere with Zima's plan."

Blak Hart: "I could've taken him." *he clenched his fists* "Every time we run, we lose a piece of ourselves. I'm done running."

Nora: "Actually, not at this time. I sent him away for now, but he will be back. He's been after us for several decades. All I can do is send him away, but eventually, he will find a way to counteract it. Wolf and I have never been strong enough to defeat him."

Blak Hart: "Why didn't you two and my parents attack him together? The four of you as a unit would've been able to destroy him."

Nora: "We tried and failed. I'm at wit's end trying to research anything that would keep him away permanently."

Cross & Rayne: "Together we can do the job."

Blak Hart: "Let's go to the office and talk about what you know so far and check out the files to see if maybe there is anything that can help us."

In one of the many chambers of hell, Amon spoke to a high-level demon. He yelled at him.

Amon: "Send someone more powerful! How could a bloodline of the Heart family, two teenagers, and an old lady defeat your hordes? This is unacceptable!"

The Demon: "You shouldn't even be here, so if you value your position in heaven, you better talk to me with respect. How dare you come to my domain and try to tell me how to handle my minions? Why don't you take care of them yourself? Oh yeah, that's right, you couldn't in the diner." *laughing* "Are you confused, fake-ass, lazy chicken-shit of an angel?"

Amon: "I could smite you. When you talk to me, remember who I am. I am your God's right-hand man, so watch what you say to me."

The Demon: "How long do you think it's going to last? You're just another plaything. You're so pathetic."

Amon: "You forget your place, demon. I am his loyal student until I draw my last breath."

Another Demon: "Are you two bitches done fighting? Mardus, why are you wasting time on him? Just get back to work!"

Mardus: "You're right, Zima. He's all yours. Have fun. I've got work to do."

Mardus opened his wings and flew off as Amon walked toward Zima and began to yell at him.

Amon: "WHAT HAPPENED UP THERE? YOU'RE USELESS! HOW COULD YOU BE BANISHED SO EASILY?!"

Zima: *laughing* "HAHAH!! I'm assuming you've never fought against the Wolf Queen before?"

Amon: "WHY DOES THAT MATTER? IT WAS YOUR JOB TO TAKE HER AND THOSE BRATS OUT!!!"

Zima: *smirking* "Relax, all in good time. I was just testing to see what they can do. Now stop your bitching and go see your master before I lose my patience."

Amon: "YOU DARE THREATEN ME? DO YOU KNOW WHO I...."

A dark voice was heard that made everything shake and shattered anyone's soul. The voice didn't echo, it devoured. Every syllable was a command etched into bone.

Dark Voice: "AMON, STOP YOUR YAPPING AND COME TO MY CHAMBERS NOW!!!!"

Amon: "Y-yes, my lord."

Zima: *sarcastically* "Ooooooo, I think he's mad." *laughing at Amon* "Get ready to bend over."

Amon: "Silence!"

The following day, at Nora's house, Nora, the teens, and Louie were in the Wolf's Den looking around for any information that would help them.

Darren: "I think I've found something. It looks like an old book that has a list of demons along with their abilities and weaknesses."

Nora: "From what Wolf told me, it updates when new and different entities are created or appear from the depths of hell once blood is spilled onto the pages. Let me see if that one might help us out. I'll call Nite to see if he has any additional information for us."

Blak Hart: "While you guys are researching, I'll see what information Ark might have. Maybe Nite and Ark could compare notes and hopefully find something that will help us."

Rayne: "I'll look online. The internet always has answers. Not always the right ones, but it has answers nonetheless."

Cross: "Well, since everyone has interesting ways of finding info, I'll keep looking around with Darren."

Blak Hart and Nora called their contacts. They told them that they needed to meet up at the house ASAP.

Nite: "Nora, I'm sorry, but I must decline. I can't meet you when Ark is there. You know that, he has no knowledge that I'm working with you."

Nora: "You two will have to come clean sooner or later. For the recent events, we need you both."

Nite: "Fine, see you then."

Ark: "I'll be there, is Nite going to be there?"

Blak Hart: "Most likely, does it matter, and who is he to you?"

Ark: "It doesn't, and he's my oldest brother. He doesn't think that I know his code name, so this will be very interesting."

Blak Hart: "Well, this will be the first time I meet Nite anyway, so what better time to meet him? We need you both for the shit that's about to go down."

Nora & Blak Hart: *both saying the same thing at the same time on separate calls* "Come by on Wednesday at 7:30 p.m. Everyone will meet in the Wolf's Den. No one will be able to listen in or attack, and if someone tries to attack, they won't survive."

Darren and Rayne took a break from the Wolf's Den and headed to the dining room. They talked about how they would take care of each other no matter what. They were about to have their first kiss when they heard someone rummaging in the rooms upstairs.

Darren: "What was that?"

Rayne: "I don't know, but I'm gonna find out."

Darren: "Wait, I'll go with you."

Rayne: "This isn't the whole 'being my hero' shit, is it?"

Darren: "Well... no, I'm just going with you, in case you need a little help, I'll be right there."

They slowly went up the stairs and heard someone throwing things around. It seemed like someone was looking for something.

Rayne: "What the hell are you doing here?!"

Rampage: *sniffing* "Another of Wolf's pups."

Rayne: "You must be Rampage then?"

Darren: "Why are you here!?"

Rampage: "Heheheh. I guess you've heard of me. As for you..." *walks toward Darren* "I should ask, why are you here, boy?"

Darren: "What do you mean, why I'm here?" *gets into a stance*

Rayne: "He's one of us, so shut your mouth!"

Rampage: "Hehe, for now. Now, let's play this game, starting with Wolf's pup. I owe your father and brother some payback." *cracks knuckles*

Rampage started charging at Rayne, and her claws and teeth grew and sharpened. Meanwhile, Darren ran toward Rampage, throwing some punches with lightning around his fists. He hit Rampage's cheek, sending him flying into the backyard. Rampage retaliated by giving Darren a swift punch to his stomach and throwing him to the side as Rayne jumped on Rampage. She clawed and bit him, and Rampage screamed in pain.

Rampage: "AAAHHHHHHH!! NOT THIS TIME!"

Rampage grabbed Rayne by her neck and started applying pressure. As she gasped for air, Rampage let her go, and his eyes changed color. He put his hands on his head and began to scream.

Rampage: "NO, NO, NO! I WON'T HURT HER!!"

Rayne: *coughs* "What is his problem? DARREN, ARE YOU OKAY!?"

Darren: *throwing up* "That bastard is going to pay."

Darren stood up and charged at Rampage. He started hitting Rampage with a storm of punches, sending him flying. Rayne jumped up and hit Darren in the back of his head.

Darren: "Ooowwww! What was that for? He tried to kill you!"

Rayne: "And he could have, but he let me go. There's something wrong with him."

Rampage staggered, standing up and grabbing his head. He screamed as Blak Hart and Cross ran to the backyard, followed by Nora. Cross and Nora ran up to Rayne and Darren as Blak Hart stood in front of them just in case Rampage attacked. Louie got in his fighting stance.

Cross: "Are you guys ok? Did he hurt you, Rayne?"

Rayne: "He tried to choke me, but his eyes changed color and he let me go. Something's wrong with him."

Blak Hart: "Check this out, guys."

Rampage started slamming his head against the pavement, cracking his mask.

Rampage: "NO, NO! I WON'T HURT ANYONE ANYMORE, GET OUT OF MY HEAD!!!!"

Cross: "What the hell, he wasn't like this before."

Blak Hart: "He sounds like he's in pain. Let's try to see if we can help him."

Rampage: "GET OUT OF MY HEAD, RAAAAAAAR!!!"

Rayne: "I feel so bad for him."

Darren: "We should still be careful!"

Rampage: *crying* "I DON'T WANT TO KILL ANYMORE, PLEASE STOP IT!!!!"

As Blak Hart walked over to help Rampage, Amon appeared and grabbed him, then disappeared. While everyone was confused about what happened, Amon appeared again. He grabbed Darren.

Amon: "Continue to stick your nose in other people's business, so for the time being, I will borrow Darren. Ta ta. See you soon." *winks and disappears*

CHAPTER 4.
SOUNDS LIKE MUSIC, DOESN'T IT?

Amon arrives in purgatory with Darren. He throws him over to Zima, and Zima drags Darren to a chamber in order to have him unlock the power of his father's bloodline. Zima has an agreement with Amon to convince Darren to convert to their side.

Amon: "You know what to do, Zima. Don't fail me this time."

Zima: "Yeah, yeah, just give the boy to me and stop your bitching."

He chains Darren up and starts chanting as a small flame, in the form of a pentagram, starts to surround Darren. He begins to scream in pain.

Darren: "AHHHHH! WHERE AM I? WHAT'S GOING ON!!!?"

Zima: "Look who's awake." *taunting*

Darren: "You're the one that tried to kill Rayne."

Zima: "Wow, you're good at this. Think about it, if I really wanted her dead, she would have been already."

Darren: "Shut up! What are you doing to me!?"

Zima: "Unlocking the gifts your father gave you."

Darren: "My father? How do you know my father?"

Zima: "Aww, your mommy never talked about me? I'm hurt, truly I am..." *acting like he's hurt, but looking at him and laughing*

Zima starts drawing symbols all over Darren's body with his blood. The blood starts to burn Darren like acid. He screams in agony as blood pours from his eyes. His eyes turn black. Zima reaches inside Darren's chest as he throws up a black, slimy substance and passes out.

Zima: "No, you fucking don't. You better stay awake. The fun just started." *slaps Darren with his other arm* "Wakey, wakey, bitch; the party hasn't finished yet."

Darren: "What do you want from me?"

Zima: "What do I want? Hehe, I want to hear you scream. Squeal like a pig, bitch, squeal!" *malicious laughter*

Zima squeezes Darren's heart, causing him to throw up and scream in pain.

Zima: "There it is, sounds like music, doesn't it?"

Zima begins to pull out a black blob and chants in a deep, distorted voice as the black mass starts to cover Darren, entering his eyes, ears, and mouth. Darren passes out from the pain. Zima tightens the chains and then shackles him, whipping and ripping his flesh until Darren passes out again. Zima wakes Darren up by ripping off his fingernails and beating him close to death.

Zima: "Look at all the fun we're having."

Darren: *coughs up the oily substance* "What did you do to me!?"

Zima: "I'm giving you a makeover, just kidding. You don't listen very well, do you?" *kicks Darren in the face* "I unlocked your true power, the one your mommy was trying to suppress from you.

It's time you join us and stop playing with those Cross brats and that bald guy."

Darren: "His name is Blak Hart, and he is my teacher." *coughs*

Zima: "I DIDN'T ASK YOU HIS NAME!!!" *kicks Darren rapidly in the stomach and face* "It's funny you call him your teacher. You never wondered why he chose you to train in the first place. He just wants to keep an eye on you. He doesn't trust you at all."

Darren: "SHUT UP! YOU'RE LYING!!! HE TRUSTS ME AND SO DOES RAYNE AND CROSS!!!"

Zima starts beating Darren inches from death, then lets several demons commence to bite and claw chunks of flesh from his body as Darren screams in pain. Lightning starts coming out of his body, burning some of the demons off of him.

Zima: "HAHA! THAT'S WHAT I'M TALKING ABOUT!! I almost lost faith in you, kid. Looks like you have some balls on you, just like your old man."

Darren: "I'M NOTHING LIKE MY FATHER!!!"

Zima: "HAHA! You're right, kid. You could be so much more than your father if you were to join us."

Darren: "I WILL NEVER JOIN YOU!!!!"

Zima: "You're stubborn, I give you that, but..." *drives a dagger into Darren's stomach and twists it* "that won't last long."

Unknown voice: *in a seductive voice* "What's going on over here? Sounds fun."

Zima turns around, and there stands a demon, looking like a human woman. She's very curvy and beautiful with black and red hair. She steps from the shadows, her voice like velvet over broken glass. She wears a transparent dress. Her lips are crimson red as she walks up to Darren and Zima.

Zima: "What are you doing here?"

The demon woman: *in a seductive voice* "I was just going for a walk, and I felt this energy out of nowhere, so I just had to find out where it's coming from. Such power in such a fragile vessel."

Zima: "Shouldn't you be in your chambers, not strolling around looking for something to fuck?"

The demon woman: "Shut up, Zima. I can do whatever I want. Plus, I wouldn't want to miss out on all the fun." *smiles* "Who do we have here?" *bites her lip*

Zima: "Don't you worry who he is, and no, you can't fuck him. Now, get out of here!"

The demon woman: *pouts her lips* "Why do you get to have all the fun? Maybe I could help you, Zima. Come on..." *she wraps her arms around Zima, then brings her lips close to his* "I will owe you one."

Zima: "No, not this time. I got punished the last time. Not happening again, bitch."

The demon woman: "But I gave you what you wanted, didn't I?" *she starts licking her lips*

Zima: "Grrr, fine, but don't get used to this. I gotta get something anyway. I won't be gone long, so make sure you're done by the time I get back."

The demon woman: "Thank you, Zima. You'll have a nice gift coming your way." *in a seductive tone whispers to Zima* "Don't worry, I'll take great care of your experiment."

Zima: "You better!" *smirks*

Zima fades away as the demon woman approaches Darren in his passed-out state. She walks around him, examining his body, and starts running her finger down his back, waking him up as he jolts up.

Darren: "AAAHHHHHH! ZIMA, I WILL KILL YOU!"

The demon woman: *seductive tone* "He left. It's just me and you..." *giggles*

Darren: "Who the fuck are you?"

The demon woman: *giggles and pulls Darren down* "Maybe if you're a good boy, I will give you my name."

Darren tries to stand, but his body is horribly wounded. The black slimy substance is leaking out of his wounds. The demon woman pushes Darren onto a torture table and sits on him, not caring about his wounds and broken bones.

Darren: *yells in pain* "AAHHHHHHH!"

The demon woman: *in a moaning voice* "Yes, scream for me, baby."

Darren: "GET OFF OF ME, BITCH!"

The demon woman: *bites her lip and moans* "Mm... mm... And you talk dirty. Let's have some fun."

The demon woman's eyes start to glow as she rips what's left of Darren's clothes. He tries to push her off, but doesn't have the strength. The demon woman ties his arms and legs down on the table, then starts biting his neck, drawing blood, making him shout in agony. She starts straddling the helpless hostage. Darren cries out for her to stop, but she just goes faster and scratches him some more. She groans loudly as she starts transforming. Her eyes turn black, and bat-like wings sprout from her back.

Darren tries to get her off him again. *He breaks the restraints.* This time, he gets her off. He tries to get up to run, but his body is too broken. The demon woman grabs Darren by his legs and drags him back to her.

The demon woman: "Hehe, where do you think you're going? I'm not finished with you."

Darren: "Please stop, I don't want this."

The demon woman: *giggles* "Awwww, you really think I care what you want? That's cute."

The demon woman steps on Darren's genitals and stomps on them with her stiletto heels. He lets out a scream of pure pain, then she chains him down and continues from where she left off, riding him harder. Darren lets out a blood-curdling scream as the demon woman chomps down on his neck, gushing blood. She whispers in his ear.

The demon woman: "My name is Lust." *she giggles*

Darren passes out from the pain and blood loss. Lust jumps off his body, unchains him, then kicks him to the side, sending Darren's body flying like a rag doll.

Lust: "I wasn't done. Wake up!" *pouting*

Lust starts kicking him in the crotch over and over, but he doesn't wake up. Zima comes back, seeing Lust kicking her toy.

Zima: "Lust, really? What did you do to him? He's out cold?"

Lust: "I was just starting to have fun, and he passed out on me. Why bring such fragile creatures here when they can't enjoy the fun?" *pouts*

Zima: *sighs* "Can I get back to work now? Our lord called for you."

Lust: "Uh!" *rolls her eyes* "What does he want now?"

Lust soars off as Zima chains Darren up and starts chanting in a deep voice.

Zima: "Your mother locked away your birthright. I'm just… unlocking the door."

Back at Nora's house, everyone is trying to find a way to get Darren back. While Nora is calling Nite, Louie is calling Ark, and Rayne is calling Gail. The three of them arranged an emergency meeting at the house. They explain what happened, then try to find a way to hold the meeting without being detected by their enemies.

Blak Hart: *on the phone talking to Ark* "We have to get him back."

Ark: "Relax, grasshopper. When Nite and I get there, we'll figure something out."

Nora: *on the phone with Nite* "Poor kid doesn't have a clue on what to do in a scenario like this one."

Gail arrives, her eyes filled with tears. She's so nervous, words can't be spoken as Nora runs to her and hugs her, explaining the current situation.

Nite: "Don't worry, Nora and Gail. When my brother and I get there, we'll make a plan on the spot."

Nora: "Ok, thank you."

Gail: "I can start using my crystals to search for his location."

Nora: "I'll assist her while you guys do whatever you can."

Hours later, Ark arrives with a blonde-haired, six-foot-five husky man in a white suit. He greets Nora in the same manner Ark greets Louie. Ark tells Louie that Nite has known the Cross family as long as he's known the Heart family.

Nite: "It's been years, Nora. How have you been?"

Nora: "Same as always, just getting older and bored. How about you?" *talks to Nite while helping Gail*

Nite: "Heavenly. I hope my brother's connect hasn't been a nuisance for you."

Nora: "No, actually, he's been very helpful. He's like another son, only he's just a little high-strung."

Ark: "Get ready. My brother is about to start trying to take over everything."

Blak Hart: "Maybe he's not focused on what you're doing. Remember, it's for the two of you to help us take care of our demon problem."

Nite: "Ark, let's get this meeting started. Nora, if it's alright with you, can we have it in the Wolf's meeting room?"

Ark: "Lou, you see what I told you. He's already starting."

Blak Hart: "Be happy that it's temporary. After the meeting, we'll have our own little game plan, aside from the one that Nite makes. What do you think?"

Ark: "Sounds like a great plan. Also, I have news that I think you're not going to like, but you're the only person I could trust with the mission."

Blak Hart: "What is the mission, and why am I not going to like it?"

Ark: "Later, Blak Hart. Later."

Nite calls everyone to the meeting room. He goes toward the board to get a plan going. Ark looks at Nite angrily and listens. Nite calls Ark to go up so they can both make the plans together.

Blak Hart: "You see, he wants you to make the plans with him."

Ark: "That's a first, but I might as well try."

Nite: "Zima is working with Amon. Amon has been seen talking to Zima. This is very alarming because Amon is an angel. He is not supposed to be on Earth, and he's not allowed to be anywhere near any demons."

Blak Hart: "How would you know that he's not allowed to be around him? What are you? An angel?" *laughs* "Anyway..."

Nite: "Shut your mouth, don't interrupt. Ark, control your pupil before I do it for you!"

Ark: "Real professional, Nite. First off, control your attitude while I do my job. Secondly, do yours and continue with the planning."

Cross: "CAN YOU TWO SHUT THE HELL UP? MY FRIEND JUST GOT TAKEN, AND WE NEED TO GET HIM BACK NOW. IF THE BOTH OF YOU CAN STOP BICKERING LIKE CHILDREN, LET'S GET THIS GOING!!!!!"

Cross punches the wall by the door and marches out of the meeting room, with Rayne following behind.

Rayne: "We can try to find him ourselves if they keep acting this way."

Cross: "Let's just listen to what they have to say. Right now, we have no way of knowing where he could be."

Rayne: "After the meeting, then we go?"

Cross: "First, let's get through the meeting, then I'll decide."

Blak Hart gives the brothers all the intel that he has. Ark starts to make the plan for the Cross family and Blak Hart to do. As for the brothers, they said they would take care of the rest.

Blak Hart: "Ark, are we gonna get Darren or not?"

Ark: "I don't think you're ready for demons of that power. You and the teens will be my distraction while Nite goes in and destroys everything in his path, giving me the diversion I need to go get Darren."

Blak Hart: "Sounds good."

Ark: "Rayne, Cross, you two and Blak Hart are going to be a distraction in order for Nite and I to save Darren. The escape route will be cleared by all of you. We don't know the condition he's going to be in. The faster we save him, the quicker we can attend to his medical needs."

Gail and Nora let everyone know that they found out the exact location. Nora reminds them that she has a separate room for medical needs. Gail also notifies them that now would be the best time to attack. They head to the woods next to the school. Right before getting there, three low-class demons are guarding the entrance. Proving that they are going in the right direction, the group charges toward the demons. Rayne and Cross transform slightly, then attack the demons full force. Blak Hart attacks the demons blocking the entrance. Now, the brothers are able to look for Darren with no interruptions.

Ark and Nite search all over for Darren. They send in Louie to try to look for him, since he has an unorthodox way of getting the job done. Outside of the chambers, Rayne, Cross, and Nite fight off hordes of demons. Group by group, they continue to fight as if they have been warriors for many years. Rayne slices the demons, Cross rips limbs off, while Nite makes sure the demons never rise again.

Nite: "Your father would be proud of you two. This is exactly how your parents started."

Rayne: "Thanks, I guess. What happened to Blak Hart? What's taking so long for Darren to be found?"

Cross: "As much trouble as we're having, I can only imagine how much trouble they're having. Hopefully, Blak Hart was sent to help find him. He's relentless when he has a job to do."

Nite: "I know he's good, but I don't know why you have so much faith in him."

Rayne: "We both have seen him get his ass whooped, then come back and somehow pull off an unbelievable upset."

Cross: "Well, shit happens to all of us. Just be patient, I know they'll find him."

Back at the chamber, Ark and Blak Hart fight their way to the torture cells. They walk past each one until they come across a cell guarded by two huge, grotesque-looking demons. The demons have electrified pipes that scorch the insides of anyone they shock. Ark charges at the first guard, and as the second guard tries to attack him, Blak Hart tackles him into the stone wall. Louie swings and misses, then the demon swings at him, striking the side of his head with the pipe, shocking him. Louie tries to get on his feet but continues to get shocked. The demon swings his claws, slicing Louie's jugular vein, causing more shocks. Louie is knocked unconscious and bleeding out. The demon laughs over and over as Blak Hart loses blood at a rapid pace. With every second that passes, Blak Hart is getting closer to death.

Second demon: "I'll continue to beat you until there's nothing left of you. No one is here to save you, Blak Hart."

While the demon beats Blak Hart, watching him bleed out, Ark frees Darren. He introduces himself to Darren and tells him to run while he holds the demons back. Darren runs as fast as he can, but he sees Blak Hart being dragged to a cell. Darren attacks the demon by shocking him with more electricity than the pipes can handle.

Darren: "Thought I wouldn't get out, did you? Now, let's see how you like to get beaten."

Darren pummels the demon into a bloody mess. He smacks Blak Hart to see if that would wake him up. Blak Hart doesn't respond. Darren shakes him, but still nothing. Ark finally reaches Darren. He picks up Blak Hart, and they head out. Before leaving, Darren takes one of the horns the demon had on his head to remind himself of the torture he endured.

Darren: "He's not getting up."

Ark: "He'll heal; the loss of blood is causing his healing to slow down. This is the worst he's ever been. I must take him to my facility to help him heal."

Darren: "No matter how his injuries occurred, he'll never be the same again. It doesn't matter how well he heals up, it's the mental injury or his ego that might cause him to change. Are you sure he'll be alright?"

Ark: "You're right, that's why we must be there to help him the best we can. We must remind him that he's not alone anymore, he has a family now. Also, remember, he is very resilient."

The three return to the group. Rayne runs up to Darren and greets him with kisses and hugs, while Cross walks up to Ark.

Cross: "What happened to him?"

Ark: "Well, he got overconfident and got his ass handed to him. If it wasn't for Darren, he wouldn't be here right now."

Nite pulls Ark to the side to have a word with him. He gives Ark the look of disappointment.

Nite: "Ark, why didn't you watch him? You should know better. You know he's too hot-headed. Why do you keep putting all your effort and time into him when he'll disappoint you in the long run? I'm not telling you that you're doing a bad job, I'm just letting you know that you should choose another champion. Because this one is going to be the death of you."

Ark: "No matter who I choose, you'll find something to complain about. I chose him because I see potential, whereas others seem to be the champion but are too stubborn to have an open mind to learn. Yes, he's hard-headed, yes, he can be a handful. But he's always willing to learn, and even though we don't see eye to eye every time, there is always a certain level of respect."

Nite: "You've been on Earth way too long, brother."

Rayne: "When will Blak Hart be back?"

Ark: "Right now, we must wait for his wounds to heal, then we'll know." *puts his head down*

Nite takes the teens back to their house. Gail is there with Nora, waiting for them. Nite explains that Louie was injured, and he would be returning once he's fully healed. Nora runs to Cross and Rayne to hug them, while Gail sprints to Darren.

Gail: "How are you feeling? What have they done to you?"

Darren: "I'm fine for now. I don't want to talk about my experience. I just want to go home and sleep. Cross and Rayne, keep me posted."

Rayne: "I will." *blows him a kiss while blushing*

Gail: "Nora, see you next time. Once again, thank you for everything."

Nora: "You're welcome, Gail. Rayne, Cross, how are you two? Did anyone hurt you? How bad is Louie?"

Cross & Rayne: "We're fine, we had each other's back. As for Louie, we don't really know. Nite wasn't being too cooperative, and Ark looked very worried."

Nora: "That's never a good sign. I'll call Ark in two hours and find out the real condition he's in, then I'll let you guys know what he tells me."

Two hours later, Ark has called in Vin to assist Louie in healing. As Vin helps clean his wounds and talks to Louie's subconscious, Nora calls Ark.

Nora: "How's Louie holding up?"

Ark: "Not good, he lost a lot of blood. All the electricity to his system burned most of his interior organs. If it wasn't for his healing, he would be dead by now. Right now, all we can do is wait. I thought his healing would have kicked in by now. You probably won't be seeing him for quite some time."

Nora: "Wow, would Hunt have survived something like this? Please keep me posted on his condition."

Ark: "Don't worry. As for Hunt," *he stops for a brief moment, then continues* "he would not have. As for Louie, the simple fact that his body is still fighting shows that there's more reason to have hope. Once his condition gets better or worse, you'll be the first person I call."

Nora: "Thank you, talk to you soon." *begins crying, since she considers Louie another one of her children.*

Vin: "Ark, even his subconscious is in shock. This warrior is going to be out of battle for a while."

Ark: "So what's the diagnosis, doctor?"

Vin: "His body is in what I call reboot. His body is shutting everything down to heal the affected areas. When he comes to, he'll be better than before." *he traces the scars* "This isn't just healing. It's evolution. Louie's body is rewriting itself—like the old guardians used to."

Ark: "How long is he going to be in this state? What do you mean 'better than before'?"

Vin: "It's hard to say, it all depends on when his body fully heals the injured parts. Louie's body is healing in a very rare way. My theory is that once his body fully heals, it will grow an unusual ability to resist electricity, and his skin will probably become thicker to prevent anything like this from happening in the future."

While Ark and Vin help Louie, Nite returns to the site where Louie got hurt. He searches all around until he finds the electric rod. As he picks it up off the ground, he is attacked by three fire demons. Nite disintegrates them with ease. He smiles at what's left of them, then walks out chuckling at the fact that the demons thought they had a chance at defeating him.

Nite: "Maybe if I find out what destroyed his body in that manner, we might be able to find a way to use their weapon against them."

Nite goes to the medical facility where Louie is, and hands Ark the electrified pipe.

Nite: "I apologize for yelling at you earlier. You have to understand, I had someone like him before, and I put all my time and effort into making him an undefeatable warrior. He then turned around and began helping our enemy. I think you know him as Rampage."

Ark: "I accept your apology, but remember, if I do need your help, I will call you. I respect your experience and what you say, but I believe he will be the one to help us when we're in need. All I ask is for you to trust my judgment and guide me in making him the warrior I know he can be. What? Rampage is a product of your training? That explains a lot."

Nite: "Alright, I will assist you only when you ask."

Ark: "Now would be the perfect time." *looks at the pipe* "I could use your help inspecting this. I see ooze seeping through the tube."

A week later, the teens are preparing for Halloween. Nora warns the teens not to be out too late.

Nora: "You three are targets, so please be back before the witching hour."

The teens let her know that they'll be back before that and head out the door. Nora has a bad feeling about how the night is going to end.

Cross: "Damn, Rayne, the makeup work you did actually makes me look like the old-school werewolf."

Rayne: "Well, you did say werewolf, not what kind of wolf."

Cross sees people possessed by demons walking around, thinking that no one sees them. He ignores them so Rayne doesn't realize. Rayne looks at Cross and notices that his eyes are changing to yellow.

Rayne: "No, not now. Can't we have a full week with no evil interference? Remember, we still have to pick up Darren. So, try to relax."

Cross: "Mom had a bad feeling all day, so excuse me for being aware. Fine, I'll try to calm down. It's just that I see how many people are possessed out here. They're trying to portray themselves as regular people. It's sickening."

Darren got tired of waiting and began walking in their direction. As he walks, he senses a few demons walking by him in human form. He acts like he doesn't notice and passes one by. When the demon turns around, he grabs him by the neck, then absorbs the demon from the human. Darren, feeling a little odd, continues on.

Darren: "I have to hurry up and get to them before they get too far from the house. Too much candy for so little time." *he says to himself*

Darren catches up with Rayne and Cross, then joins them in their mission for tasty treats.

Rayne: "Hi, it's about time."

Rayne sees that Darren's eyes are dilated due to recent events. She begins to worry.

Rayne: "Darren, is everything ok with you?"

Darren: "I'm fine, just eager to see you and what the night has in store for us."

Cross: "Dude, since when? You usually try to rush us through the night to head back home."

Darren: "When people have traumatic experiences, some mature faster than others. Sometimes they change how they act or think."

Cross: "Ok, it wasn’t for all that. I’m just stunned that you want to be out longer than usual, that’s all." *a little skeptical about Darren's actions, feeling like he's hiding something*

Darren: "My bad, I haven't been sleeping much lately, and I'm still worried about Blak Hart. Have any of you heard any updates about him?"

Cross: "My mom has been keeping in contact with Ark."

Back in heaven, the angels are meeting to talk about the recent events that have happened.

Head angel: "I see that Blak Hart was injured for being overzealous. Any comments on how the teens handled themselves?"

Third angel: "He just needs to be taught how to survey the area and plan."

Angel trainee: "He needs more training; after that, he should be fine."

Second angel: "Brothers, what kind of training does he need?"

An unknown angel: "He should just be left to die. Why help a knuckle-headed human with no consideration for his group? If it wasn't for Darren and Ark, he would have rotted in hell where he belongs anyway. By the way, who are Ark and Nite? Who assigned them to the teens and Blak Hart?"

Head angel: "It is none of our concern. Our job is to oversee and make sure things don't get out of hand."

Second angel: "Every mortal should have a teacher or a guardian angel."

An unknown angel: "Not everyone is worthy of having an angel, and definitely, Blak Hart is one that doesn't deserve one."

Third angel: "What is your issue with that mortal? Must I remind you that it was his parents that helped you survive a mission?"

Angel trainee: "He has done more to help us in our cause than you have in centuries."

An unknown angel: "Seems like everyone has higher expectations for Blak Hart than they do for me, so if that's the way you guys want to think, then so be it. Don't ask me for anything."

Third Angel: "Not like if we did, you would actually do work anyway."

In the medical facility, Vin continues to work on Blak Hart. The road to Louie's recovery has been slow. Vin checks on his vitals when suddenly Louie wakes up from his coma.

Blak Hart: "Where am I? Why am I here, and what happened?"

Vin: "You are at a top-secret medical facility. You were severely wounded in battle, then Nite and Ark brought you here. Do you remember who I am?"

Blak Hart: "You are Vin, correct? And sorry, I don't remember ever being in this facility. Are the twins okay? And was Darren rescued?"

Vin: "The teens are doing fine. Darren and Ark were the ones who saved you."

Blak Hart: "How bad am I, Vin? Don't hold back, tell me the bad news."

Vin: "Actually, you're healing. Considering how severe your injuries were, it's an improvement. You've been out of commission for a while, but you shouldn't make too much movement. Your vitals aren't strong enough. If you do too much, you could regress."

Blak Hart: "So I can't leave until I'm fully healed? By the way, how long have I been out?"

Vin: "Correct. The slightest action or a highly emotional state can cause your body to reverse back. There's no guarantee that you'll come back again from it. You've been here for a week."

In town, the teens continue on their mission for sweets when Rayne sees the silhouette of several apparitions lurking around. Cross sees them and tells Darren to brace himself—they are about to fight.

Darren: "What are we fighting? If it's those mist demons, they're not strong enough to harm us."

Cross: "How do you know that? Is that something you learned when you were captured?"

Darren: "Ha, ha, very funny, Cross. It's just something I know. How? I don't know. Now, that beast over there..." *points on the roof of one of the houses* "he has no reason to be here. He's looking directly at us. If he's here, that means there are more close by."

Cross: "Okay, what do you suggest? Stay and fight or run?"

Darren: "Take Rayne, I'll hold them off."

Rayne: "I'm not going anywhere. You saved Louie, and now you think you can handle anything? Get real, Darren. Together, we can have each other's back."

Darren: "I took down a demon that Blak Hart couldn't by myself. I really don't need anyone to help me."

Cross: "You're full of shit. Everyone needs someone." *Darren grabs Cross* "What are you doing?"

Rayne: "Let him go! Is this how you treat friends?"

Darren: *Lets Cross go* "I... I... don't know why I did that. Did I hurt you, Cross?"

Cross: "What's the last thing you remember?" *Now Darren's actions prove his theory—Darren has been affected by the incident.*

Darren: "Talking about the mist demons, then everything went black."

At the medical facility, Vin has to call Ark. The condition of Blak Hart has taken a turn for the worst. He is getting ready to explain to Ark how this was possible. He picks up the phone and dials the number. Ark picks up.

Ark: "Hi, Vin, what's the good word?"

Vin: "Salutations, I don't have any good news."

Ark: "Why, what happened?"

Vin: "Louie's heart rate keeps dropping, his blood keeps clotting, and I think you should get here as soon as you can."

Ark: "I'll be there right after the meeting."

Vin: "Sounds good... oh my..." *Sound of beeping in the background, then flatline*

Ark: "What's going on over there? Vin... hello... I'm on my way."

Ark gets to the facility. Vin runs to Ark. Ark sees that Vin looks shocked.

Ark: "Vin, what's going on?"

Vin: "It... it's Blak Hart. He's gone. One minute his body was dying, the next he's gone."

Ark: "Do you know where he went?"

Vin: "I have no clue. What do we do?"

Ark: "I'll take care of it. Stay here, just in case he returns. I'll go and look for him."

At the teens' Halloween walk, the teens' worry for Darren grows.

Rayne: "How can we help you, Darren?"

Darren: "You can't. I gotta go."

Cross: "Why? We can help you."

Darren: "No, stay here. I have to go."

He speeds off, losing the teens and starts chasing the beast on the roof. He is trying to follow the being known as a Tiktik, a shapeshifter that can transform into animals, usually a giant bird. It got its name from the sound it makes when it's about to hunt.

Rayne: "DARREN, COME BACK!"

Cross: "Rayne, let's go get him before he gets himself killed."

Rayne: "How are we going to catch up to him?"

Cross: "Just run as fast as we can and hope he doesn't catch up to that creature."

Cross and Rayne start running after Darren as fast as they could. Rayne gets Darren's scent heading toward the school. Their way is blocked by Nite.

Nite: "Where are you two going?"

Cross: "Nite, get out of the way. We are going to get Darren."

Rayne: "What are you doing here anyway?"

Nite: "Protecting the Cross family, what else?"

Cross: "Okay then, come with us to get Darren."

Nite: "No, leave Darren, head home."

Rayne: "You're joking, right?"

Nite: "No, I'm not. Head home."

Cross: "Nite, I'm going to warn you just one more time, please move out of the way."

Nite: "Michael, this is not a game, head home now!" *Cross marches toward Nite.*

While at the school, Darren catches up to the Tiktik and begins charging at it with lightning surrounding him. He attacks the Tiktik with his lightning punch, but the beast is unfazed. The beast darts at Darren, and in a blink of an eye, it claws and bites him with amazing speed and power, knocking Darren down. Before Darren can act, the Tiktik grabs him by the neck and digs its claws in, starting to squeeze. Suddenly, a flash of flame reaches out and grabs the Tiktik by the arm, snapping and breaking it like a twig, releasing Darren from its grip. He falls to the ground and sees a black silhouette, then calls out.

Darren: "What the fuck, who are you!?"

The silhouette turns around with glowing red eyes and walks toward Darren. As the moonlight hits the silhouette, Darren calls out again.

Darren: "Blak Hart, w-what are you doing here?"

Blak Hart: *in a deep, roaring voice* "GET OUT OF HERE NOW!!!"

Darren: "I can help..."

Blak Hart: "I WASN'T ASKING, LEAVE NOW!!!!!!!"

Blak Hart's eyes glow a fiery red as flames surround him. Darren starts running with much more speed than before, catching up to Cross. Darren and Rayne see Cross grab Nite by his jacket. Darren stops to catch his breath.

Rayne: "Darren, are you okay?"

Darren: *panting* "I... I... saw B... B... Blak..."

Cross: "Easy, Darren, calm down. You're not making sense."

Rayne: *holds Darren's face* "You're so pale. What did that thing do to you?"

Nite: "I will heal your wounds."

Cross: "Now, you want to help him? What was all that 'leave him' shit before!?"

Nite: "I was trying to keep you two safe."

Cross: "By letting my friend get hurt? Great job, Nite." *Claps sarcastically*

Rayne: "Can you two shut up and help me with Darren?"

Nite heals Darren while trying to calm him down emotionally.

Nite: "Did I hear you right? You said you saw black. Black what?"

Darren: "I said I saw Blak Hart. He's the one who saved me from the Tiktik."

Cross: "What?" *thinking Darren was seeing things*

Rayne: "How? He's still in the hospital."

Darren: "Just telling you what I saw."

Nite calls Ark to find out if Louie is still in the facility. Vin picks up the phone.

Vin: "Salutations, Nite, how can I help you?"

Nite: "I was wondering if Louie is still in there with you."

Vin: "Well, he was, then he disappeared. Ark is looking for him."

Nite: "When did this happen? And when you hang up, call Ark, tell him that Louie is near the teens. Darren said Blak Hart saved him from a Tiktik."

Vin: "Several hours ago. The unusual part of the whole situation is he was dying. Then, when I was talking to Ark, he disappeared. Nite, hello? Nite?" *The sound of the dial tone is heard.*

Nite asks Darren where he saw Louie. Darren tells him that Louie was on the roof at the end of the block. He destroyed the Tiktik with little to no effort at all.

Nite: "I'll go find Louie. You three should go home."

Cross: "Let us help you. Why are you being so stubborn? When we worked together, we were able to get a lot accomplished."

Nite: "No, I don't want any of you to get hurt the way Blak Hart did." *He looks down at his feet, with a depressed look on his face.*

In the distance, a voice can be heard. The teens and Nite turn around. When they look, they see Blak Hart walking toward them in a hospital gown.

Blak Hart: "Hey guys, what are you up to?"

Cross: "What the fuck are you doing here? You're supposed to be in the hospital."

Blak Hart: "I sensed you guys were in trouble, so I just came to help out and have some candy. Damn, I forgot my clothes." *Embarrassed because of his lack of clothes as the breeze blows his butt, making it seem like a front cape.*

Ark: "Where were you? By the way, here's your clothes." *He throws Blak Hart a bag of clothes.*

Blak Hart: "I'm here, obviously."

Ark: "Have you been here the whole time? You're supposed to be at the facility."

Blak Hart: "I got the feeling the teens were in danger, so I headed over and got here just in time. Don't worry, Ark. I'll go back for further tests."

Ark: "Alright, don't stay out too late. Vin is worried that your body might crash."

Blak Hart: "Will do. Alright guys, see you soon. In case of anything, just call me and I'll be there."

Rayne, Darren, Cross: "Hope you get better."

Blak Hart: "Me too."

Ark, Nite, and Louie leave the teens at the house, then head to the medical facility.

Blak Hart: "How did we get here?"

Ark: "Let's just say it's some form of magic, I call it faith."

Nite: "Why go to the teens when you're not at full health?"

Blak Hart: "Without counting Ark, they're the only family I have. If I feel they're in danger, I'd give my life to save them."

Nite: "Louie, sometimes you amaze me with your way of thinking." *Looks at Ark with a smile.*

Ark: "Now, do you see why I do what I do, brother?"

Nite: "I see. He reminds me of a younger version of you." *chuckling*

Vin: "Where did you go? You had us worried."

Blak Hart: "I apologize, Vin. I left to save the teens."

Back at the house, Nora asks the teens how their night was. They look at each other, and then Rayne begins to tell the events that transpired. When she's done, Nora asks Darren to come to her. She looks deep into his eyes and sees that his soul has something dark that wasn't there before.

Rayne: "Mom, what's wrong?"

Nora: "Whatever happened during his capture has added something extra to his soul. I don't know exactly what it is, but when I find out, I'll let you know."

Darren: "Mrs. Cross, what did you see? You seem worried."

Nora: "Don't worry, I'm fine. Whatever is wrong, I'll find a way to take care of it."

Nora heads down to the Wolf's Den to find information on what's going on with Darren. After hours of searching, she finds a journal written by Alpha, Wolf's father. It speaks of a group that were descendants from the unborn. This group's origin goes as far back as the creation of Earth. The legend states that close to the destruction of Earth, the descendant of the unborn will rise to assist the new leader of Hell and his guard to take over the heavens and the universe.

As Nora continues to read, she finds out that Alpha has battled several of them in his missions. She begins to read about one of his missions involving an unborn, and he starts off:

Alpha: *"Today, December 25, 1813, Nite has given me a mission to go investigate reports of churches getting destroyed by Zima. There are rumors that Zima has been working with a member of the Gier family. They're known for having the seed of the unborn. If it is woken up, the seed gives the host amazing power of chaos, demonic, and angelic abilities. I have to get there fast before anything happens. I met up with my contact, and he told me that he saw a man walk into a church up the mountain. Some other man was going after him. He described the man as a huge, bulky man with a bald head. He was smoking a cigar, and he had a heart on the back of his jacket with a scar on his face. It seems like my old friend Sparta is on this hunt as well. This should be very interesting. Hopefully, I can get to Sparta before he gets to Zima and his champion. If not, this could get very ugly. I'm heading up the mountain, and I'm sensing a lot of demonic energy and the stench of burning flesh everywhere. I'm climbing the mountain when I notice some trees that have handmade burns. Sparta must be close, and by the looks of it, he's not happy."*

Flashback:

Alpha transforms into a Timber wolf and runs up the mountain. He follows the scent of Sparta until he finds a cave that appears recently made. He heads into the cave, sniffs the air, and smells blood ahead of him. Alpha tries to get to where the blood scent is coming from when he's attacked by a huge figure consumed in flames. Alpha jumps back and transforms back to his human form, wearing his trench coat with his hood up and the Cross-family crest on the back. He draws his sword and charges at the figure, hitting it with the back of the blade. Then, the figure grabs the blade, stops, and speaks in a gravelly deep voice.

Huge figure: "Azrael Cross, is that you?"

Alpha: "It's Alpha when I'm out in the field, Sparta."

Sparta: "No one is here, relax. So what brings you here, old friend?"

Alpha: "The same reason you are, more or less."

Sparta: "Ah! The unborn seed."

Alpha: "Nite sent some of the info."

Sparta: "Ark sent me the same thing. The only thing he asked me to do was assist whoever was sent here and not destroy the area."

Alpha: "Can't blame him due to what happened last time."

Sparta: "I know, but I'm not like that in every mission."

Alpha: "Let's talk about it later. Right now, let's do the job."

They continue to journey deeper into the cave. Suddenly, two demons appear: Coron and Pagas. Two demons that were angels cast out of Heaven during the war. Their upper halves are angelic with huge black bat wings, while the lower halves are goat-like. These demons usually work with Amayo and Zima. Together, they cause havoc and carnage wherever they go. The demons jump out and attack the two hunters.

Coron and Pagas: "Why are you invading our sanctuary? Your penalty for trespassing is death."

Sparta: "Alpha, how about we show these guys what happens when good meets evil?"

Alpha: "Nah, I think we should show them what happens when they threaten us." *His nails begin to grow, revealing sharp claws for hands.*

Alpha and Sparta hit back with combination moves. Alpha hits Pagas, then transforms, while Sparta hits his opponent with a fiery fury of punches, changing his body into a humanoid flame. Coron tries to escape and runs further into the cave. Alpha claws at Pagas, piece by piece, until each body part is torn off. By the time Alpha realizes that Sparta is gone, his enemy is shredded into pieces. Covered in black blood, Alpha once again searches for the scents. He finally gets the scent of Sparta and follows it. Trails of blood and black ooze are splattered all over the walls. The walls have scorched hand marks that lead to a cavern with a ten-foot drop. He jumps down, following the trail of charcoaled rocks and boulders. The smell of iron and the trickle of ooze are all around. Once he gets to the bottom, there is a river of black ooze.

Alpha: "Sparta!" *in a deep, raspy voice* "Where are you?"

In the distance, he hears what seems to be yelling and fighting. He follows the river of ooze and sees Zima fighting Sparta, with Coron burnt to ash on the ground. He charges over to help when Amayo hits him with Coron's humerus, causing Alpha to stagger back. Alpha turns in the direction he was hit from; Amayo jumps into the ooze. Alpha looks in Sparta's direction and sees multiple demons holding him down as Zima stabs him repeatedly in the stomach. Alpha zooms to his area and begins fighting the demons away. This is what Sparta needed to retaliate, and he uses his attacks toward Zima. Sparta takes out his sword and swings it at Zima. Alpha transforms back and helps Sparta. The two continue to attack until Amayo emerges from the ooze. He grabs Sparta by his leg and tries to drag him into the ooze. As Amayo pulls, Alpha takes out his sword once again and uses one hand to drive the sword into the ground to prevent Amayo from pulling him in. Then, he uses his other hand to throw a ball of fire in an attempt to burn the arm, but narrowly misses. Zima uses his nails as razor blades and slices Sparta's neck, having him bleed out rigorously. Amayo slithers back into the ooze, accompanied by Zima.

Alpha: "Sparta, hold on. I'll try to save you."

Sparta: "Don't worry about me. Complete the mission." *Sheaths his sword.*

Alpha: "I'm not leaving you here for them to have as a trophy."

Sparta: "I'll heal. By the time you complete the mission, I'll be fully healed. Just get the job done. I'll be right here."

Alpha: "Fine, don't go anywhere."

He follows the river to the end and sees several small beings getting out of the ooze. He takes his sword out and starts decapitating the small beings until all that is left of them is a pool of blood and parts. More oozy beings get out and attack him.

On the other side of the river, Amayo comes right back out of the ooze, licking the trail of blood coming from Sparta. He goes to Sparta and drags him into the river with little to no fight. Sparta's healing hasn't been curing his body as fast as he thought. He tries to fight Amayo, but his efforts just anger Amayo more. Amayo picks him up and throws him into the ooze. Sparta tries to get out, but Zima pulls him back in. With one of his gasping breaths, he yells.

Sparta: "Alpha, I need your help now!" *Gasping for air.*

On the other side, Alpha hears his friend's cry for help. He runs over there as fast as he can, but it's too late. He sees Zima punch Sparta back into the ooze as Amayo drags him in.

Alpha: "Sparta, don't give up; I'll get you out!"

Zima: "Too late. His lifeless corpse will be suitable for our experiment."

Amayo: *Popping back out of the ooze* "Don't fuss. We have enough demons for the both of you. I have a special legion just for you."

Sparta tries one more time to fight his way out. With his last strength and breath, he gets out, only for Zima to grab his neck and break it. The snapping sound of his neck sends shivers down Alpha's spine. Amayo picks Sparta up. Before he can take him to the ooze, Alpha hits Zima, sending him flying across the river. He darts over to Amayo, but Amayo slides back into the ooze. Alpha looks down and sees Sparta's motionless body lying there. He holds him.

Alpha: "We've been through worse and fought stronger things and survived. Don't give up. Listen to the sound of my voice, come back."

He starts giving Sparta CPR. No response. He tries again, nothing. Out of frustration, he hits Sparta in the chest. Suddenly, Sparta chokes and begins to throw up some of the black ooze. Alpha pats Sparta on the back, making sure he throws up all the black ooze.

Alpha: "Let it all out. You don't want that in your body."

Sparta: *Coughs* "Thank you for helping me. Yeah, like I really want the liquid in my body." *Laughs and coughs because of the pain.*

Alpha: "Don't thank me yet. This mission isn't over. How's your neck?"

Sparta: *Snapping his neck back in place* "Very true. Let's finish this. Next time you want to kiss me, make sure you buy me dinner." *Laughing & coughing.*

Alpha: "Asshole." *Laughs.*

Sparta stands up and blows his nose, pushing some of the black ooze. They walk to the river as the ooze that Sparta threw up slithers back into the river.

Sparta: "You sure we have to go in that shit?" *Watching the ooze that was in his nose slither back from where it originated.*

Alpha: "Do you have a better idea?"

Sparta: "Sometimes I wonder if you're trying to kill me."

Alpha: *Hehe* "If I was trying to kill you, you would know. Now stop complaining and jump in. Remember to surround yourself with your flames."

Sparta: "Yeah, yeah, just don't get lost in there."

Alpha's hair turns snowy white as ice forms around his body. Sparta's body begins to steam up as his beard turns to flames. The flames begin to surround him. Both jump in and notice that the ooze is its own realm. They hit the bottom and then realize they can breathe. In the distance, they see a figure talking to what seems like eyes staring at them. Then a voice is heard.

Voice: "Zima, you have company."

Zima: "About time you two showed up."

Voice: "When everything is done, give me what's mine and don't take all day!"

Zima: "Just let me work."

Alpha charges at Zima, sword in hand, ready to kill Zima. Alpha is suddenly grabbed by the ooze and stopped in his tracks.

Zima: "Did you really think it would be that easy? Tsk tsk."

Alpha: "Do you think this could really keep me from ramming my blade through that black mass you call a heart?"

Zima: "Ouch! Alpha, that really hurts. Here, I thought we were friends, even after killing my BROTHERS!!!"

Zima surrounds himself with the black ooze and turns it into a spear, then aims it at Alpha's heart.

Alpha: "Did I hit a nerve?"

Zima: "I will finish you here, Alpha. It was great knowing you."

As Zima gets ready to throw the spear, Sparta appears behind him and hits him with a punch consumed in lava, sending Zima flying. Sparta rushes Zima while airborne and slams him to the ground. He starts punching him into the ground, then Zima uses the ooze to throw Alpha at Sparta, knocking Sparta off of him. They both go flying as Zima stands up and begins chanting, causing the ooze to surround him again.

Zima: "I WILL AVENGE MY BROTHERS!!!!!!!"

Alpha and Sparta stand up and prepare themselves for Zima's attack when a huge figure manifests from the black ooze in front of Zima and speaks.

Huge figure: "Zima, let me avenge your brothers. Plus, I want to see what I can do with my new powers."

Alpha: "Kevin Gier, I assume."

Huge figure: "That's my human name. As you can see, I'm not human anymore. Call me Mezun."

Alpha: "Sparta, get ready. This guy's aura is going crazy."

Sparta: "What is he now?"

The huge figure appears behind them and whispers.

Mezun: "Your death."

Mezun smacks Alpha and Sparta into the air and starts hitting them from all sides. He uses the ooze as tentacles, smacking Alpha and Sparta all over. Out of nowhere, Sparta throws a fireball at Mezun. He absorbs the flames and throws it at Alpha. Alpha blocks it with a shield made of thick ice. Mezun sends several tentacles at Alpha, not realizing that Sparta has made a sword of fire. Sparta slashes the tentacles as Mezun teleports behind Sparta, kicking him out of the way. Alpha jumps at Mezun, plunging his blade into Mezun's head. He isn't fazed. He smiles at Alpha.

Mezun: "Haha, is that all you got!?"

Alpha: *Breathing heavily* "You're a pain in my ass."

Alpha starts chanting, and his sword starts freezing Mezun from the inside out, turning him into a statue of ice. Then Alpha calls out to Sparta.

Alpha: "SPARTA, YOU'RE STILL ALIVE!!!"

Sparta: "Haha, very funny. Yes, I'm alive."

Alpha: "I need you to get him out of here. I don't think that will hold him for long. If we get him out of here, we might have more of a chance sealing him away in our temporary cell until Nite gets there."

Sparta picks up Mezun as they head out of the river of ooze. Right before they get out, Mezun manages to get movement in his limbs and tries to free himself from Sparta's grip.

Sparta: "Stay still, you slimy piece of shit."

Alpha: "Hold him still so I can freeze him again."

Sparta: "What the fuck do you think I've been trying to do?"

Sparta holds Mezun long enough for Alpha to refreeze him. While being refrozen, Mezun keeps trying to fight. After all his efforts, he ends up frozen and is taken to the Warriors' stronghold, where Alpha and Sparta interrogate him with Nite and Ark watching. Mezun is being held down by enchanted chains with holy symbols carved into the ground. Alpha steps forward and starts the interrogation.

Alpha: "Mezun, why are you working with Zima and the unborn?"

Mezun: "HAHAHA! Why else would I work with them? For the POWER!!!"

Sparta: "Well, that worked out for you, didn't it?" *Laughing.*

Mezun: "SHUT UP!!! YOU GOT LUCKY!!!!!!"

Sparta: *Walks up to Mezun* "I can finish you right now, you piece of shit!"

Mezun: "HAHAHA, TRY IT! IT TAKES A BIG MAN TO ATTACK SOMEONE WHILE THEY'RE RESTRAINED!"

Sparta starts hitting Mezun with devastating blows. Then Ark steps up and catches Sparta's fist.

Ark: "Enough, Sparta. We need him alive."

Mezun: "HAHAHA, HE COULDN'T KILL ME EVEN IF HE WANTED TO!!!"

Sparta: "YOU SON OF A BITCH!!" *Gives Mezun a punch, knocking him to the ground.*

Alpha: "Sparta, I need you to stop letting him get to you."

Sparta: "He's just so annoying!"

Alpha: "Ignore him or it can possibly influence your frame of mind."

Sparta closes the cell, then joins the others in the meeting room. Unexpectedly, a loud crash comes from the dungeon. Sparta, Alpha, Ark, and Nite run to see what the commotion was about. They see Mezun escaping with the help of Zima. The four try to stop them, but their efforts are foiled when the two disappear in black ooze and smoke. Then a flash of light shines the way a huge angel would. He takes Mezun and Zima by the throat and slams them to the ground. Nite and Sparta rush out to see; the angel calls out to Nite and Ark.

Huge angel: "NITE! ARK!"

Ark: "Sensei Zechariah!"

Zechariah is the warrior angel, physically buff, long blonde hair, and extremely tall. He teaches all angels in the art of battle—everything from fighting styles to battle strategies.

Zechariah: "How could you let them escape!?"

Alpha: "Who's the big guy?"

Sparta: "He seems to be Nite and Ark's sensei."

Alpha: "About time someone rips them a new one."

Zechariah turns to Alpha and Sparta, then begins yelling at them, causing them to shake nervously. He yells, and Zima and Mezun drop to the ground. They start to crawl as another light flashes. A teenage angel charges out, ready for battle. He sees Zechariah yelling at Alpha and Sparta, so the young angel charges at them, giving Alpha a punch to the stomach and giving Sparta a kick to the head. But neither of them moves. He grabs the young angel, telling him to stop.

Zechariah: "STAND DOWN! YOU'RE ATTACKING THE WRONG PEOPLE!"

Young angel: "But you were yelling... I thought..."

Sparta: "Awww, he hits pretty good for a young angel."

Young angel: "What's that mean!?"

As they talk back and forth, Zima and Mezun manage to summon some help: Amayo, and a huge demon manifest from a cloud of smoke. The young angel notices them and rushes for the huge demon, hitting him in the chest. The huge demon grabs the young angel and calls out to Zechariah.

Huge demon: "Zechariah, IS THIS YOUR NEW STUDENT?"

Zechariah: "LET HIM GO, RAIMUS!!!"

Raimus: "HAHA, easy, old friend. Just needed the distraction."

Zima, Amayo, and Mezun disappear in a cloud of smoke. Alpha charges to go after them but is stopped by Raimus throwing the young angel. Zechariah rushes at Raimus and hits him with an open palm to the stomach, sending him flying. The huge demon flies back, giving Zechariah a devastating uppercut.

Raimus: "Tsk tsk." *As he waved his finger.* "Zechariah, you're not even trying."

Zechariah: "Grrr, why are you doing this, Raimus!?"

Raimus: "Hehe, that's the real question, my old friend."

Zechariah calls Ark and Nite to fight against Raimus. Nite rushes the demon with a sword in hand, attacking him and missing as Ark hits him with a barrage of punches. Alpha jumps up and immediately chases after Raimus. The young angel leaps up; he tries to hit the huge beast. Raimus catches his punch and throws the young angel at Alpha. With a wave of Raimus's hand, a circle of fire surrounds him.

Raimus: "Another time, old friend. Next time, I won't hold back."

In a tornado of fire and smoke, Raimus gets swallowed into the circle of fire.

Updated reports state sightings of Mezun in various countries. There are additional reports that he has a family. His whereabouts and the location of his family are unknown. The case is currently open.

Nora puts down the journal in shock.

Nora: "Poor Darren. Information like this would confuse him. I must find a way to make the unborn gene in Darren return to a dormant state. But how will I do it?" *She thought.*

Cross: "Mom, are you alright? Darren and Rayne told me you didn't look like your cheerful self."

Nora: "I'm just worried about Darren, that's all. I think he's hiding his emotions. He doesn't want us worrying, but he's scared and doesn't want to bother us. What do you think about the way Darren is acting?"

Cross: "He's always been that way, so it seems almost normal to me. Have you heard anything concerning Louie?"

Nora: "He's just making sure that when he comes back, he's at full strength. He blames himself for what happened to Darren."

A week later at the facility, Louie walks to Vin.

Vin: "You are all set. What else would you like to know?"

Louie: "Who are all the other people and what is wrong with them?"

Vin: "Their information is classified, just like yours."

Louie: "So I can go back to active duty, or do I still have to let the action pass me by?"

Vin: "Like I stated previously, you are ready to resume your missions. Before you leave, I have something I believe might help you with your missions and more."

Louie: "What is it? Whatever it is, I'll appreciate it."

Vin: "Ark has a vehicle waiting for you at the area where you'll be appearing. Weapons and other information will be found in the trunk. As for what I have for you, it's something that was your grandfather's." *He pulls out a box.*

Louie: *He opens the box and takes out a leather black and red trench coat.* "Thanks, Vin. I'll thank Ark when I call or see him."

Vin: "Be very careful with the coat. Not only is it a fashion item, but it also has a special enchantment protecting the user. It enhances their strength and protects from physical attacks. This was the only thing found from your grandfather. His code name was Sparta."

Just to test the theory out, Blak Hart calls Ark to ask him about the reports of the Pine Barrens in Suffolk County. Ark tells him the information and warns him.

Ark: "Check it out first, then after you report what you found, we'll plan your next move."

Blak Hart: "By the time I wait for you to come up with a plan, I'll be done already. By the way, thanks for the tactical SUV."

Ark: "My pleasure, be patient. We don't know exactly what you're up against. Blak Hart..." *Dial tone* "Hello, Blak Hart! Damn it, I hate when he does this."

Blak Hart gets into his new toy. He decides to call it his War Tank. After he gets over the excitement of his new toy, he goes to investigate how accurate the reports were. He gets to the woods and hears multiple growls and yelling. He creeps slowly and sees Wendigos feeding on civilians, demons assisting Wendigos, and Collin Dirksen directing them all.

Blak Hart: "What the hell is going on here?"

Blak Hart finds an area to observe without being seen. He climbs a tree nearby to watch from a safe distance. Collin seems to be injecting demon blood into the Wendigos, causing them to turn into some hybrid of both. They don't look easily controlled. Then he sees a demon putting a collar-like object around their neck. The beast slashes the demon and starts devouring it as a smaller Wendigo approaches the one killing the demon.

Blak Hart: "That little one is going to get destroyed."

The little Wendigo walks on all fours toward the bigger one. With one swift strike, it takes the head off and starts devouring the bigger one, leaving the other Wendigos shaking in fear.

Blak Hart: "Damn, this got interesting fast. Hehe."

Blak Hart continues to survey the area, noticing the little Wendigo parading around the others as a sign of dominance. Collin tells the little one to go to him and begins talking as he would to a regular person. Not too far from that location, a dark black mass of sludge-like being was also watching. It was trying to blend in with the shadows. When it catches a glimpse of someone watching the Wendigos from the trees, it signals Collin to send one of the Wendigos to get the intruder. Collin tells the little Wendigo to send one to do his bidding. Suddenly, the Wendigo sprints toward the tree. Blak Hart looks down, seeing the Wendigo heading in his direction. Blak Hart jumps from one tree to another, over and over, trying to evade it. He runs out of trees and jumps down, preparing himself for battle. The Wendigo jumps down, only to get punched in the ribs and sent flying five feet away. The Wendigo charges back at Blak Hart, but Louie realizes that the beast is coming toward him with a few more. The creature waits for the others to join him.

Blak Hart: "What the fuck is happening? Wendigos aren't creatures that usually work as a pack. Collin must be pretty powerful to be able to control these beasts."

Once they gather up, they charge at him. They attack Blak Hart with skilled hits and cuts. He retaliates with several punches, then catches one and rips its head off.

One by one, Blak Hart decapitates the Wendigos with his blessed sword. The last two surround him. Right before they attack, he sheaths his sword, takes out his double-barrel pistols, and shoots their heads with precision.

The little Wendigo arrives with more to attack. The little one growls, and several of them move slowly toward Blak Hart. The Wendigos appear to be measuring him. Once again, he is surrounded. The warrior holsters his pistols.

The first one charges at him. The second one jumps toward him, and the others attack him randomly. He blocks the first one's hit, then grabs it by the throat and crushes its windpipe, causing it to spit out black blood.

The second one is hit so hard that its eyes pop out of its sockets, blood and brain matter splattering all over Blak Hart. The others continue to attack him, knocking him to the ground, clawing and biting him at the same time.

Blak Hart gives out a loud yell, then jumps back to his feet and starts going primal. He bites one, tearing the flesh off of the Wendigo. His fangs begin to grow, his eyes glowing red with yellow. His body bulks up, and he takes on the frame of a high-level demon. He attacks the Wendigos, tearing off heads, ripping limbs off of them, then consuming them as if they were a tasty treat. He puts his hands through one and begins eating its intestines.

While he's eating, the ones left flee. Once he's done, the little Wendigo is going to charge but is called back by Collin. Blak Hart hears a sound coming from the shadow next to the tree. He sees the shadow with its red eyes glaring and smiling at him. Collin yells to Blak Hart.

Collin: "Now you've done it. Here's something not even you can defeat." *He looks at the shadow* "Mezun, he's all yours now."

CHAPTER 5.
ARK, A PLEASURE AS ALWAYS

Mezun: "Collin, you idiot!" *Looks at Blak Hart* "Blak Hart, you're pretty strong. Not as strong as your grandfather, but you'll have the same fate he had."

Blak Hart: "How did you know my grandfather?"

Mezun: "I was the one who ended his life. I guess I'll end his bloodline now."

Blak Hart takes out his sword and tries to attack the demon. He swings, cutting through Mezun's ooze-like body, but it regenerates instantly.

Mezun: "Cute attack. I hope that's not the best you've got. I'd like to have a little workout before training these worthless Wendigos."

Blak Hart: "Maybe, or maybe not. You'll just have to find out."

Mezun sends a tentacle of slimy black ooze at him. Blak Hart turns his body on fire. The tentacles try to wrap around him, but they burn up and turn to ashes.

Mezun: "Now that's more like it. I can't say I was surprised; the fire is a trait your family has."

Mezun conjures six groups of fire demons to destroy Blak Hart. As the demons walk out of the dark mist, suddenly, Ark appears.

Mezun: "Ark, a pleasure as always."

Ark: "My, my, my, you don't look a day over 90."

Mezun: "Flattery before dying, that just bought you an additional ten seconds."

Ark begins fighting off the demons, tossing them aside like garbage at Mezun. Blak Hart helps by fighting the fire demons with ease. As Mezun prepares to attack, Nite jumps down from the trees, landing on Mezun. While Nite fights, Ark knocks out Blak Hart, then carries him to the car. As Ark drives Blak Hart away, Nite calls upon two allies to assist him: Spaz and Poena. Poena sees Mezun running away. He decides to hunt him down without the knowledge of Nite or Spaz. He chases him everywhere until he gets blown back by an explosion. The explosion was a diversion to get Mezun away from Poena.

Back with Nite and Spaz, they ignore the fact that Poena left and begin to fight the fire demons. One after another, Nite and Spaz take them out. Nite hits them like a piñata while Spaz takes his time. His attacks are precise and very calculated, and he seems like he's enjoying it; in actuality, he's just practicing for a bigger battle.

Nite: "Spaz, that’s it; they stopped moving. You can stop attacking."

Spaz: "Just practicing. You never know when the real threat arises."

Nite: "When it does, we’ll be ready. Until then, just do as you're told and go back."

Spaz gives Nite a dirty look and disappears. Nite looks around for Poena but doesn’t see him. He starts walking into the woods, looking for him.

Back at the Cross household, Ark arrives with an angry Blak Hart by his side. He tries to reason with him, but Blak Hart doesn't want to hear it.

Blak Hart: "Why did you feel you had to knock me out? You could've told me to stop!"

Ark: "I saw you were about to be in danger, and I know you wouldn't stop until your last breath."

Blak Hart: "You could've given me the benefit of the doubt. Don't assume I'll do something without trying other options."

Ark: "I apologize. Maybe you're right. I'll try other options before doing certain things. I guess I should take a dose of my own medicine."

Blak Hart: "Finally, I accept your apology."

Nora walks up to Louie and hugs him while Cross, Rayne, and Darren run to him. After hugging Louie, Nora converses with Ark.

Nora: "Is he ready to battle?"

Ark: "He's ready, but he needs a whole lot of training. He's already being targeted by class S++ level demons, and I can protect him to a certain point. Can you help me train him without telling him who I truly am?"

Nora: "Why don't you tell him the truth? He can handle it, but if you think he's not ready yet, then your secret is safe with me."

Ark: "Thank you. I'll tell him when he has achieved the beginning of his destiny."

Nora: "I think we might have an issue with Darren. I looked into his eyes, and I saw a void. The activity that took place when he was taken has tainted his soul. Is there anything I can do to make that darkness dormant?"

Ark: "Not at the moment. My brother and I are trying to find ways to help him. Just try to keep positivity surrounding him. Notify his mother to let her know the possibilities of what can happen with her son. We were hoping that there would be more time, but I guess we must work quicker to find a remedy for him."

Nora: "I'll call her later on. That way, when she comes to pick up Darren, she could come inside to talk."

Nite: "We'll talk about it later on. I'm busy looking for Poena."

Ark: "Try to follow his aura; there's no way to hide that unless he has help to do it."

Nite: "That's what I've been doing. Then the trail seemed to have vanished."

Nora talks to Gail about Darren. Gail tells Nora that she'll be there right after work. Hours later, Gail arrives at the house. She's greeted by Nora.

Nora: "Hi, Gail, come in. You're going to need this." *Hands her a cup of coffee.*

Gail: "Hi, Nora, what exactly is going on with my son?"

Nora: "After Darren was rescued, I looked into his eyes and saw a glimmer of light in the darkness where his soul was."

Gail: "Haven't you spoken to Nite?"

Nora: "Not lately, but I spoke to Ark. He told me that he's assisting Nite to try and find a way to keep whatever Darren has inside dormant until he can find a way to free Darren from it."

Gail: "Is there anything I can do?"

Nora: "I asked the same thing. I'm looking through Wolf's old files and even checking his father's journals to see if I can find anything. I suggest you try to do the same. Look through the files that Gil might have had."

Gail: "Sounds like a plan."

Nora: "Remember, Gail, you don't have to go through this alone. I'm here to help you whenever you want."

Gail: "Thanks, Nora, I appreciate it. If I find anything, I'll let you know."

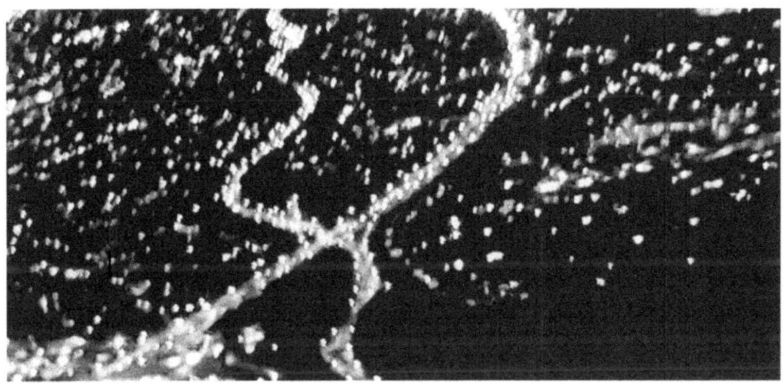

Way beyond the pearly gates, the angels called another meeting as they looked down toward Earth. This time, it has nothing to do with Blak Hart and the teens. This session involves the rebellious angels: Amon and Raimus.

Head Angel: "Before we decide the penances for them, is there anyone that has anything to say as to why they shouldn't?"

Second Angel: "As far as I can see, everyone at this table agrees that they should be punished for their actions."

Third Angel: "I think this should have happened a while ago, but now, with all the evidence against them, they can't deny it."

Angel Trainee: "They must know by now that they are going to be hunted. Are we going to send Blak Hart and the teens to fetch them for us, or do you have other plans in the works?"

Head Angel: "That information will be released after we have sentenced them."

Angel Trainee: "Are we going to cast them to Hell or Purgatory?"

Second Angel: "First thing that has to happen is they need to have their wings clipped. Just to start off their eternal torture."

Third Angel: "Then they should have their powers stripped from them, along with the gift to heal."

Head Angel: "They all sound like excellent ideas. I will go to the king, then I will return with the verdict."

Back at the **Cross residence**, Blak Hart is talking to the teens about a mission that involves **Collin's** capture.

Blak Hart: "It's time to bring him in and hand him over to Ark."

Cross: "What's the rush? What's the sudden urgency to get him?"

Blak Hart: "He was in the Pine Barrens. He was leading the Wendigos around. He was controlling the head Wendigo to do his bidding."

Darren: "Do you think we're strong enough to take on this mission? Remember, you just got out of the hospital."

Blak Hart: "I have faith that we can tackle this mission. We just have to formulate a plan and execute it."

Rayne: "Where is this all going to take place?"

Blak Hart: "The place where he feels the most confident: the school. To his knowledge, no one would try anything while he's posing as a teacher."

Cross: "We have to make sure that the plan involves keeping him away from the students. We don't need anyone getting hurt except for him."

Blak Hart: "Exactly. The mission is simple, but it will be dangerous. Lure him to me. The rest of you will handle his lackeys, giving me the chance to capture him and hand him over. Once he's handed over, I'll return to help you out—if it's needed at that time."

Cross: "Why are you so sure that we can handle this?"

Blak Hart: "You are, after all, Wolf's kin. Darren has the cool electricity and speed on his side. So why wouldn't I think that you are capable?"

Rayne: "What kind of being are we going up against? Not like it matters; just saying, it would be nice to know so we can prepare accordingly."

Blak Hart: "That's the right attitude to have. The issue is that it could be several different things. It could be Wendigos, mist demons, possessed school staff, shadow demons. Prepare for all, because we don't know which kind we'll be up against. Don't be surprised if he has one of each there."

Cross: "Mist demons, shadow demons—we can handle them. As for the Wendigos and the possessed school staff, we need training on how to deal with them."

Blak Hart: "With Wendigos, they are strong and heal quickly. The only real way to kill one is to burn them to a crisp or cut their head off. Possessed people? They must be captured, tied up, then an exorcism must be performed to release them from the hold of whatever demon has them in their grasp. If it was involuntary, it will be easier; if they accepted it, then that would take more work."

Rayne: "How the fuck are we supposed to exorcise the people?"

Blak Hart: "As for the possessed..." *looks at Cross* "Your mom can train you in how to properly extract the demon from its host. I'll be there too so I can get a refresher course. And as for the Wendigos, I'll show you how to handle them. Exorcising has been in our families for generations." *He looks at Darren* "We can go to your mother, and she can do a protection spell. That spell will help us greatly with this mission."

Darren: "My mom doesn't do none of that. She just works, cooks dinner, then watches her soaps."

Blak Hart sees Gail in the living room and calls her over.

Blak Hart: "Gail, have you told Darren about your incredible gifts?"

Gail: "No, I haven't."

Blak Hart: "I think now is the time. We'll be going on a mission within the next several days, and we're going to need your help."

Gail: "How soon are you all going to battle?"

Blak Hart: "In three days. That gives us enough time to train, effectively plan, and equip ourselves."

Gail: "I'll talk to him today."

Blak Hart: "Thank you, Gail. I know this was last minute, but you know how this job is. And as for the rest of us, our training begins in an hour after dinner."

At the four corners (Arizona, Colorado, New Mexico, and Utah), Amon is preparing to fight the angels. Before the battle, he holds his own mini meeting with Mezun and Mardus.

Amon: "Get more of my fallen brethren to assist us."

Mezun: "What are you expecting from this battle? If you win, great things will happen for us. If you lose, we're fucked."

Mardus: "That's the reason he wants the other fallen angels, dumbass."

Mezun draws multiple symbols on the ground, surrounding a pentagram with his blood. He begins to chant. Slowly, several beings surrounded in dark smoke rise from the circle. As they come out, they look at Mezun.

The beings: "Why have you conjured us?"

Mezun: "Amon and I have called upon you to assist us in battling the angels."

The beings: "What gives you the right to bring us here?"

Amon: "I, Amon, the right hand of our dark lord, command you to."

The beings: "Don't make us laugh. Why would a rejected angel, who couldn't cut it, be the dark lord's right hand?" *in a deep, hair-raising tone.*

Amon charges toward one of the beings and uses his power to disintegrate it, in an act of dominance.

The beings: "How are you trying to show dominance using angelic powers? You're a walking contradiction, you fool."

Mezun slides over to Amon and nudges him.

Mezun: "You do know that they're right. If you have any, use your dark powers."

Mardus: "He uses whatever power he chooses; he has nothing to prove to you. You just have to follow orders before your fate is the same as the one he destroyed."

The beings: "Did you truly get permission from the dark lord to call us for this particular fight?"

Amon: *Scratching his head* "Of course. Whatever I need, he told me to get the job done."

The beings: "Alright then. After the battle, we'll ask him to make sure."

Mezun goes up to the being who stood up to Amon and makes it combust from the inside out, causing it to turn to ash.

Mezun: "Anyone else that doubts us, step forward. If not, then do as you're told!"

At Gail's household, she is trying to explain to Darren about her powers.

Gail: "I have dabbled in multiple arts of magic. After your father passed, I stopped, until recently. I helped Blak Hart with a spell for one or two of the missions you guys were in."

Darren: "Why didn't you tell me about this years ago?"

Gail: "I didn't want you to be involved in this. I did this for your safety. I don't want you to be upset with me, but I should have realized that this is part of our family's gift and curse."

Darren: *Voice turns deep and raspy* "Bitch, you knew damn well that this bullshit would hound me because of you."

Gail: "Darren, what's wrong with you? Why are you talking to me like this?"

Darren: *Voice returns to normal* "What are you talking about? I didn't say anything. I'm still trying to wrap my brain around you doing anything supernatural."

Gail: "Are you sure you don't remember saying anything?"

Darren: "I would know if I said something. Why are you acting so weird, Mom?"

In Cross's yard, Louie is training with Cross and Rayne. Nora brings them drinks to hydrate them. They continue training on how to handle battling Wendigos and how to capture a possessed person. As the teens train, Louie walks to Nora and asks her several questions.

Louie: "Nora, when are you going to let the twins know that your injuries have healed and you've been secretly helping us?"

Nora: "I don't think they're ready."

Louie: "They are warriors. Remember, they are, after all, children of warriors themselves. Don't underestimate them. They can handle more than you give them credit for."

Nora: "I know. It's just that I... I'm not ready. Please don't tell them until I'm ready."

Louie: "You have my word." *He hugs her.*

The phone rings. It's Gail on the other end. She tells Nora about the incident that happened to her with Darren.

Gail: "He ended up cursing me out when I told him about my abilities and why I kept it from him."

Nora: "Is he back to normal?"

Gail: "He is, but the way his voice was when he spoke to me... he was so heartless and cold."

Nora: "If anything else happens, call me, and I'll be there as fast as I can."

As Gail proceeds to talk to Nora, she notices a silhouette of a figure on a treetop watching her. She realizes that it looks hunched over. She tells Nora what she sees. Nora asks her to describe it. Gail tells her it's difficult to see full details, but all that can be seen is that the figure looks humanoid. It's holding onto the branch like a monkey, but its eyes glow a bright red as it watches her every move.

At that moment, Darren goes out to toss out the trash. The figure looks at him and jumps down. Darren looks in the direction of the figure and prepares to fight. The figure rushes toward him with lightning speed. As it ran, it passed a light coming from inside the house. Gail finally sees what it is. The figure is a Wendigo, and though it’s small for a Wendigo, it has a very muscular build. Gail looks at the face—it resembles her dead husband, Gil.

Gail runs out to deter the Wendigo from going after Darren. She begins to chant a repelling spell. The beast stops and laughs at her, then moves forward toward her.

Gail: "Darren, go inside. I'll take care of this."

Darren: "Mom, I can help you."

The beast looks at Darren and smiles. Gail chants again, sending a ball of energy that hits the Wendigo. The beast turns back to Gail. Suddenly, it starts getting bigger. The muscle tone becomes extremely muscular. The Wendigo is now larger than any of the others he’s been around. Simply out of nervousness, Darren shoots an electric bolt directly toward the Wendigo. The Wendigo is unfazed, enraging the beast.

Gail: "Darren, don’t do anything else. You’re going to make it angry."

Darren: "Mom, trust me. I know what I’m doing. Leave it to me. I’ll save us both."

The monstrous Wendigo turns its attention to Darren and quickly heads in his direction. Gail chants stronger, speaking in Latin. She blasts the Wendigo with a thick ray of blinding bright light, causing the beast to flee. As it runs away, it stares at Darren.

Wendigo *in a deep tone*: "This isn’t over, Gail. I’ll be back for our son, not you or anyone will stop me from showing him how to turn into his true form!"

Gail: "Not in my lifetime, Gil. Not in my lifetime!"

Darren: "Did I hear that correctly? Was that my father? I thought he was dead. You told me he was dead!"

Gail: "Whatever that is, that's not your father anymore. He's just a shell of his former self."

Back at the four corners, Amon and the rest have been awaiting the arrival of the angels for several hours now. As the beings begin to get ready to go back to where they came from, Mardus yells as a ray of light comes out of his chest, turning him to ash.

Amon: "Where are you, you fuckin' cowards? Come face us, you poor excuse for a so-called angel!"

All of a sudden, thunder is heard as multiple bolts of lightning come down from the heavens. The Archangels, Michael, Raphael, and Gabriel, appear with their battle gear ready to contest all evil in their presence.

Archangel Raphael: "You rang, bitch? Don't stay quiet now, you had plenty to say a couple of minutes ago."

Archangel Michael: "Relax, brother. No need for foul language. Remember, we don't have to lower ourselves to their standards."

Archangel Raphael: "You take the fun out of everything."

Archangel Gabriel: "I bet he's gonna run! I'll get him."

Archangel Michael: "Brothers, attack at will; no demon shall survive the fury that heaven hath for thee."

Amon: "What the fuck was that?"

Archangel Gabriel: "Raphael, what did he say?"

Archangel Raphael: "He said 'fight.' Well, what are you waiting for?"

Gabriel swings his mighty sword at the mist demons. Raphael flaps his wings and flies on his way to Mezun, while Michael charges at Amon. The battle rages on for over five hours. Gabriel is beaten and battered but manages to defeat the demons. Michael has Amon on the run, while Raphael is beating Mezun senselessly. Another burst of light comes from the sky. Raimus grabs Amon and Mezun. In another burst of light, they vanish.

Archangel Raphael: "Damn it, I guess Lucifer needed his bitches back."

Archangel Gabriel: "We had them. This just can't be."

Archangel Michael: "I agree, this didn't go well, but we will not give up until he is in our grasp."

At Cross's house, Nora is worried about Gail and calls for Louie. Louie runs to Nora's side.

Louie: "Nora, what's wrong?" *in a worried tone*

Nora: "Gail was on the phone with me when she saw a figure on top of a tree watching her. She yelled at Darren to go back inside, then the phone went dead. I was wondering if you could do me a favor and go over there to make sure they're alright. From the sound of her voice, it sounded like something was about to go down."

Louie: "No problem, I'll go check up on them. Once I get there, I'll call you and let you know what I see."

Nora: "Thank you, Louie. I greatly appreciate it. And please be careful. I don't know what she saw."

In the blink of an eye, Louie changes his clothes, gets in his car, and speeds over to Gail and Darren. He arrives at the house to find Gail and Darren arguing.

Blak Hart: "What's going on? Nora told me that something was here. Are both of you okay? And why are you two arguing?"

Darren: "Did you know? Huh? Did she tell you about him?"

Blak Hart: "What the fuck are you talking about, kid? Gail, what's he ranting about?"

Gail: "The figure that was here was my ex-husband."

Blak Hart: "I thought he was dead. How could that be?"

Darren: "Don't cover up for her. If you knew, you should've told me."

Blak Hart: "Damn it, Darren, shut the fuck up! I just found out that he's not dead. Stop acting like a bitch and listen to what your mother is trying to tell you. She thought he was dead too. Why else would she try to hide that from you? Hiding something like this wouldn't help anyone, especially you." *Clenching his fists but keeping his anger under control.*

Gail: "Yes, I did hide the fact that I have special gifts, but something of this nature—I'd have no way of hiding this from you. To my knowledge, he was dead years ago. How that thing resembles your father is beyond me."

Darren: "I don't understand. How do neither of you know that he's still alive? And on top of all that, what has he changed into now?"

Gail: "He is now a form of a Wendigo. Not the typical kind either. As far as I know, Wendigos do not speak and usually don't leave survivors."

Blak Hart: "That Wendigo... was it smaller than a regular Wendigo but muscular as hell?"

Gail: "How did you know?"

Blak Hart: "Oh shit, I saw a small, muscular Wendigo leading others to do Collin's bidding. If I'd known it was your ex-husband, I would've let you know. Is he still small?"

Gail: "No, he's now much larger than any Wendigos I've ever seen."

Blak Hart: "Was he that huge when you first saw him, or did he change right in front of you?"

Darren: "He changed after my mom hit him with an energy ball."

Blak Hart: "Damn, so that means your energy ball made him stronger somehow. He was difficult to fight before; I can only imagine how he could be now." *Rubs his head.*

Blak Hart calls Nora and explains the whole story. She tells him to pass the phone to Gail. Gail tells the whole ordeal in detail. Nora tells Gail that she'll be right over. Nora calls Blak Hart on his cell.

Nora: "I was wondering if you could come and pick us up to go to Gail's house."

Blak Hart: "Sure, I'll be there in five minutes." *Hangs up the phone, then he turns to Gail* "Gail, I'll be back; I'm bringing Nora and the twins."

Gail: "Okay, see you then."

Deep in the Pine Barrens, Collin yells at the huge Wendigo, Gil. Gil looks at Collin and growls at him.

Gil: *In a deep, evil tone* "You may have some power, but nowhere near mine. So watch your tone, Collin."

Collin: "You must still listen to me. If it wasn't for me, you'd be a mindless beast like the rest."

Gil: "I could've had any warlock do that favor for me. You should be thanking me for having other Wendigos and demons following you. Don't you ever forget that without me, you'd have been a sacrifice for your dark lord!"

Collin: "Yes, I understand; please accept my apologies." *in his thoughts* "I'll make you think twice before you ever try defying me again. Not you, nor anyone else, will make that mistake."

Somewhere in Hell, Raimus is talking to Amon and Mezun about the previous battle. He starts off by yelling at Mezun.

Raimus: "Mezun, what the fuck were you doing to help Amon against Michael? Amon, what about you? How could you let Michael wipe the floor with you?"

Amon: "How can you talk? You were nowhere near the battle. At least we had the balls to face them."

Raimus: "I'm smart enough to know which angels to fuck with and which ones to stay away from. When it comes to Michael or Raphael, I know it's best to keep my distance."

Mezun: "I am the only one who would help him out. No one else was brave enough to."

Raimus: "Next time, let's plan before going out half-assed and getting our asses whooped like a bunch of humans. As stupid as it may sound, we must train."

Amon: "Funny, how you act so much like Michael."

Raimus runs at Amon, punching him and stomping on him. Raimus finally stops when Amon is a bloody mess.

Raimus: "I've fought demons and angels that make you seem like an infant. Don't you ever call me Michael. You ever make this mistake again, you won't have to worry about Michael because I'll kill you myself. Is that clear, you wannabe demon?"

Amon: "Damn, Raimus, why does it have to be like that? What happened? Can't you take a joke?"

Zima: "Oooh, Amon has a boo-boo."

Amon: "Zima, where have you been?"

Zima: "None of your damn business. Where have you been? Oh yeah, getting your ass whooped by Michael and Raimus, for being a dumbass bitch."

Amon: "Are you forgetting who I am?"

Zima: "I guess you're forgetting who does more work for the dark lord. And you still haven't figured out the hierarchy here. He would take my word over yours. I've been loyal to him from day one. Unlike you, you don't get your way; you're quick to change allegiance in a blink of an eye."

Amon: "You're trying my patience, Zima. Don't make me show you what I can do."

Zima: *He speeds up to Amon and has his hands on Amon's throat* "The same way you showed Michael, Raphael, and Raimus, right?" *laughs* "Do me a favor, keep your mouth shut before I get my brother here to keep it closed permanently. So Raimus, you're with him; why? Mezun I can understand, but Amon; really? Amon, remember, I could have ended your pathetic existence and allowed you to live. Please give me a reason to finish you off." *releases his hold*

Raimus: "I didn't choose. He was selected. What about you? What are you up to?"

Zima: "Once again, none of your business. I'll see you soon. Hopefully, you get better help instead of comedy relief."

Blak Hart arrives with Nora, Rayne, and Cross to see Darren and Gail. Gail greets everyone, and the twins go to see Darren as Blak Hart stays outside and calls Ark.

Blak Hart: *leaves a message on Ark's voicemail* "Ark, it's me. Call me as soon as you get this. Something interesting was just uncovered. So try to return the call ASAP. I'm at Gail's house with Nora and the twins. I have information about Darren's father. He's not dead. He is alive and not human anymore. He has somehow turned into a Wendigo. Not to mention, he is much larger than your average Wendigo, and he can actually talk." *The beep of the voicemail ending the recording.*

Cross comes out to see what Blak Hart is doing. Rayne and Darren come out to join them.

Cross: "What are you up to?"

Blak Hart: "Trying to notify Ark and update him on the new findings about Darren's father."

Darren: "So, you really didn't know anything about my father? Wow, my bad. Due to my mom withholding things from me, I figured everyone is probably hiding things from me."

Blak Hart: "Everyone has secrets. Not all secrets have to be unveiled. Everyone, including you, has things that no one else knows about. So, if I have something that involves you, I can promise you one thing: I'll let you know I have info. Now, if I can tell you, I will. If I can't, I'll honestly tell you that I can't."

Darren: "So you're keeping secrets from all of us."

Blak Hart: "Yes, I am. But it doesn't involve any of you. So why should I have to reveal it?"

Darren: "So, how are we supposed to trust you?"

Cross: "He could've killed you several times and didn't. Not to mention, he's saved you multiple times. Well, if you still feel that he can't be trusted, you don't have to be involved in anything we're doing."

Rayne: "Do you two have to be so hard on him?"

Darren: "Don't forget, I saved him too."

Rayne: "So, does that mean you don't trust me either?"

Darren: "That's something different..."

Rayne: "No, it's not. If you can't trust people in my family, then that means you don't trust me."

Rayne walks up to Darren, holds his hand, and looks into his eyes.

Rayne: "I know you're going through a lot, but turning against your friends isn't going to help the situation get better. We need each other if we're going to survive the madness we're born into."

Cross: "She's right. We need to stop arguing about nonsense and try to help each other in critical times."

Blak Hart: "We must be able to trust each other if we're going to be working together. As for secrets, the only ones I am withholding are the things that I've done and I am not proud of. If anyone needs me, I'll be in the dining room." *He walks away.*

Darren: "I trust you two. It's him that has me feeling a bit defensive."

Cross: "Why? What has he done to make you feel that way? From what I've seen, he's been treating us with respect so far. As for me, he hasn't given me a reason to suspect him of anything shady."

Darren: "I have a weird feeling. I can't explain it. I just do."

In the dining room, Nora, Gail, and Louie talk about the possible reasons why Gil would come back for Darren.

Gail: "Years ago, he was trying to have something enter Darren's body in order to join whoever was going to dethrone the king of Hell. That's when I caught him, and I thought I killed him. Then, I packed all my things along with Darren's and left. When I came back to pick up the rest of my things, I thought someone smelled his rotting flesh and removed the body. The one thing I couldn't figure out is why there was so much blood. When I left, there wasn't that much blood on the ground. Now, seeing his condition, I can only imagine who he must have eaten to turn into an unusual version of a Wendigo."

Nora: "Maybe he's back to finish the ritual and end your life for doing whatever you did to him, interrupting him in the ritual."

Louie: "Or he probably knows something of what might have happened to Darren when he was captured, which could make it that much easier in completing the ritual."

Gail: "Come to think of it, everything we said combined would make sense. He was probably waiting for the right time to awaken what he started. With the recent events, now would be as good a time as any to complete it and have him by his side."

Louie: "The only thing we can do now is wait for Ark to return my call. While I'm waiting, Nora could call Nite to see if both of them together can give us a solution on how to go about this with little to no injuries or fatalities."

Nora: "Sounds like a temporary plan. I'll call Nite right now."

Louie: "Just a heads up, Darren is having trouble trusting us because of the things that have been hidden from him. I think the best move for him would be for us to sort of keep him in the loop without giving up all the information. What do you think?"

Nora and Gail: "We agree. Some information is better than hiding it completely."

Louie: "I think he at least deserves that much. I can sense he feels different toward me. I believe it's because of what happened to him when he got captured. Gail, has he been acting a little different since his return?"

Gail nods as a tear slides down her face.

Louie: "I didn't mean to make you feel bad. I apologize."

Gail: "It's alright. We were arguing before you got here, and he said some hurtful things. I know he's not exactly himself, but it still hurts nonetheless."

Before Louie goes back outside to join the teens, he calls Nora over as he looks at Darren and sees black static electricity sparks being emitted from his body. Nora walks over.

Nora: "Yes, Louie?"

Louie: "There's something I have to tell you. I can't tell the others."

Nora: "You know you can tell me anything."

Louie: "Well, here it goes..." *sighs* "I remember the night I ended up in the coma."

Nora: "What happened?"

Louie: "When I got jumped by the demons, they had help from someone we were there to save."

Nora: "What exactly are you saying?"

Louie: "While the demons were shocking me, I almost got up to attack. Then, Darren showed up. I thought he was going to help me, but instead, he killed one of the demons and tried to finish me off." *Nora is shocked and horrified* "I was so hurt and confused, and everything faded to black. I didn't try to do anything because I figured that maybe he was in a fight-or-flight response and attacked anything in his way, but to hear afterward that he saved me with Ark makes me think he might have been affected more than anyone can imagine. After what Gail said, it confirms it. What do you think I should do? Should I tell Gail and the twins or just tell the oddball brothers, Nite and Ark?"

Nora: "We can tell Nite and Ark, but we have to find the proper words to say. Because if it's not worded properly, they'd probably try to take him, then Gail would get involved."

He agrees and walks over to Darren to see if he's feeling well when Darren turns to him and begins to tell Louie how he truly feels.

Darren: "Get the fuck away from me. I don't know how everyone can trust you, but don't expect shit from me."

Louie: "What the hell is your problem?"

Darren: *begins to rant* "You are! What the hell makes you think that you are qualified to help us, or as you call it, save us from evil?" *laughing in an eerie tone* "You went in to rescue me, and what ended up happening?" *laughing* "I had to save your sorry ass."

As Louie stared Darren down, the dark static surrounded Darren, and he starts sending electric shocks at Louie. Louie tries to control his own anger while deflecting it and attempts to get Darren's mind frame back to normal. Cross talks to Darren, but is quickly dismissed. Rayne is trying to reason with him, but he ignores her. Darren looks at Louie and continues on.

Darren: "You think you're so powerful? Take a good look at me; what the others couldn't get done, I will." *thinking to himself* "I will kill Blak Hart." *his eyes turn black*

Darren unexpectedly darts toward Louie, but before anyone can react, a blur passes by, taking Darren with it, too fast to be seen.

Blak Hart: "What the fuck was that? Did anyone see where Darren went?"

Cross: "All I saw was a blur, how about you, Rayne?"

Rayne stands in shock, and Gail is screaming and crying. Nora tries to console her.

Nora: "We'll get him back. All we have to do is find out where Gil is hiding."

Blak Hart: "Last time I saw any Wendigos, they were deep in the Pine Barrens—training."

Gail: "Then that's where we're going. Whoever doesn't want to come, I'll understand."

Cross, **Nora**, & **Rayne**: "I'm going."

Blak Hart: "I'll head over there first to scout the area, find him, and see what we're up against."

Nora: "When you call us, we'll be about a block away. Just let us know which way we should go through."

Blak Hart: "Sounds good."

As Blak Hart walks to his car, he calls Ark again. This time, Ark picks up.

Ark: "So, what happened exactly?"

Blak Hart: "Gil took Darren. Nora, Gail, and the twins are going to help me get him back. This is the reason why you should pick up your phone when I call!"

Ark: "I was on a mission; I just couldn't."

Blak Hart: "Well, from my previous message, how can you help us with that and this situation?"

Ark: "I'll meet you at the Pine Barrens. Don’t go in without a game plan."

Blak Hart: "That’s why I’m going to survey the area, then call them with routes to get in if he’s there. Once they get here, I’ll tell them the rest of my plan."

Ark: "So, what's your plan?"

Blak Hart: "I’ll make one once I see what we’re dealing with."

Ark: "Well, about time, my student finally learns." *sarcastic and proud tone*

Blak Hart: "Now is not the time, Ark."

Blak Hart gets into the car and drives to the Pine Barrens. A block away, Ark is waiting. Blak Hart walks over to him.

Blak Hart: "Since you’re here, how should we handle this?"

Ark: "The way you said before—see who’s there, how many, and which one is the weakest link. Then, figure out which entrances are best for them and the quickest exits."

Blak Hart: "So, what are we waiting for? Darren's sanity—the little that's left—is depending on us."

Ark goes up to the treetops to have a better view of the area, while Blak Hart sneaks in on the ground level to get a closer look and quietly take out any enemies that could alert the others. Ark notifies Blak Hart on the two-way headset that there are two enemies coming up on his right side. Without hesitation, Blak Hart pulls out a pistol with a silencer and shoots both in the head. Blak Hart looks up and sees four Wendigos jumping from tree to tree in Ark's direction. He notifies Ark, then shoots two in the temple. He calls Ark back as Ark kills the rest.

Blak Hart: "You're welcome."

Ark: "I had them, but thank you anyway."

They continue looking through the woods until they find a small shack near the exit.

Blak Hart: "Do you see what I see?"

Ark: "Yes, I do. Let's slowly go and see if Darren is in there."

Blak Hart: "Don't you notice something? No one is guarding it. This seems too easy. Check and see if there's anyone else in the vicinity. I have a funny feeling about this."

Ark: "Alright, bitch, I'll check."

Blak Hart: "Damn, Ark, watch your fuckin' language." *smirks*

Ark: *laughing* "Just get to work, smartass."

Blak Hart: "I got it."

As Blak Hart approaches the shack, Ark notices something heading toward the shack, moving like Darren.

Ark: "Hold on, Blak Hart, there's something heading your way."

Blak Hart: "From where and how fast?"

Ark: "It's headed right to you, up on your left."

Blak Hart turns to see a black figure jumping from branch to branch, moving with lightning speed. Blak Hart gets behind the tree for cover, the black figure lands in front of the shack. Blak Hart quietly radios Ark.

Blak Hart: "What do you see?"

Ark: "It looks like a teenager; it's just standing there."

Blak Hart: "Why is it here, then?"

The figure starts sniffing the air like a hound looking for a scent.

Ark: "It's looking for something."

Blak Hart: "Do you think it knows I'm here?"

Ark: "Doesn't look like it, but just in case, hold your position."

The black figure sees Blak Hart and charges at him. A bright light scares off the figure. Blak Hart tries to find the source of the light as Ark lets Blak Hart know that the figure went into the cabin. He looks at Ark.

Blak Hart: "Where did that light come from?"

Ark: "Don't worry about that right now. Call Gail."

Blak Hart calls Gail.

Blak Hart: "I'm here with Ark. We found a shack almost at the end of the woods. As far as we can see, a figure is in there. If you guys want, come to the cabin; wait outside and make sure that no one goes in. Ark and I are going in to find Darren."

Gail: "Alright, we'll be right there. If you find him, don't forget to call me."

Blak Hart: "You'll be the first person I call."

Ark and Blak Hart carefully walk into the shack, which has only one floor. They search around the entire house, but the figure is nowhere to be found.

Blak Hart: "Where did he go? You told me that son of a bitch went in here."

Ark: "He probably vanished."

Blak Hart: "Or he probably went into a hidden compartment. You should look around for a basement door or a hatch that blends in with the floorboards."

Ark: "Not a bad idea. I taught you well. I'll search the kitchen."

Blak Hart: "Are you hungry? If you are, I doubt you'll find any food. I'll go look in the living room."

Ark: "Just because I chose the kitchen doesn't mean that I'm hungry. Well, maybe a little, but that's not the reason I chose it. Don't get mad that I thought of it first." *smirks*

Blak Hart: *laughing* "Don't take it seriously, loosen up a bit, it was just a joke."

Ark: "Now is not the time." *laughs* "If you picked the living room to relax, those sofas are infected with ticks." *chuckles*

Blak Hart calls Ark over to the living room, pointing to the center of the room. Several of the boards look similar in material, but the way they're positioned looks off. Ark pulls the wooden planks by the edges to find a trap door. He pulls it open, and the stench of rotten flesh and feces engulfs the air.

Ark: "I opened the door, after you."

Blak Hart: "You should be going in first, remember I found it. This stench is nothing. Try being behind you doing a recon mission after you've eaten tacos, then talk to me."

Ark: "Try being around you when you haven't showered in a week, that's a horrid smell, my friend."

Blak Hart laughs as he flips him off and goes down the steps. Ark follows right behind him. He turns on his flashlight, illuminating the walls inside the tunnel. He realizes that on the ground are drops of dark black ooze. He shows Ark.

Ark: "I think the figure left us a trail. Seems like the residue of one of the creations of the unborn."

Blak Hart: "Those are still around? I thought they were taken care of centuries ago."

Ark: "The last reported sighting was by your grandfather and Cross's grandfather. They were hunting Darren's grandfather. So you see, it's in Darren's bloodline."

Blak Hart: "I don't have that information. But that makes sense." *thinking of Darren's recent change* "I'll have to check those files once we get back."

Ark: "Look through your father's files. He has some of his father's journals and files. In there, you might find something that can help Gail out."

Blak Hart: "Once I get back, I guess it's research time."

They continue to follow the trail of ooze, and then Ark sees the figure running. They chase after it only to find Mezun chanting. Darren is chained to the center of a pentagram, surrounded by red and white candles. The figure stands next to Darren, and with every chant, it slowly transforms into what appears to be Darren. Ark bolts over toward Mezun, as Blak Hart goes to rescue Darren. Ark throws a mighty blow, sending Mezun flying to the wall. Blak Hart frees Darren, but he's contested by the figure, which has now taken a form that replicates Darren. The figure swings at Blak Hart, then Louie retaliates with his fist covered in flames, attempting to hit the figure. The figure retreats, pulling Mezun with him. Ark makes an effort to chase after Mezun.

Blak Hart: "Leave him for now. We got what we came for."

Ark: "Next time, he won't be so lucky. Check you out, giving well-thought-out orders."

Blak Hart: "Well duh, I have an exceptional teacher. Anyway, don't worry, next time will be his last. For now, let's make sure Darren is taken care of."

Blak Hart carries Darren as they head out of the tunnel. He yells for Gail. She runs over as Blak Hart puts him on the ground. Rayne accompanies Gail.

Gail: "Thank you both for saving him from his father."

Blak Hart: "I don't think that was his father, but you're welcome either way. He should have a nice little knot on his face, all thanks to Ark." *smirks*

Gail: "Thanks, Ark. At least this will teach him a lesson."

Rayne hugs Darren and starts to cry as she holds him. She whispers.

Rayne: "Why does this keep happening to you, why? I can't lose you too, what are they doing to you?"

Blak Hart: "Rayne, it's ok; he's with us."

Rayne: "No, it's not ok, as long as they keep coming for him!"

Gail hugs Rayne to calm her down as Darren starts to wake up.

Darren: "Mom? Rayne? What happened? Why am I here?"

Blak Hart: "Look who's up. Wish you were awake while I was carrying your ass." *sarcastic tone*

Rayne clenches onto Darren as tears run down her face. Darren hugs her.

Darren: "It's ok, I'm fine." *glares at Louie*

Cross runs up to them and calls out to them.

Cross: "Yo, is everything alright over there?"

Rayne stands up and slaps Darren as she yells at him.

Darren: *feminine scream* "What was that for?"

Rayne: "You need to stop thinking you can fight everything; you're going to get yourself killed!!!"

Darren: *sobbing* "Rayne... I'm sorry."

Cross turns around and heads back to Gail's car.

Cross: "Hey, look, the car lights are on." *awkward laughter*

Rayne: "Cross, why did you leave?"

Cross: "I'm gonna get the car ready!"

Rayne: "Why are you so weird?"

Cross: "Why are you so weird?" *mimicking tone*

Rayne: "Stop being immature and get your ass over here."

Cross: "You're not going to slap me, are you?"

Rayne: "I'm thinking about it."

Cross: "I'll stay in the car."

Ark taps Blak Hart on the shoulder with a puzzled look on his face.

Ark: "Does this happen often?"

Blak Hart: *trying to hold in his laughter* "Sometimes it's worse."

Darren gets back on his feet. He walks to the car slowly and quietly, then sits in the back seat with his head down. Rayne turns to Gail.

Rayne: "Maybe I shouldn't have hit him."

Gail: "He needs some tough love right now. Whether he likes it or not, he will be just fine. Just give him time."

Rayne: "I hope so..."

Cross sticks his head out of the window and yells out.

Cross: "Are we leaving or not?"

Blak Hart: "We should head home."

Ark: "Alright, I have some research I have to do. I will call you guys later."

Blak Hart: "Cross, come on and hop in my car."

Cross jumps out of the car and walks with Blak Hart to his car.

Blak Hart: *turns to Gail* "Be careful driving. In case of anything, have Rayne call me."

Gail nods and heads to the car with Rayne, Nora, and Darren. They then drive off. Blak Hart sits in his car, talking to Cross as they drive away.

Blak Hart: "How are you holding up with all this?"

Cross: "There's something big happening, and it's making me nervous."

Blak Hart notices that **Cross**'s legs are shaking.

Blak Hart: "That bad, huh?"

Cross: "Yeah..."

Blak Hart: *takes out a pack of cigarettes* "Here, take one."

Cross: "I don't smoke."

Blak Hart: "Hehe, really? Who are you fooling? I've seen you smoking outside the high school. Just take one."

Cross: "I guess I'm not that good at hiding things." *takes one and lights it, then takes a long drag*

Blak Hart lights a cigarette as he pulls up to a 7/24 convenience store.

Cross: *exhaling a cloud of smoke* "What are we doing here?"

Blak Hart: "I'm getting some coffee and I'm running low on cigs. You want coffee?"

Cross: "Well, since you asked so nicely, I'd enjoy a cup of coffee."

Blak Hart: "Why do you say it like that?"

Cross: "What do you mean?"

Blak Hart looks back at Cross, then turns around laughing as he heads inside to get the coffee.

Cross: "Lou, tell me, how do you do it? How do you stay calm when everything's falling apart?"

Blak Hart: "Training, training, and you, Nora, and Rayne, and maybe Darren. If I lose my shit, one of you can get hurt—or worse. So, I have to think clearly and keep myself in check so everyone can get home safe and sound."

Meanwhile, in Hell, Mezun and the Darren doppelganger are met by Amon and Zima.

Zima: "So, how did it go... DAMN! What the fuck happened to you, Mezun?"

Mezun: *growls* "It was that damn, Ark."

Zima: "I remember when he used to do that to me, but he got you good." *laughs*

Amon: "Zima, shut up."

Zima laughs even harder.

Amon: *sighs* "Did the ritual work?"

Mezun: "See for yourself."

Out of the shadow walks the doppelganger, who smirks at Amon.

Amon: "It wasn't completed?"

Mezun: "It would've been if those brats didn't show up, but he's a lot stronger now."

Amon notices that the doppelganger doesn't stop staring at him.

Amon: "Why does he keep looking at me like that?"

Zima: "He sees through your bullshit." *laughs at Amon*

Amon approaches the doppelganger.

Amon: "So, what is your name?"

The doppelganger tilts his head at Amon in a questioning manner.

Mezun: "He can't speak; he's incomplete. It's all primal senses."

Amon: "Well, I got you a gift."

Amon snaps his fingers; an armor rises from the lava below.

Meanwhile, at the Cross household, Darren is sulking in the Wolf's Den as Louie and Cross walk down. Louie nods at Cross, and then Cross walks to his father's desk. He takes out the files of Darren's parents, along with his grandfather's files, and gives them to Louie.

Louie: "If you still don't trust me, here." *in an angry tone*

Louie slams the files on the desk in front of Darren as they both leave him in the room alone.

Darren reads the files of his family, along with his father and grandfather's files. He begins reading about Gail helping the Cross and Heart families with her abilities of sorcery, telekinesis, premonition, and electrokinesis. She assisted them in numerous missions and battled alongside them for the greater good. There were several occasions where she helped them capture Gilbert Gier and nurse him back to health. As Darren continues to read, he begins to imagine what he's reading.

Gilbert awakens in a cell and starts to yell.

Gilbert: "Where the hell am I?!"

Frustrated and confused, Gilbert starts banging on the door with all his might until his hands bleed. He hears a soft voice coming from the other side of the door.

Soft voice: "You need to calm down. You're going to hurt yourself."

Gilbert: "Who's there? *in a confused state* Get me out of here!"

Soft voice: "If you calm down and relax, you could be let out."

Gilbert: "Do you know who I am? Let me out!!!"

Soft voice: "Let me know when you're ready to calm down."

The sound of heels can be heard walking away. Gilbert keeps hitting the door until he passes out. Hours later, the cell door opens, and Gilbert is brought to a room filled with bright lights. Gilbert is placed on a chair and chained down. He starts to wake up and is blinded by the lights.

Gilbert: "What the fuck is going on here?!"

A man's voice responds to his question.

Man's voice: "You know why you're here. Where is Mezun?"

Gilbert: "Hahaha!!! Like I would tell you!"

A man wearing a red and black trench coat walks up to Gilbert and punches him in the face, almost knocking him out.

Man in the trench coat: "Don't make me beat it out of you."

Man's voice: "Hunt, why did you hit him that hard?"

Hunt: "He was looking at me funny and wasn't cooperating. Wolf, stop acting like you don't want to beat the shit out of him."

Wolf: *sighs* "If you keep hitting him, we won't get the info we need."

Hunt: "Fine, I'll hit him softer next time."

Two women walk out from the back room.

First woman: "Hunt, are you behaving?"

Hunt: "Come on, WQ, I'm always on my best behavior." *smirks*

WQ: "Yeah, sure you are." *sarcastic tone*

The other woman who came down with WQ walks up to Gilbert, she starts looking at him and checking his wounds. Gilbert begins to reawaken and sees a beautiful woman with long black hair and red lipstick checking his injuries. Hunt realizes that Gilbert is waking up and puts his hand on her shoulder.

Hunt: "Be careful, Enchantress."

Enchantress: "He won't hurt me, after that last punch you gave him."

Wolf: "You never know; he is a dangerous man. Don't underestimate him."

Enchantress looks at Gilbert and smiles as he blushes and looks away.

Enchantress: "He doesn't look too dangerous to me."

WQ: "Looks can be deceiving. He is the son of Mezun."

Enchantress: "That's why? His father is a monster. Killing his wife in front of his son, then forcing his son to do his dirty work. He has suffered a lot."

Gilbert sits up straight with his head down, his hair covering his face.

Gilbert: "What do you know? No one knows what I've been through... the beatings, the torture, the pain I had to deal with... Alone..."

Enchantress: "Gilbert..."

Hunt: "Then tell us where your father is."

Wolf: "The faster we know where he is, the faster we can stop him. Then he'll never harm you again."

Gilbert: *with an evil grin* "You can't stop him. No one can."

WQ: "There's always a way."

Enchantress: "Please, just tell us where he is; you don't have to suffer anymore."

Gilbert looks into Enchantress's eyes and notices that they're crystal blue. He blushes and puts his head down once again. He whispers.

Gilbert: "Silence is key..."

Hunt: "What did you say?"

Gilbert: "Silence is key..."

Hunt: "What the fuck is that supposed to mean?"

Enchantress notices that Gilbert has a tattoo around his neck that wasn't seen before.

Enchantress: "WQ, Wolf, check this out."

Wolf and **WQ** walk over and see the tattoo.

Enchantress: "Did he have that when you found him?"

Wolf: "No, not that I can remember. Hunt, do you remember seeing this tattoo?"

Hunt: "No, he didn't have it when we got him."

Gilbert: "It's one of my father's ways of making sure I don't fuck up, even if I knew where he was. I couldn't tell you."

Enchantress: "He put a spell on you so you wouldn't tell anyone his work or his hideouts? Does it hurt?"

Gilbert: "Such a caring father, huh..."

Wolf: "Damn, is there anything you can tell us?"

Gilbert: "I can't tell you anything, but I can show you something. So could you take these chains off of me?"

Hunt: "HAHA, yeah like that's going to happen."

Wolf: "Hunt, take the chains off."

Hunt: "What?! Are you fucking kidding me?!"

Wolf: *staring at Hunt with glowing yellow eyes* "What else are we going to do? What's your plan, huh?!"

Hunt: "Could we take that tattoo thing off of him?"

Gilbert: "Without killing me? Good luck with that. Haha."

Hunt: "FUCK!"

Wolf: "Take the chains off!"

Hunt: "Can you ask nicely?"

WQ: "Really, Hunt?"

Hunt: "Yes, really. I have feelings too, you know." *smiles*

Enchantress: *giggling*

Gilbert: "Are they always like this?"

Enchantress: "More than you would think." *smiling* "Let's get those chains off of you. Just be happy they're in a good mood."

Enchantress takes the chains off Gilbert as he looks at her, blushing and with a puzzled look on his face.

Gilbert: "Th... thank you."

Enchantress: "You're welcome." *blushing and smiling*

WQ hits Wolf and Hunt on the back of their heads.

Wolf & Hunt: "Ouch, what the hell was that for?!"

WQ: "You two need to stop acting like kids and focus on the mission."

Wolf: "She's right, Hunt."

Hunt: "Yeah, but did you have to hit that hard? Damn!"

WQ: "Would you prefer if I called the Huntress?"

Hunt: "Noooo, please no."

Gilbert: *chuckling* "You guys are strange."

Hunt: "Who took the chains off of you?"

Enchantress: "Well, while you two were busy having a lovers' quarrel, someone had to."

Gilbert: "You guys ready to head out?"

Wolf: "You didn't say anything about leaving."

Gilbert: "We have to go back to my old house. I'm sure you can find whatever you're looking for there."

Hunt: "You really think we trust you enough to just follow you? What if it's a trap?"

Gilbert: "It's your only choice, take it or leave it."

Wolf: "Sounds risky."

Hunt: "Why can't you just show us on a map where your house is and we can go?"

Gilbert: "Because without me, you won't get in. Now, the more time we waste talking about it, the less time we have to find my father. He's always on the move, so let's go."

WQ: "We have to go."

Wolf: "Let's go then."

Hunt walks up to Gilbert and grabs him by his shirt, pulling him up and looking him in the eyes.

Hunt: "If you are leading us to a trap or you try to run away, I will make sure you can't run ever again." *smirking*

Gilbert: *smiling* "Sure, whatever. Are you done? You're just wasting time."

Wolf: "Let's go."

Wolf: *sniffing the air cautiously* "Everyone stay alert. This place reeks of death. Keep your guard up."

Hunt: *cracking his knuckles* "Yeah, thanks for the heads-up, genius." *he smirks, trying to ease the tension*

Enchantress: *shivering, clutching Gilbert's arm* "This place feels wrong. What happened here, Gilbert?"

Gilbert: *his voice lowers as he watches the team carefully* "It's a place where dark things happen, where lives were lost, and where they stayed. This is where my father's twisted legacy began. And I'm afraid, it's not over yet."

The mansion is dark, filled with the smell of decay. The atmosphere is heavy, and the silence is deafening as they make their way further into the house.

Wolf: *grimly* "I can't shake the feeling that whatever killed your mother is still here... watching us."

Gilbert: *his face hardens* "If my father is still lurking around, this place is ground zero for whatever madness he's planning next. We need to move quickly, but carefully."

They reach the back room where Gilbert had disappeared. Enchantress tries to stay close to him, though the strange aura of the mansion is unsettling.

Gilbert: *pauses, speaking quietly* "Don't follow me further. I have to face what's in there alone."

Wolf: *stopping him* "Gilbert, we stick together on this one. You don't have to do this alone."

Gilbert: *turning to face him, his expression cold* "You don't understand. The last time someone came in here with me, they didn't leave. I can't let that happen again."

Enchantress: *placing a gentle hand on Gilbert's arm* "You don't have to carry this burden by yourself anymore. We're with you."

Wolf: *nodding* "Whatever it is, we're in this together."

Gilbert hesitates, then nods reluctantly. He leads them into the room, and the banging noise they heard earlier becomes louder, more frantic. The team enters cautiously, weapons ready.

Hunt: *whispering* "Whatever is in here, we better deal with it fast."

As they enter the room, the sight that greets them is both chilling and disturbing.

Gilbert: *with a grim tone* "It's here."

A figure stands in the center of the room. At first glance, it seems like a person, but the twisted, demonic energy emanating from it is unmistakable. Its eyes glow with an unnatural light, and the air around it seems to warp. The figure speaks in a deep, guttural voice.

Figure: "You think you can stop it? It's already too late."

Wolf: *unsheathing his weapon* "What the hell are you?"

Gilbert: *his voice cold, betrayed* "That's my father's creation. A soulless husk, part of the ritual he tried to finish with me."

Hunt: "What do you mean, part of the ritual?"

The figure laughs, a distorted sound that chills them to the bone.

Figure: "Gilbert's blood is the key. His father made sure of it. His essence... his soul... is linked to this place. And you, foolish mortals, will pay the price."

Wolf: "Not if we stop you first."

Without warning, Wolf charges at the figure, but it's faster than expected, vanishing into a blur of shadow. Hunt and Gilbert exchange a look, then move to flank the figure, with Enchantress casting a protective spell around them.

Gilbert: *clenching his fists* "This ends now."

The figure reappears behind Wolf, slashing at him with claws that seemed too sharp to be real. Wolf spins around, narrowly dodging the attack, and he slashes at the figure in return. The creature howls, retreating back into the shadows, only to reform in front of Gilbert, who steps forward with a determined look.

Gilbert: "You want me? Then come and get me."

The room seems to grow colder, the air heavier as the figure charges again. The team fights back with everything they've got, but each strike they make is met with a swift counterattack.

Hunt: *grunting* "This thing's too fast! We need to cut it off from its power source."

Wolf: *while fighting* "Gilbert, do you know how to kill this thing?"

Gilbert: *eyes narrowing* "It's tied to the bloodline... to my bloodline. The only way to destroy it is to sever the connection it has to the Gier legacy. But I don't know how to—"

Enchantress: *rushing forward, focusing on a spell* "I think I have an idea! Stand back!"

She chants an incantation in an ancient language, drawing symbols in the air with her hands. The figure hesitates for a moment, but then charges toward her with terrifying speed. Gilbert jumps in front of her, and as the figure slashes, Gilbert holds up his hand, a surge of dark energy pulsing through his veins.

Gilbert: *shouting* "I won't let you destroy everything I've ever known!"

A blinding light erupts from Gilbert's body, and with a final, guttural scream, the figure disintegrates, turning to dust that scatters across the room. The energy in the air dissipates, leaving the room eerily silent.

Enchantress: *exhales in relief* "Is it... over?"

Gilbert: *breathing heavily, his face pale* "For now... but this isn't the end. My father... he's still out there, and so is the curse he left behind."

Wolf: "Then we keep fighting. But for now, we get out of this hellhole."

Hunt: *laughing* "Man, I thought we were done for."

WQ: "We're not done. But we will be... when this is all over."

The team moves through the mansion, each member still reeling from the fight but knowing that more challenges lie ahead. Gilbert, still haunted by his father's legacy, knows that his journey is far from over.

CHAPTER 6.
HE SEES GILBERT

Hunt goes to the back room, and he sees Gilbert ripping and punching a painting. Hunt yells at him.

Hunt: "What the fuck are you doing!!!!"

Gilbert looks at Hunt with glowing red eyes and a black mist floating around him. He's clenching his fist, driving his nails into his hand, causing his hand to bleed black blood.

Gilbert: "It's my house, why does it matter to you?"

Hunt: "What's the point of attacking a painting, are you that weak?" *grins*

Gilbert walks to Hunt and gets in his face, smiling a devious smile.

Gilbert: "You want to see how strong I am?"

Hunt: "Hahahahaha!!! Are you trying to threaten me, I can't really tell?" *still laughing*

Gilbert: *still smiling* "Or I could pay that wife of yours a visit. Hehe."

Hunt grabs Gilbert by the neck and lifts him off the ground. Gilbert just keeps smiling as Hunt starts to squeeze down on his neck. Hunt's eyes start glowing yellow and red.

Hunt: "How dare you threaten my wife, I will break your fucking neck, you piece of shit!!!!"

Enchantress walks in and sees Hunt choking Gilbert. She calls out to Wolf and tries to pull Hunt off Gilbert, but she can't pull them apart. Wolf rushes in and yells out to Hunt.

Wolf: "Hunt, put him down!"

Hunt: "Why, he threatened my wife. He's better off dead!"

Wolf: "We need him alive, Hunt. He is just trying to get you to put him down!!!"

Hunt slams Gilbert to the floor, cracking the area he landed on. Then Hunt walks outside, lighting a cigarette. As he inhales the smoke, he walks away to calm down.

Hunt: *thinks out loud, lighting another cigarette* "I know I shouldn't have snapped. But when he mentioned my wife…"

Wolf: "Not like that." *sighs while shaking his head*

WQ walks into the back room. She sees Gilbert on the floor gasping for air and Enchantress helping him up. A piece of the painting on the floor catches her eye.

WQ: "Wolf, look." *pointing*

The painting seems to be a family portrait of Gilbert and his parents. On his face appears to be a nasty bruise.

Wolf: "That's probably why he destroyed it, bad memories."

WQ: "Life couldn't have been easy, being him."

Wolf: "You feel bad for him?"

WQ: "Why, are you jealous?" *teases him*

Wolf: "No, why would I be?" *blushing*

WQ gets close to Wolf and whispers in his ear.

WQ: "You're a horrible liar." *chuckles*

Wolf blushes as he walks up to Gilbert and Enchantress.

Wolf: "You alright there, Gilbert?"

Gilbert: "Yeah, I'm fine. Did he have to throw me like that?"

WQ: "You threatened his wife. What did you expect exactly? Be happy he let go."

Gilbert: "It was a joke; I don't know who his wife is." *rubbing his neck*

Enchantress: "Bad joke and wrong person to tell it to." *putting her hand on her forehead*

Gilbert looks at Enchantress. She blushes and stands up. She walks to where the painting is.

Gilbert: "Here's the entrance where my father keeps his notes in the room behind the wall."

Enchantress: "How do we open it?"

Gilbert puts his hands on the wall and speaks in the unknown language. The door opens, dust and cold air bursting from the room.

Gilbert: "There you go."

Wolf and WQ walk into the room. They start looking through papers and notes.

Wolf: "Take everything. Any info can help. Hunt! Get in here."

Hunt: *in a gravel voice* "I'm coming, I'm coming."

Wolf: "So you're coming?" *laughs and points at Hunt*

Hunt walks back into the house, shaking his head, and gives Gilbert a dirty look as he goes into the hidden room. Enchantress notices another piece of the painting on the floor that had Gilbert's mother on it. She picks it up and walks up to him.

Enchantress: *looking at the picture of Gilbert's mother* "She's beautiful. She must have been so loving and caring."

Gilbert acknowledges that she's looking at the picture. He reaches into his pocket, takes out the picture of him and his mom smiling, and shows it to Enchantress.

Gilbert: "She had a smile that could light up a whole room. She always knew how to make me smile even when I didn't want to." *tears roll down his cheek*

Enchantress: "You love her and miss her a lot, huh?"

Gilbert: "She was my world. My mother was always there for me. When my father would beat me, she would pull him off me and heal my wounds. She would fight with him to leave me alone, but… that never ended well…"

Enchantress: *getting closer to Gilbert* "Oh no, it must have been a nightmare for you."

Gilbert: "Not when she was here. She taught me that no matter what darkness my father put me in, there is still light. Then, he took her from me…" *his eyes start to glow, and his teeth sharpen.* "I didn't even get the chance to say goodbye."

Enchantress hugs Gilbert, then kisses his cheek. Gilbert is speechless and blushing. He hugs Enchantress.

Enchantress: "I'm sorry you had to deal with all that. It isn't fair." *Tears roll down her cheeks.* "You poor thing."

Gilbert: *puts his head down* "Why are you apologizing? You didn't do anything. Don't cry." *wipes her tears away.*

Enchantress: *looks up at him and blushes,* "No one should have to deal with all that. It's not fair."

Gilbert leans forward as Enchantress leans up, their lips inches from each other. They can feel each other's breath on their lips as they touch, but they are interrupted by Wolf.

Wolf: *coughs,* "You two ready to go?"

Enchantress: "Wolf, sometimes I just want… ughhhh!!!" *stomps off*

Wolf: "What did I do?"

Gilbert: "Great job, Wolf. Great job, let's go."

Wolf: "What did I do?"

Hunt: "Ha ha ha! What are we going to do with you?"

WQ: "You could be so clueless sometimes, I swear."

Wolf: "What did I do? Someone tell me." *confused*

The team heads to the front door. Suddenly, a huge explosion sends everyone flying. The house crumbles down, dust and debris covering where the house was. Through the dust, a huge figure can be seen.

Hunt: "Who the hell is that?"

Wolf: "I don't know, but we're about to find out."

Gilbert jumps up and starts looking for Enchantress in a frantic manner, throwing debris everywhere as the huge figure walks toward him. Enchantress calls out.

Enchantress: "Gilbert, watch out!"

Gilbert looks up and sees Enchantress under a beam. He runs to her and starts lifting the beam as the huge figure stands behind him. The figure speaks to him in a deep voice.

The Huge Figure: "What are you doing helping them, my son?"

Gilbert turns pale, then he begins to tremble, hearing his father's voice. He turns slowly to see his father towering over him. Gilbert nervously replies.

Gilbert: "F-father, y-you're here?"

Mezun: "Of course. I knew you would fuck up and bring them here. You are so predictable, Gilbert."

Mezun slaps Gilbert, sending him flying. Then he teleports and slams him down to the ground. Wolf charges at Mezun with Hunt as WQ helps Enchantress up. Wolf kicks Mezun off of Gilbert, shortly after Hunt starts shooting at Mezun. Wolf gets Gilbert away. Hunt then holsters his pistols and smirks at Mezun.

Hunt: "It's just me and you, old man," *cracks his knuckles*

Mezun: *laughing,* "Is that supposed to mean something, boy?"

Wolf drops Gilbert off by Enchantress and WQ.

Wolf: "WQ, let's go. Hunt can only hold him for so long."

WQ: "Enchantress, will you be ok?"

Enchantress: "Yeah. Is Gilbert ok?"

Wolf: "He's fine."

Enchantress: "Alright, go help Hunt."

Wolf and WQ run to help Hunt, who's attacking and retaliating. Wolf jumps up and punches Mezun in the face, sending him airborne as WQ slashes him with her daggers. Hunt rushes Mezun and punches him back to the ground, creating a crater.

Hunt: "Are you guys almost done? I can only dance for so long before I get tired. Hurry up and get Gilbert out of here!"

Wolf: "We're trying, hold him a little bit longer."

Hunt: "Fuck it, I'll finish him off now!"

Mezun grabs Hunt by the neck, then holds him up as he commences to squeeze. Mezun stands up and sends tentacles to Wolf and WQ, pinning them down. He calls out Gilbert.

Mezun: "Gilbert, get your worthless ass over here now!!!"

Gilbert stands, then walks over to Mezun. Slowly, Enchantress grabs Gilbert by the arm to try to stop him. She looks into his eyes as she starts to cry.

Enchantress: "Please, Gilbert, don't go."

Gilbert: "I have to..."

Enchantress: "No, you don't. Come with us. Please, don't leave."

Gilbert hugs Enchantress, kisses her passionately, then walks away. Mezun teleports behind Enchantress and smiles.

Mezun: "Well, who do we have here? Such a beautiful thing."

Enchantress starts to shake nervously and falls to the ground. Mezun walks toward her and reaches for her. Gilbert teleports in front of Mezun and slaps his hand away. Mezun yells at him.

Mezun: "Out of my way!!"

Gilbert: "No, leave her out of this!!!"

Mezun: "I'm not going to tell you again!!!"

Mezun raises his arm and swings with all his might. Gilbert catches his hand and stares at Mezun, his eyes glowing red. Black mist emits from his body as his fangs come out.

Mezun: "Hahahahaha!!! Are you turning on me now?"

Hunt jumps up and rushes Mezun, punching him in the jaw and sending him sliding back.

Hunt: "Forgot about me?"

Gilbert turns to Hunt and starts hitting him with a flurry of punches, sending him upward. He rushes over to Hunt in midair, then slams him powerfully into the mountain. He lands next to Mezun.

Mezun: "That's my boy!!!"

Enchantress calls Gilbert.

Enchantress: "Gilbert, nooooo, stop! You don't have to do this!!"

Wolf gets up and charges at Gilbert. He starts kicking and punching him as WQ grabs Enchantress, taking her to safety. Mezun begins pounding Wolf as he tries to hit Gilbert with a finishing blow.

Wolf smacks into a tree. Immediately after, Gilbert teleports next to Wolf and slams him to the ground. Hunt drops from the mountain, surrounding himself in flames. He punches Gilbert into the dirt.

Hunt: "I knew you couldn't be trusted." *He continues hitting Gilbert, making his head go deeper into the soil.*

Mezun turns to Hunt and shoots tentacles at him. Hunt stops hitting Gilbert and retaliates by throwing fireballs at Mezun, but they're swatted away. He taunts him.

Mezun: "You have so much talent, yet all that is wasted on your cause." *He zooms over to Hunt and hits him.*

As Mezun hits Hunt repeatedly, he looks in Enchantress's direction. He throws Hunt while Wolf runs to Enchantress. Gilbert finally gets up from the ground and tackles Mezun into a tree.

Gilbert: "I told you to leave her alone. Torment me all you want, but she is off limits."

Mezun: "Love makes you weak. Kill her and join me, Gilbert."

Gilbert looks at Enchantress. He lowers his head with a depressed look on his face. He mouths to her that he's sorry, then turns to WQ.

Gilbert: "WQ, please protect her for me."

WQ: "I will. What are you going to do?"

Gilbert: "Don't worry."

Gilbert grabs Mezun and takes him away. They disappear in the dust. WQ and Enchantress look at each other, wondering.

Enchantress: "Where is he going? He was going to stay with us."

WQ: "You do know that he did what he did to save you."

Wolf: "Where is he?"

WQ: "He left, and he took Mezun with him."

Hunt finally gets up and looks at Wolf with anger in his eyes.

Hunt: "I told you he couldn't be trusted. This is bullshit. When I find him, I'm going to kill him."

Wolf: "We'll find him together, don't worry. Next time, we'll make sure he doesn't escape."

Hunt: "Oh, he won't escape. I can guarantee that he won't be doing anything when I'm done."

Darren puts the files down, sits back, and takes a breather. So many thoughts cross his mind, questions ranging from how his parents actually got back together to why his grandfather is still around. Gail walks in and sees Darren confused and overwhelmed.

Gail: "Is something wrong?"

Darren: "How long were you planning to keep this from me?"

Gail: "I was afraid to tell you your father was an evil man who could never leave his past behind. You don't deserve to grow up like he did."

Darren: "Why did you marry him? How could you be with someone like him?"

Gail: "I was stupid and lovestruck, but in the end, I'm happy. I was with your father because he gave me the best thing in my life..." *She looks Darren in the eyes.* "He gave me you."

Darren looks into his mother's eyes and starts to cry. He hugs her tightly as she hugs back. She whispers in his ear.

Gail: "I will protect you with everything I've got."

Darren: *He looks in his mother's eyes again.* "No, it's my turn to protect you. You've been through enough."

Gail: "Darren..."

Darren: "Mom?"

Gail: "Yes, sweetheart?"

Darren: "How did you and Dad get back together after the mission in the files?"

Gail: "It was two months after that mission. Nora and I went out for a ladies' night to recover from the missions done in those months. We were going to go to the bar for a few drinks when I got a call from one of my clients. She wanted me to bring the group to do an investigation. Nora and I looked at each other and decided to do the job without the guys. As we left, I noticed a man with hair covering his face and a dark blue hood standing outside the bar. He seemed to follow me with his eyes, but I continued on my way. I asked Nora if she saw him, but when I went to point him out, he was gone. So I just ignored it and we rushed to my client's house."

Gail begins to get flashbacks of that night.

Nora: "So, what's the mission?"

Gail: "My client said she will explain when we get there. I don't think it will be too difficult."

Nora: "I just need a break from the two knuckleheads."

Gail: "They're not that bad."

Nora: "You're joking, right?"

Gail: "Well, Albert is always busy with Aggy, and Will is always training or trying to impress you."

Nora: *Blushing.* "He can be a real sweetheart, but sometimes I wish he would just relax."

Gail: "Awww, you're worried about him."

Nora: "Shut up!" *She blushes and giggles.*

They pull up to an old cabin deep in the woods and start heading to the front door.

Nora: "This place is kind of creepy."

Gail: "Well, Kate is a bit of an oddball, and her husband Jeff isn't all there himself."

Gail knocks, and a tall, slender man answers the door. He has long hair and a beard, dressed like a lumberjack. Gail waves as he greets them.

The tall slender man: "Oh, hi Gail, it's been a while. How are you?"

Gail: "I'm good, Jeff. This is Nora. I don't believe you two have met."

Jeff reaches out with his skinny hand and fingers.

Nora: "Hi, Jeff." *She shakes his hand.* "It's nice to meet you."

Jeff: "Likewise. Come on in, Kate is in the kitchen."

They walk in and head to the kitchen. They see a short female with long brown hair, freckles, and glasses, cooking with a black cat on her shoulder. Gail walks in and hugs her.

Kate: "Hey, Gail, sweetie, it's been too long." *She hugs her back tightly.*

Gail: "Yes, it has." *She grabs Nora.* "This is Nora. She will be helping with the mission."

Kate: "Any friend of Gail is a friend of mine." *She hugs Nora.*

Nora: "It's great to meet you finally." *She hugs her back.*

Gail: "So how are things, Kate?"

Kate: "They're good, but they've been kind of busy. A lot of the wood nymphs have been getting killed or wounded out here. I've been healing the ones I can."

Gail: "Is that why you called me?"

Kate: "Yes, I need you to kill whatever is doing this."

Nora: "What do you think is doing this?"

Kate: "Well, I wasn't sure until a few minutes before I called Jeffy. He was getting firewood when a young wood nymph ran out of the woods with a huge Duwende (a demon hitman that can change its form. Unlike a shapeshifter that's limited, it has no limit on what it can change into) running after him, snarling and already covered in blood."

Gail: "A Duwende? But they don't normally hunt like that."

Kate: "It didn't look like a normal Duwende. It seemed feral, and it was acting like it was burning from the inside or something. Black smoke was coming out of its mouth, and its eyes glowed a deep blood red."

Nora: "How did Jeff get rid of it?"

Jeff walks into the kitchen quietly and stands in the corner.

Kate: "Oh, Jeffy, you must be hungry."

Jeff nods his head slowly up and down as his stomach growls.

Kate: "Sit at the table, and I will give you your food. You can explain to Gail and Nora your side of the story."

Jeff sits at the table quietly as Gail and Nora join him.

Gail: "So Jeff, how did you get rid of the Duwende?"

Jeff: "Well, as the young nymph hid behind me, I told the Duwende to stop, but it just kept growling and clawing at the air. As I picked the nymph up, the Duwende charged at me, so I had to use my light, and I pushed it back. It started going crazy and ran back into the woods."

Nora: "The light? Flashlight or candlelight?"

Gail: "Oh yeah, I forgot you're half-angel."

Jeff: *Nods.*

Nora: "Wait, you're part angel?!"

Jeff: "Yes. My father was an angel, my mother was human."

Nora: "So if you're half-angel, and your power affected the Duwende, that means it had a demon in it."

Jeff: "Well, I saw a man come up this way a month ago. He made me feel uneasy and smelled like sulfur. He asked me if there was a lake for him to get some water. I pointed him to the nearest one and haven't seen him since."

Gail: "Maybe it was a demon hiding in a human's body."

Jeff: "I could take you to the lake if you want."

Gail: "Sure, but you should eat first. Nora and I will look around the house to see if we can find anything."

Nora: "Where was this attack?"

Jeff: "Around back by the firewood."

Kate walks in with some cookies and cakes.

Kate: "Here you go, honey. Enjoy your food."

Gail: "We'll be out back looking around to see if we can find anything."

Kate: "Ok, be careful. And if you need anything, just yell."

Gail: "You know me, I'm always careful."

Nora: "Really?"

Gail: *Kicks Nora in the leg.* "Who asked you?"

Nora: "Ow! What the hell?"

Kate: *Giggling.* "You two are funny."

Nora and Gail head to the back of the house and start investigating. Nora notices there's blood on a nearby tree and calls Gail over.

Nora: "Gail, I found some blood over here, but it doesn't smell like normal blood."

Gail: "Should I test it?"

Nora: "Yeah."

Gail goes into her bag and takes out a small bottle of holy water, then sprays some on the blood. It starts to bubble and smoke.

Nora: "That's demon blood then."

Gail: "Is it possible that the Duwende could have ingested some demon blood? That’s what is causing them to act this way."

Nora: "It is a theory. We need to check out that river."

They went to the river and saw what seemed to be an animal's footprint next to the water. They slowly get a closer look at the print. Gail sees a vision of how the beast reacted after drinking from the river. She saw through its eyes that it was hard to focus, enraging it. Gail talks to Nora.

Gail: "Nora, check this out. This is the footprint of the Duwende. I got a vision from it; I saw what it saw and felt. It drank from the river."

Nora: "So, how do we cleanse the river?"

Jeff: "I can purify the river. It'll take me a while, though." *In a low, soft tone, startling both the women. They both scream.*

Gail & Nora: "Don't sneak up on us like that! You almost gave us a heart attack."

Jeff: "Sorry, Kate hates it when I do that to her, too. I just wanted to make sure I would be here in case you needed me. Luckily, I was." *He smiles weirdly.*

Gail: "It's fine, just let us know you're with us. You know how things can get out here."

Nora: "What would you like us to do while you're doing the cleansing?"

Jeff: "Please, guard me. Make sure that I don't get attacked once I start the ritual. Once I start the process, I won't be able to protect myself."

Gail & Nora: *In unison.* "Consider it done."

Jeff begins to speak in an unusual language as his eyes and hands begin to glow. The water begins to illuminate, appearing like the sun itself was in it. Suddenly, in the bushes across from the river, the leaves and branches start to move.

Nora: "Did you hear that?"

Gail: "Yes, I did. Let's just stay here. If anything comes out of it, then we'll handle it accordingly."

Out of the bushes jumps the Duwende that was around earlier. Nora and Gail run toward it to keep it away from Jeff. As they charge toward it, the beast dodges their attacks and lashes out with massive hits of its own. It knocks Nora into the bushes, then grabs Gail. Right before the Duwende begins to bite Gail, a hooded figure jumps down from a tree and hits the beast. The beast drops Gail and turns its focus on the hooded figure.

Hooded figure: "Are you alright?" *Looking at Gail.*

Gail: "You're the guy from the bar."

Nora can be heard from the bushes.

Nora: "I'M FINE! WHO ARE YOU?!"

The Duwende bites onto the hooded figure's shoulder and starts to shake its head, but the figure lifts the Duwende by the throat and slams it to the ground. He starts punching it deeper and deeper into the dirt until the head explodes and nothing is left. Even after the Duwende dies, he keeps punching until Gail yells for him to stop. The hooded figure stops and looks at her, then starts walking toward her.

Just as he gets close, Jeff finishes the cleansing. He heads in the hooded figure's direction, but Jeff appears different. He looks extremely buff and is about to hit the stranger when Gail yells.

Gail: "He saved us!"

Jeff: "Sorry, I thought he was going to harm you. I sensed an evil presence."

The hooded stranger senses Jeff's power and sprints to Jeff, but Gail gets in the way. The moonlight hits the stranger's face, and Gail realizes it's Gilbert. She notices a horrible scar around his neck.

Gail: "Gilbert, is it really you?"

Gilbert: "Yes, it is."

Gail: "What happened to your neck? Did your father do that to you?"

Gilbert: "No, I had to do it to get away from him. I had to scrape off the tattoo to escape and find you, no matter what it took."

Nora: *Looking at the scar.* "Wow, didn't that hurt?"

Gilbert: "I did whatever I had to do to be near Enchantress."

Nora: "Aawww, that's so sweet."

Gail: "Don't give him a swelled head." *She chuckles.* "Oh yeah, by the way, Jeff, this is Gilbert. Gilbert, this is my friend Jeff."

Gilbert looks at Jeff with disgust, but still shakes his hand. While they are shaking hands, their hands begin to steam. Gilbert pulls his hand away almost immediately. Gail walks up to Gilbert and asks:

Gail: "Are you ok?"

Gilbert: "I'm fine. I've been thinking about you ever since the last time I saw you."

Gilbert and Gail begin to talk amongst themselves while Nora walks over to Jeff.

Nora: "What was that all about?"

Jeff: "I don't like it. He has some demon blood in him. He's not good for her. I hope she is very careful around him."

Nora: "She can handle herself. But I do agree, I don't think he's right for her."

Then Gail finishes telling Darren as she blushes.

Gail: "He was a sweet, sweet man back then. We kept talking and got engaged. Then, you finally got married, and then you were born. If he stayed the way he was when you were born, I know you two would have been inseparable."

Darren: "So what happened to him? What made him change?"

Gail: "He let the darkness get the best of him. He didn't know how to control it. He tried to handle it alone. I told him to let me help him, but he just didn't want me to see him at his worst. Then you saw how that turned out."

Darren: "Yeah, I know. Wish I could've known him the way he was. By the way you describe him, he sounds awesome."

Gail: "Me too, baby, me too." *She looks down, and her eyes begin to water.*

Just as Darren and Gail are hugging, Louie comes in with Cross.

Louie: "Gail, may I have a word with you? It involves the previous mission."

Gail: "Sure, what is it?"

Louie: "I was wondering if we could talk in the living room."

They walk to the living room and begin to talk about the mission. Nora joins in the conversation.

Nora: "Just out of curiosity, how would you feel if we had to kill Gil and his father?"

Gail: "Years ago, I thought I killed Gil. But I guess I was wrong. So if it has to be done, so be it. He's not the man I fell in love with."

Louie: "How do you think Darren will handle it?"

Gail: "He's a strong boy. He'll bounce back."

Nora: "Yes, he is. At least you won't be alone in dealing with this anymore, Gail."

Gail: "I know, and I appreciate it, but you know me. I don't like being a nuisance."

Nora: "It's never a nuisance. This is what family does for each other."

Somewhere in hell, Mezun, Zima, and Amon are plotting something.

Zima: "I think I've found the perfect way to destroy the group from within."

Mezun: "And how exactly are you going to do that? Is it their family?"

Abruptly, a female walks in and asks Zima if it's time.

Zima: "Not yet, my dear, be patient. Your time is coming very soon."

Unknown female: "When? I am eager to do my mission. My revenge is all I want."

Zima: "It's a process. You'll get your opportunity. Just wait until I tell you. Now leave before I get someone else. GO!"

Mezun: "What the fuck was all that about?"

Zima: "I have many plans; she is just one of my pawns to do my bidding, just as Amon is your pawn."

Amon: "I am no one's pawn; I am..."

Mezun: "You are nothing. The voice you heard was Zima imitating the Dark Lord. So shut up, or we'll call Archangel Michael and Archangel Raphael to come get you."

Amon: "I left heaven on a promise that I would help reign with the Dark Lord. It's too late to go back now. I'll fall back."

Zima: "My, my, how the unfaithful are quick to jump ship when they don't get their way. Aawww, what happened?" *In a sarcastic tone.* "Did I crush all your hopes and dreams?" *Evil laughter.*

In heaven, the archangels are training potential angels to take their place if anything were to happen. The head angel, Michael, walks over to the angel trainee, Guardian Angel Isaac.

Archangel Michael: "Your archangel training begins today. If anything were to ever happen to Raphael, Gabriel, or me, you would be next in line to protect."

Guardian Angel Isaac: "Understood. I am honored that you have chosen me to become next in the ranks."

Archangel Michael: "This is the reason why I had you sit in our meetings and participate in them, giving you some experience when the time comes to take over."

Archangel Raphael: "Haven't you noticed that no other angel has been in the meetings with us? Only a select few are worthy of this ranking or title."

Guardian Angel Isaac: "Are there going to be any other angels that might get promoted?"

Archangel Michael: "By the time you have finished your training, out of the potentials, you shall choose which one you'll train to be our successors."

Back in New York, while Nora, Cross, Rayne, Gail, and Darren take a trip to the southern state, Louie decides to go back to New

Jersey to try to look through his father's files. On the trip to Jersey, Ark calls him.

Ark: "Louie, what are you up to?"

Louie: "Just going to Jersey to see what information I can find in my soon-to-be office."

Ark: "Your father's old office?"

Louie: "Yes. Is there something I should know about it?"

Ark: "The reason your parents bought this place was because it is in the center of all paranormal gates. I would suggest that you choose a time when you're not alone."

Louie: "I can handle whatever comes my way."

Ark: "There's still a lot you don't know. Don't be so stubborn. At least wait until I get there."

Louie: "It's going to take me about two to three hours to get there, so depending on where you're coming from, you should be there before me. Which is nothing new, 'cause wherever I go you're already there."

Ark: "Alright then, I'll see you at the house."

Louie: "Would you like me to pick up anything for you?"

Ark: "No, thank you. See you then. Actually, could you pick up some pie?"

Hours later, Gail pulls over to a diner on the side of the highway so everyone can stretch their legs, use the facilities, and eat food. Nora and Rayne head to the restroom, while Cross and Darren go into the restaurant to find a seat. Gail looks around and then realizes that a car parked two lanes away has been following them for the past five miles. She calls Nora to let her know what she has just seen.

Gail: "Nora, when we were on the highway, remember the car I asked you to watch?"

Nora: "Yes, why do you ask? I saw the car get off at the exit ten exits ago."

Gail: "Well, when you're done, I need you to verify that this isn't the same car."

Nora: "No problem. Why are you so paranoid? You do know that the things that have been happening to us are supernatural. Everything else we can handle with no problem at all."

Gail: "I know. Just want to make sure that we get to enjoy this vacation without any incidents or emotional rollercoasters."

Rayne walks over to join the boys, and Nora joins Gail. She begins to calm her down when Gail points at the car.

Gail: "Since the vehicle arrived, no one has gotten out of it."

Nora: "I see why it has you nervous, but your paranoia has great senses. That car is the same one. I recognize it from the dent on the right side above the wheel well."

Gail: "I think we should go inside and order the food to go."

Nora: "I agree. I'll get the food as you and the kids go back to the car. Explain it to them as you wait for me to return. If things seem like something is about to happen, honk the horn twice."

Gail nods her head, and they get their plan underway. The teens ask Gail why they can't eat here, then she explains the situation.

On the highway in New York, Louie is stuck in traffic on his way to the George Washington Bridge. He calls Ark to let him know his current status. Ark laughs at him and tells him that he'll be waiting for him in the office.

Three hours later, Louie finally gets to the house. He walks in and checks to make sure no one else is inside. He moves the rubble around until he finds the doorway to Hunter's Lair. Louie calls for Ark out loud and waits for a response. After a minute and a half, he doesn't hear anything. He clears off the rust and debris from the lock and then opens the door. He walks down the stairs and hears someone humming. Once at the bottom of the steps, he sees Ark sitting down, reading a file.

Blak Hart: "Here's your pie. Whatcha doin'?"

Ark: "Reading a file on me. Just wanted to see what info they had on me and to see if it is accurate."

Blak Hart: "Well, is it?" *He wonders.*

Ark: "It's on point, but this information will be given to you when it's time."

Blak Hart: "After all this time, you're still withholding info? Really, Ark?"

Ark: "Soon enough, grasshopper. Soon enough, you will know what's in the file and more. Just have to be patient."

Blak Hart: "Is that the reason why you're here? How did you get down here without using the secret entrance or the door in the kitchen when there was debris in front of it?"

Ark: "How I did that is kind of explained in the file, which you will get after a specific mission."

Blak Hart: "So give me the mission now so I can do it."

Ark: "That mission is for the future, not now. All will be revealed in time."

Blak Hart: "It's always the future! What do I have to do to prove that I'm strong enough?! Fine, I'm going to the liquor store while you think of more shit you want to hide from me."

Ark: "Stop acting like a bitch. Go get something to drink so you can shut up."

Blak Hart: "So what do you want to drink? Maybe some verbal diarrhea to cleanse the palate." *Joking.*

Ark: *Smiling back.* "No, thank you, wise-ass. I would like an Italian drink that has a sweet almond flavor. I believe the bottle is square-shaped."

He walks to the store when a physically fit, short beautiful and slightly tanned complexion, dirty blonde-haired woman wearing a jean jacket, white tee shirt, and black jeans bumps into him.

Blonde-haired woman: "Oh, sorry, I was looking at my phone."

Blak Hart: "It's ok, it's my pleasure. I'm Louie. Who is the lovely angel that graced me with her presence?"

Blonde-haired woman: *Blushing.* "Catherine. Wow, handsome and witty." *She giggles.*

On the road, Nora and Gail are trying to decide whether to go to Virginia Beach or Florida.

Gail: "So what are we going to choose, Virginia Beach or Florida?"

Nora: "How about we go to Virginia Beach first for a few days, then we can go to Florida."

Gail: "I think that would be a great vacation."

As Gail continues to drive, she sees the car again in the rearview.

Gail: "Nora, it's back! What should we do?"

Nora: "Floor it. Let's park somewhere to find out who dares confront us."

Gail speeds to the next rest stop, but the car doesn't follow. Nora and Gail look at each other in confusion.

Gail: "Why didn't that person follow us?"

Nora: "I don't know. How far are we from Virginia Beach?"

Gail: "About an hour and a half away."

Nora: "Then let's stay the night at this hotel to collect our thoughts. Then we'll continue in the morning."

Gail agrees. They go to the front desk to get a room for the night and head upstairs. Nora keeps looking out the window, but there is no sign of the vehicle anywhere.

Darren: *Thinking while looking out the window.* "If my father couldn't fight it… what chance do I have?"

Hours go by, and everyone is enjoying themselves. They decide to go to the gift shop on the first floor of the hotel. Rayne decides to relax in the room while everyone else goes. Once Nora, Gail, Darren, and Cross get into the elevator, Rayne hears a knock on the door.

Rayne: "Did you guys forget the key?"

She walks over to the door and opens it. No one is there. She thinks nothing of it and closes the door. Rayne sits back down to continue watching television when it happens again. She tries to ignore it, but the knocks keep on until she yells.

Rayne: "Stop knocking on the door, I'm not opening up!"

Suddenly, it stops. A couple of minutes go by, then the knocks turn into banging. She yells and screams, but to no avail. The banging is so strong that the door seems like it's about to burst open.

In the gift shop, Cross senses something wrong with his sister. He lets Nora know he's going to check on Rayne and rushes off.

Gail and Nora continue browsing when Nora suddenly sees someone.

Gail: "Nora, what's wrong?"

Nora: "I just saw a man in one of the aisles. He looks very familiar… not in a good way either."

Gail: "Do you know where you might have seen him?"

Nora: "I'm trying to figure it out. Since we've been having a stalker issue, I think we should be extra aware of our surroundings."

Gail: "I agree, but we don't know who they are or what we're looking for. Where's Darren?"

Nora: "He said he's going to join the twins."

In the hotel room, Cross is doing all he can to calm Rayne. She seems to be having a nervous breakdown. Cross hugs her. Moments later, there's another knock on the door.

Cross: "Who the fuck is knocking on the damn door?"

Darren: "Dude, it's me, Darren. Damn, what's your problem?"

Cross: "Some asshole has been fuckin' with Rayne while we were downstairs. When she opened the door, no one was there."

Darren: "That's fucked up. Who would dare do something so childish?"

Cross: "I just hope they try again, but we both know they probably don't have the balls to do it again."

At that exact moment, the doorknob starts turning. Since they can't get in, they begin banging on the door. This time, a deep dark voice is heard.

The voice: "Let me in, kids. Don't make me break the door down."

Cross: "Over my dead body, asshole."

The voice: "That can be arranged. Learn to choose your words carefully, boy."

Darren: "Do you really think your threats frighten us?"

The voice: "Hope not, then it wouldn't be fun. Your grandfather and father are pretty powerful. If you join me, you could be ten times more powerful than both combined."

Darren: "Now, why would I want to join you? I don't know who you are." *Sarcastically.*

The voice: "If you let me in, you'll be able to find out who I am."

Rayne: "Never! You would have to get through all of us before you could get him."

The voice: "You kids make such interesting proposals. My, oh my, how ever could I pass it up?" *Sarcasm.* "Just let me in so we can talk. As for not knowing who I am, my name is Julius. Now that we know each other, why not just open the door? I could've broken it down a while ago, so take that into consideration."

Cross: "If you could've done it, then do it already."

Julius: "Is that an invitation?"

Cross: *Slowly transforming, he begins to speak in a growl-ridden voice.* "If you get in, don't think we're your welcoming party."

Julius: "It wouldn't be fun if I didn't get at least a little resistance."

Meanwhile, in the gift shop, Nora follows the stranger around the store. Gail advises her not to, but Nora keeps stalking him. She loses him, then Gail sees him leaving. The stranger is clearly visible, with a light complexion, black hoodie, denim jeans. Gail calls Nora over to the window. They watch as he gets into a jet-black pickup with tinted windows and speeds off.

In the room, the teens' adrenaline is at its highest.

Cross: *Looking at Darren and Rayne.* "When he comes in, you two be ready to back me up."

Julius: "Oh no, they have a plan." *Sarcastic tone.*

Cross walks to the door, unlocks it, and steps back.

Julius: "Ah, good choice, kids. Now we can have a nice conversation." *He starts to open the door slowly.*

As Julius opens the door, Cross jumps and grabs his face, slamming him into the wall. He starts punching and clawing at him. Julius kicks Cross off and retaliates with an arsenal of kicks and punches, sending him back into the room. Cross looks up at Julius as all the wounds he inflicted heal instantly. Julius licks blood dripping from the side of his mouth.

Julius: "How lucky am I, huh? I get to fight a werewolf. It's been so long."

Cross: "What the fuck are you?"

Julius: "HAHAHA! Come on, I know you can smell it on me."

Cross: "You smell like blood and death. So... are you a vampire or something?!"

Julius: "Ding! Ding! Ding! Give this pup a biscuit. Correct, kiddo. Now let's finish this game."

Cross jumps up and starts ripping and slashing. Julius does the same. Cross bites down on Julius's arm and rips it off, then slams him to the ground.

Julius: "I'm so impressed. You were able to catch me off guard. I would clap, but you kind of ripped my arm off."

Cross: *Enraged.* "Why are you still talking?!"

Darren and Rayne run out. Rayne uses her hydrokinesis to push water at Julius, while Darren sends a blast of lightning at the same time, accidentally shocking both Cross and Julius. Julius fights through the current, grabs his severed arm, and runs to a nearby window. He smiles at the kids.

Julius: "Ta-ta for now, kids. You're way too much fun to kill. Now let's promise to play again when you're more experienced."

Julius jumps out the window. Right before hitting the ground, he turns into a swarm of bats and fades into the night.

On the first floor, Gail and Nora finish shopping and head up. When they reach their floor, they see blood on the hallway floor, the wall in front of the door dented, and Cross's body steaming.

Nora: "What happened here? Are you all alright?"

Cross: "We just had a run-in with a vampire named Julius. No matter how much I hurt him, he would just heal up and keep talking crap."

Rayne and Darren come out and explain everything in full. Gail runs up to Darren and hugs him so tight he can barely breathe. Nora sits them down to check for injuries. She notices that Cross's cuts and bruises are already almost healed.

Nora: "I'm going to the laundry room in the hallway to clean up the pants. Hopefully, I can get the blood off the jeans." *She looks at Gail, Rayne, Darren, and Cross.*

She walks down the hall. Nora hears footsteps several feet behind her. She turns, sees nothing, and continues on. In the laundry room, while she's washing clothes, someone sneaks in. She senses a weird presence and decides to call Cross.

Nora: *Whispering.* "I think someone is here with me. Please come quickly without causing a commotion."

Before Cross arrives, Nora turns and sees the man from the gift shop. In a blink, he stabs her. Nora looks down at the blood pouring from her stomach. She grabs the guy by the shirt and slams him against the wall. She hears Cross's voice as he rushes to help.

Nora: "Forget what I said! Go back and protect the others. I can handle this imbecile."

Cross: "You don't have to do it alone. I can help you."

Nora: "I know, sweetie. Just stay in the hallway. If I need help, I'll yell for you, ok?"

Nora commences questioning as she hits him in his pressure points.

Nora: "Who sent you?" *She hits him.* "What do you want?" *She hits another point that paralyzes him.* "What is your main objective?"

She glances at her wound, it's healing rapidly. The stranger looks shocked. A stab like that should have killed her, yet she's beating him senseless. Cross, impatient, rushes in and recognizes the man.

Cross: "That's the substitute teacher, Collin Dirksen!"

While Cross is talking, Collin recovers and vanishes.

Cross: "Where did he go? As far as I know, he wasn't that powerful."

Nora: *Upset tone.* "I told you to stay in the hallway next to the room! This is why! I was taking care of it!"

In New Jersey, Blak Hart and Catherine exchange phone numbers. She smiles at him and walks away. Blak Hart smiles back, then runs to Ark.

Blak Hart: "I met someone. She's gorgeous."

Ark: "What about her appeals to you?"

Blak Hart: "Everything. She has beautiful blue eyes, dirty blonde hair, a tanned complexion, and the sweetest voice I have ever heard."

Ark: "Slow down, Casanova. Did you get her name, or did you just stare and drool?"

Blak Hart: "Ha, ha, very funny." *Slightly offended.* "Of course, I got her name, and her phone number. So ha."

Ark: "Still didn't hear her name."

Blak Hart: "Her name is Catherine."

Ark: "What, no last name? Just Catherine?"

Blak Hart: "Her last name is Structer. Why does it sound familiar?"

Ark: "Nah, just bustin' your balls." *Laughs at his protégé.*

Blak Hart: "Ballbuster. Anyway, I think we should try to find anything we can on Collin Dirksen. He's been causing problems long enough. What do you think?"

Ark: "I agree. Try to find everything you can. I'll search for any additional info so we can get him to where he belongs."

Back at the hotel, Cross, Nora, and the rest of the gang are trying to compose themselves from the night's events.

Cross: "Mom, I'm sorry. I was just trying to help."

Nora: "I know, hun. Next time, do as you're told."

Rayne: "Cross said you were bleeding. How did you heal?"

Nora: "I am able to heal so quickly because I am a half-shapeshifter. Just like werewolves, we can cure a specific area while the rest of the body defends itself."

Gail: "I think we should get out of here. While I'm driving, Nora, could you call Nite to let him know what happened?"

Nora: "Then Nite could contact Ark so he could keep an eye on Louie."

They leave the hotel, get into the car, and begin their journey to Virginia Beach. Nora calls Nite. The phone rings twice, then hangs up. She tries again, but there is no response.

Nora: "Cross, call Louie and tell him everything that occurred. He needs to know that he has to watch his surroundings."

Cross: "Ok, ma. I'll call him and let him know."

Cross calls Louie, tells him everything, and warns him of a possible ambush.

Blak Hart: "Thanks for the heads-up. You want to know something funny? Ark and I are currently looking up all the info on Collin. Eerie, isn't it?"

Cross: "Creepy, but cool."

Blak Hart: "Just keep me posted on your vacation. If you need me down there, just call."

Cross: "Same here. Keep us posted."

Blak Hart: "Enjoy it for me. See you when you guys return, or I'm called to kick some ass."

Blak Hart hangs up, then tells Ark about the warning and everything he was told.

Ark: "Let's search for Collin and all possible vampires fitting the description."

Ark and Blak Hart look in one file, then another. Several hours later, the files on Collin are complete, and a game plan has been made to hunt him down and end his schemes. As for the vampire, no one fits the description given.

On the road, Gail finally gets to Virginia Beach. During the entire ride, there was no sign of Collin or the car. Everyone begins to plan what they want to do first. The teens want to go to the beach and dive in the water. The adults want to go to the spa and enjoy a nice hot bubble bath.

As the teens are at the beach, Collin has an eye on them from one of the food carts on the boardwalk. He gets on his phone and calls someone, telling the person on the other line where he is located and what Nora, Gail, and the teens are doing.

The person on the other line: "Just keep the adults separate from the teens. I'll notify you when to head back home."

Collin: "Wolf's widow and I already had a confrontation. Their son recognized me. So, what do you suggest?"

The person on the other line: "If it happens again, kill whoever gets in your way."

The teens are diving, racing out to sea. Everything is going smoothly until a huge wave pushes Cross farther out. The undercurrent carries him beyond the swimming parameters. He struggles to swim back, but his arms are getting heavier and heavier. Suddenly, something grabs his legs, pulling him under. He can feel himself slowly fading away. When he comes to, he looks around and realizes he's somehow in a cave.

Back at the beach, Rayne and Darren are frantically looking for Cross. They try to contact him, but he left his phone on the beach with his towel.

Rayne: "Cross! Cross, where are you?"

Darren: "He never goes anywhere without his phone. I hope he's fine. I think you should call your mom, she'll know what to do."

Rayne: "I should be able to sense him. I can feel he's still alive, but… he's confused. I don't know why, I just know he is."

She calls Nora to inform her of the incident.

Nora: "How did you lose him?"

Rayne: *Worried and scared.* "We all went swimming, then a strong wave came by, so the lifeguards told everyone to get out of the water. When we came out, we were looking for him and even called him. He left his phone with his towel. I sense him. The issue is that it doesn't feel like he's close by. The feeling I'm getting is that he's confused and anxious to get out of somewhere."

Somewhere in the cave, Cross is trying to find his way out. The only problem: every possible escape route is submerged in water.

Cross: "I am a great swimmer. The only problem I'm having is that I don't know how far into the sea I am."

He starts to concentrate. Since he's a twin, he tries to send a telepathic message to Rayne. He can sense that she's too nervous to fully understand him. He tries again, this time to Nora. He sends the message differently, attempting something he's never done before — creating a beacon, so that if she doesn't get it, another person with similar senses can pinpoint his location.

At the Four Corners, Zima and Amayo are speaking to Amon about Collin.

Zima: "Who sent Collin to follow the Cross family?"

Amon: "Someone has to keep tabs on them."

Amayo: "The further away from us they are, the better our plans will work."

Zima: "If we don't tell you to do something, don't just make things up. By doing that, you can cause them to get in the way of our plans."

Amon: "I thought—"

Amayo: "That's the problem. You thought. Zima, just let me tear him apart while he stays alive, as I eat his organs."

Zima: "No, not yet. He can still serve us. Next time you fuck up, Amayo will do with you what he chooses."

Amon: "No, please!" *On his knees, begging.* "I have served you to the best of my ability. If you grant me more power, I'll be better able to serve you. So, what's with the female that was talking to you last time?"

Zima: "That is none of your business. First of all, she's too much of a woman for you. Second, she has been more useful to me in a day than you have in the past year."

At the Heart property, Ark and Louie are done searching. They compare notes and look for a pattern of activity through the decades.

Louie: "At least we know he's somewhere in the area around Nora and Gail. Maybe I should head over there and see if I can track him down."

Ark: "I think it would be best if you try to find Cross. I was notified by Nite that the teens were on the beach, and he disappeared."

Louie: "I'm on it." *Worried for Cross.*

Ark goes outside, and Louie follows to say something — but then watches as Ark disappears. Confused, he's about to walk to the car when Catherine jumps in front of him.

Catherine: "Hey, you never called me."

Louie: "I was busy working, sorry. I didn't forget about you. Believe me, I wouldn't be able to forget you, ever."

Catherine: "It's fine. I thought you were blowing me off."

Louie: "Why would I do that? I was going to call you when I got back from my business trip."

Catherine: "What line of work are you in?"

Louie: "I'm a bounty hunter."

Catherine: "Sounds dangerous. How long have you been doing it?"

Louie: "Since my parents died."

Catherine: "Wow, sorry to hear about your parents. How long has it been? How have you survived on your own for so long?"

Louie: "Thank you. It's been twelve years, give or take a year or two. I just picked up a few tricks of the trade from my father. He always prepared me for 'just in case' scenarios."

Catherine: "Why was he preparing you at such a young age?"

Louie: *Defensive.* "Well, my parents were always prepared for any and all possibilities. That's just the way they were. I'm sorry, I have to cut this conversation short. I have a plane to catch."

He begins to walk toward the car. She runs up to him and kisses him on the cheek.

Catherine: "Don't forget to call me when you get back."

Louie: "Don't worry, I'll remember." *He grabs her, pulling her close to his chest, and gives her a long, passionate kiss.* "You're not a person I want to forget."

Catherine blows him a kiss as he gets in his car and drives off. She touches her lips, slightly hot and bothered. As he drives, all he can think about is how good that kiss felt — from the scent of her perfume to her soft, warm lips that gave him goosebumps.

At the beach, Nora and the teens nervously try to think of ways to find Cross. A young man, about 18 or 19, approaches them. He has tanned skin, black hair, a medium build, brown-yellow eyes, and stands about 5'11".

The young man: "I'm sorry to overhear that you might have lost someone in the sea. By any chance, can you describe his appearance to me?"

Rayne describes Cross to him. The young man excuses himself and runs off. Darren looks around.

Darren: "Where did he go? I thought he was asking so he could help us out."

Gail & Nora: "Hopefully, he went to start looking."

In the cave, Cross uses his senses to find an exit. Suddenly, he feels an evil presence watching him. He calls out.

Cross: "Who's here? Why did you bring me here?"

The water in the middle of the cave begins bubbling. A male figure pops his head out.

The male figure: "Are you Cross? Your family is worried sick, hoping for your return. Can you hold your breath for a long period of time?"

Cross: "Who are you? Why did you bring me here? Why do you want to know if I can hold my breath?"

The male figure: "The longer you can hold your breath, the quicker you can get to your family. I didn't bring you here. Right now, I'm going against my family code by showing myself to you."

Cross: "So you're here to help me?"

The male figure: "That is correct. Are you ready?"

Cross: "Ready as I'll ever be. I can hold my breath for at least five minutes, so we can submerge every five. How about you? Do you have any scuba gear?"

The male figure: "I am going against my clan by saying this… I am a water breather."

Cross: "So you're a mermaid?" *Shocked.*

The male figure: "No, no, no." *He chuckles.* "Mermaids are female. Males are mermen. But we prefer to be called water dwellers. Before we leave, you must promise me you'll keep my secret."

Cross: "No problem. (There's probably a file on the water dwellers anyway.)" *He thinks.* "Before we go, what's your name?"

The male water dweller: "Suma. Let's leave it at that for now. We'll meet again."

Cross: "How would you know that?"

Suma: "It's a hunch. A lifetime is long enough to see someone more than once."

Cross: "I agree."

Suma takes Cross's hand and tells him to hold his breath. They swim as fast as they can to the shore.

Cross: "How did we get here so fast?"

Suma: "Sea creatures can swim much faster than any land dweller could ever imagine. I believe your family is over there." *He points to the lifeguard station.*

Cross looks over and sees them. Just as he turns to thank Suma, he's gone.

Cross: *Thinking out loud.* "No one is going to believe this. Maybe they would… If we're able to deal with things people consider myths, then I guess they might believe me. But what pulled me into the cave? He said he didn't do it… so what was it?"

Cross runs over to Nora and Rayne. They all hug. She bops him on the back of his head.

Cross: "What was that for?"

Nora: "For scaring the crap out of me. Don't you ever do that again."

Rayne: *Punches his arm.* "That's for driving me crazy — sensing you but not being able to reach you."

Cross: "Sorry. You guys are acting like I meant to get lost. I was pulled under by something. What it was, I don't know."

Gail: "Next time, make sure you're close to everyone."

Cross: "I was. Like I said, something pulled me under and took me to a cave."

Nora: "Do you know where?"

Cross: "No."

Darren: "So how did you get back?"

Cross: "Someone named Suma helped me get to shore."

Rayne: "Can you describe him?"

Cross: "Sure. Tanned skin, black hair."

Nora: "That's the young man who asked me about your description. Where is he now so I can thank him?"

Cross: "He showed me where you were. Then, when I turned around, he was gone."

Back in Jersey, Louie is racing down the highway to help Nora when he receives a call from Ark.

Ark: "Go to Shepherdstown, West Virginia. There are reports of Collin being there."

Louie: "What about Cross?"

Ark: "He got help from a teenage merman. He's safe."

Blak Hart: *Thinking.* "A merman? I guess if ghosts and goblins are real, it's not a stretch that sirens and mermaids are too."

Four hours later, Blak Hart finally arrives in Shepherdstown. He looks around for a place to rest. He spots a small hotel in the center of town. After parking, he opens the trunk to grab his weapons for the hunt. Then he walks into the hotel.

After getting his room key, he notices a demon in the lobby talking to someone. He rushes over, pulls out his pistol, and shoots the demon. The woman with the demon screams and runs. He chases her into a corner. With nowhere to run, she turns around, shocked.

My life is filled with heartache and pain, the struggle is shared by few but felt by many. Blood and carnage are a basic part of my life, many are the cause, but few will survive. This is why we punish those that deserve it, while we help those that are in need. Call us demons or call us angels; it really doesn't matter, for we're simply doing whatever will help the greater good.

Blak Hart

There is light and dark, good and evil, yet most choose to do what come easy instead of working hard to enjoy the fruits of their labor…

Archangel Raphael

CHAPTER 7.
FLORIDA

He sees the woman covering her face. He walks up to her and begins to ask her questions.

Blak Hart: "Why were you talking to a demon?"

She had nothing to say. She just kept shrugging her shoulders.

Blak Hart: "What's your name?" *getting annoyed*

Still no response. She points behind him.

Blak Hart: "What are you pointing at?"

He turns around and observes another demon coming out of the ground. He turns back around; the woman is gone. The only thing left is the smell of her perfume.

Blak Hart: "Fucking lady made me aware of an attack but gave me the slip. Her perfume smells familiar. I just can't pinpoint where at the moment."

The demon charges at him, and once it gets close, it jumps at Blak Hart. Unfazed by its actions, he puts his hands through the demon's chest. He throws it to the ground, revealing that he grabbed the demon's heart in one swift motion. He drags the carcass outside and torches it to ash, then walks back into the hotel as if nothing had ever happened. Still baffled by the perfume, he sits on the hotel bed to think.

Blak Hart: "I really wish Catherine was here. Her beautiful smile, her soft, sensual voice, the smell of her..." *Then it hits him* "That's where I smelled that perfume from, but that couldn't be her.

She's miles away from here. That woman didn’t look like her. Fuck it, maybe the long drive has me imagining things. I'll rest up and hunt later on."

Hours later, Blak Hart gets a phone call. He looks at the phone. It’s Catherine. He goes to answer it, but he's too late.

Blak Hart: "She should've waited a little longer." *He calls her back. It rings, then she picks up.*

Catherine: "Hello."

Blak Hart: "Hi, is everything alright?"

Catherine: "Sure, I was just wondering if you were back yet."

Blak Hart: "I'm still working. When I get back, I'll make sure to call you."

Catherine: "I'm sorry if I'm bothering you."

Blak Hart: "It's no bother. I was actually thinking of you."

Catherine: "Really?" *playing coy* "What were you thinking?"

Blak Hart: "I was thinking of your beautiful smile and the smell of your perfume. By the way, what's the name of it?"

Catherine: "Why? It’s Antique Love by Tamara Beth Hamilton."

Blak Hart: "I just wanted to know. I'll call you when I get back. I have to rest up for a few. It’s going to be a long several days."

He hangs up the phone and thinks out loud.

Blak Hart: "She called shortly after the whole ordeal. Pretty coincidental. If it was her, why didn’t she come to me? And what is she doing here? Nah, I'm probably overthinking it. I'll just rest up for at least an hour, then Collin, here I come."

Before going to sleep, he calls up Cross to see how he's doing.

Blak Hart: "Hey, how are you feeling?"

Cross: "Confused and tired. Where are you?"

Blak Hart: "Pretty close to you, I'm in West Virginia."

Cross: "Come by, so you can enjoy the vacation too."

Blak Hart: "I would love to, but I'm working. I'm chasing Collin. This time, he's not getting away."

Cross: "Need a hand?"

Blak Hart: "Would be nice, but this mission I have to do alone. Ark already stated that Collin has to be stopped, no matter what I have to do to get it done."

Cross: "Wow, Ark must really want him."

Blak Hart: "It's more that we both want him off the streets so he doesn't hurt anyone else ever again. By the way, I met someone."

Cross: "How did you meet, and how does she look?"

Blak Hart: "I'll explain all that when I finish this mission. Then I'll meet you in Florida."

Cross: "Sounds good. I'll see you then. Call first to make sure you know where we're at."

Blak Hart: "Will do. I'll talk to you later."

In Virginia Beach, Cross takes a nap while Rayne answers her phone.

Rayne: "Hello?"

Luna: "Hey, so how's your vacation going?"

Rayne: "It's okay. We're getting ready to head toward Florida."

Luna: *extremely excited* "That's great!"

Rayne: "Why?"

Luna: "I'll be going to Florida too."

Rayne: "Cool, where in Florida are you going to be?"

Luna: "Wherever you and the charming Cross are going to be."

Rayne: *squeals with excitement* "Here you go again, did you ask my mom yet?"

Luna: "I was wondering if you could ask her for me?"

Rayne: "I'll ask her. When are you going to be there?"

Luna: "Tomorrow night, that's when my flight arrives in Tampa."

Cross walks up to Rayne and asks her who she is talking to.

Rayne: "One of your biggest fans."

Cross: "I have a fan?!"

Luna hears Cross's voice over the phone and gets quiet. Rayne teases her a little. Cross grabs the phone and speaks in a heroic voice.

Cross: "Hello there, it's Michael A. Cross speaking. How is my biggest fan doing?"

Rayne tries to hold in her laughter, letting little bursts out. Luna responds in a shy tone.

Luna: "H-hey Cross, I'm good. How are you?"

Cross's face turns bright red, realizing it was Luna, and he starts to stutter.

Cross: "L-L-Luna, I-it's you. Hehe. H-H-Hi, I-I'm good."

Rayne bursts out laughing and falls to the ground.

Rayne: "HAHAHAHA! Your face is so red! What happened to the confident hero voice?" *still laughing*

Cross: "Shut up, Rayne!"

Luna: "I'm heading down to Florida, so I will be hanging out with you guys."

Cross: *blushing and sweating from the heat he's feeling* "That's g-g-great. Can't wait to see you. It's been a while."

Luna: "Yeah, it has. Umm, I gotta go. Tell Rayne I will call her later. Could you relay the message to me, please?"

Cross: "Sure, no problem, Luna. Anything for you."

Cross blushes so much his face gets even redder. *He slaps his forehead with his palm.*

Luna: "Hehe, thank you, Cross. See ya soon."

Cross: "See you soon." *still blushing*

Cross hangs up the phone, then throws it at Rayne.

Cross: *embarrassed and angry* "WHY DIDN'T YOU TELL ME IT WAS LUNA?!"

Rayne: *in a superhero voice, mimicking her brother* "What's the fun in that?"

Cross throws himself on the bed and lands face-first on the pillow. He starts thinking about Luna, still blushing, while Rayne sits next to him.

Rayne: "She likes you, I hope you know that."

Cross: "How do you know?"

Rayne: "It's obvious. From what I see, you like her too." *teasing Cross*

Cross: "Maybe." *blushing* "So what?"

Rayne: "I just think that's so cute." *has a huge smile on her face*

Cross: "What did she say about me?"

Rayne: "Stuff. Whatever she tells me is between us, not you. Why do you need to know?"

Cross: "I'm just curious, that's all." *changes the subject quickly* "So, you remember years ago when Darren first came back?"

Rayne: "Yeah, what about it? I knew there were things you hid from me."

Cross: "I had to. At that time, you wouldn't have been able to understand or believe me."

Rayne: "You assumed that I wouldn't. You never gave me a chance."

Cross: "Sorry. It was hard for me to explain what exactly happened. It wasn't until Louie came into our lives that I was able to make sense of it all."

Rayne: "That's messed up, how you would tell him before you told me."

Cross: "I only told him because of his experience with the unexplainable. I figured he would help me make sense of it all. Anyway, I'll start with that moment when I was in the middle of the high school class." *begins to remember, flashing back to that time*

Cross: *sighs and thinks* "That was the last time Rayne and I saw Darren. I heard something about his dad dying the following day. I hope he's okay after all that happened. Not that far from that event, our dad..."

Cross: *sighs and thinking* "It's always something, isn't it? My little sister dating the wrong kind of people and a sick mother at home... It's not like I have anxiety or suffer from depression. Doesn't help much either."

Class ends as Cross tries to cut his next class, but he's stopped by his sister.

Rayne: "Seriously, again? You're gonna try to go without me? Either you let me go with you, or I tell Mom."

Cross: "Nooo!! Come on, I'll buy your favorite candy, like old times."

Rayne: "I'm not a little girl anymore, Michael."

Cross: *laughs* "Look, I just need to get some air, okay? Just stay in school."

Rayne: "Fine, but you're going to owe me one."

Cross: "Yeah, yeah."

Cross walks to the double doors at the end of the hallway and exits the school. Once he's out of the school, he pulls out a cigarette and a lighter. As he tries to light it, the lighter slips from his hand and slides to the middle of the street.

Cross: "Fuck! Just my luck."

He looks both ways. Once it's clear, Cross goes to pick up the lighter when a white vehicle turns onto the street, speeding through. The driver stops just in time before hitting Cross. Cross angrily slams his fists on the hood of the car.

Cross: "Hey!! School speed limit, ASSHOLE!!!"

The driver steps out of his car.

Driver: "GET THE FUCK OFF THE STREET, YOU DUMB FUCKING BIT—"

The driver pauses as he and Cross look at each other and realize that they might know each other.

Cross: "Darren?"

Driver: "Cross?"

The two smile and laugh loudly at the situation.

Cross: "Damn, how long has it been? You changed a bit."

Darren: "You look hairier." *chuckles* "Just kidding. Shouldn't you be in class?"

Cross: "Yeah, you should talk. Mr. 'I never show up to my class.'"

Darren: "Fair enough. How do you know that?"

Cross: "You're in half of my classes and always a no-show. Soooo yeah, that's kind of obvious."

Darren: "Well, let's just say I know something about the Principal that he wouldn't want the public to know." *smiles* "So I don't have to worry about the grade as much as the rest."

Cross: "Oh, what do you have on him? By the way, some of my grades are slipping in this one class... Sooo if you can kindly help."

Darren: "Alright, alright. Turns out he's fucking a student. Not something he wants the papers. Anyway, get your ass in the car already."

Cross: "Nice car, how can you afford this? Sorry about denting your hood."

Cross enters the passenger's side door, then Darren drives off.

Darren: "Well, you know, working non-stop to help my mom keep a roof over our heads."

Cross: "I know the feeling, even though she can't be happy about you cutting all your classes."

Darren: "The less she knows, the better. She doesn't know about the blackmail either. In her eyes, I always show up to school and get good grades."

Cross: "That's not good. She'll kick your ass if she finds out."

Darren: "Yeah, she probably would. How about you? Does your mom or Rayne know about your cutting class?"

Cross: "No, and I plan to keep it that way. I don't want to stress her out even more and Rayne's just been a handful."

Darren: "Is she alright?"

Cross: "She's fine. Just doing a lot of stupid shit that's stressing me out—picking fights at school and picking idiots for boyfriends."

Darren: "Your secrets are safe with me, you know that."

Cross: "What kind of work do you do?"

Darren: "Sorry, Cross, I can't really talk about it."

Cross: "Does your mom know what you do?"

Darren drives quietly, ignoring Cross.

Cross: "Can you at least tell me it's nothing dangerous where we're headed?"

Darren: "My place. I don't have to be at work right now. I work later on tonight."

Shortly after, Darren and Cross pull up to the house. Gail steps out the front door.

Gail: "What are you doing home so early? It's not like you to cut class. Do you want me to kick your ass? Get back to school!"

Darren: "I know, I know, mom. You wouldn't believe who I ran into at school."

Cross: "More like who almost got run over. Hi, Ms. Gier. It's been a long time."

Gail: "Cross? I remember when you two were kids. I'm sorry that Darren and I haven't been around, especially after what happened with your father and my husband dying. It wasn't easy for us to get things back to normal—or what normal can be considered."

Cross: "It's fine, Ms. Gier. I understand that you both needed time to heal."

Gail: *smiling* "Thank you for understanding. Darren, what did Michael mean when he said almost ran him over? Should I be concerned?"

Darren: "A... no... Cross was just joking... We'll be in my room if you need us."

The teens enter the house as Darren gives him a look.

Cross: "What?"

Darren: "Asshole!"

Cross laughs as they walk down the hall to Darren's room. Upon entering, Cross notices the differences between before and now: holes in the wall that look like they were punched in and written on, dirty clothes all over, along with the smell of something rotting.

Darren: "Welcome to hell, Cross."

Cross: "Darren, I... I hate seeing you live like this. It must have been hard on you after losing your father. You shouldn't have had to go through that alone."

Darren: "H-hey, don't start getting soft on me now... I mean, your old man died shortly after mine, right? Not like I could hold it against you. You were going through the same thing. I'm sure Rayne and you distanced yourselves from the world too, like we did."

Cross: "Thank you for understanding."

Darren offers Cross a cigar, but Cross refuses.

Cross: "I don't smoke."

Darren: "Riiight, 'cause you almost ended up a hood ornament on my car for a fucking lighter. Who are you trying to fool?"

Cross: *laughs* "Alright, you got me."

They light their cigars and take a moment of silence to enjoy them.

Darren: "Still... something doesn't feel right about my father's death."

Cross: "What do you mean?"

Darren: "I mean, from what I remember, he was pretty healthy. I was very young back then, but I remember the last day I saw him alive. He called me to the basement. There were all these weird symbols on the ground. I can't explain it, but my dad wanted me to stand in the middle of one of them and..."

Cross: "Then what?"

Darren: "He asked me to chant in a strange language that I don't recall, then... I saw something that couldn't be explained. Some kind of dark figure watching me. It's strange though, it was like I knew who that already was."

Cross: *sighs* "It's not uncommon to see strange things that are unexplainable. Trust me, I know."

Darren dismisses what Cross tells him, then continues.

Darren: "I wasn't joking, Cross."

Cross: "Neither was I."

The teens look at each other with serious expressions. Both silently agree without speaking. Darren breaks the silence.

Darren: "Either way, I know there's more to the story than my mother's not telling me."

The smoke from the cigars travels through the house. Gail smells it and then marches to the room.

Gail: "You better not be smoking that shit in my house again, Darren."

Darren: "Shit, Cross, put your cigar out."

Gail walks into the room. She looks directly at her son, then at Cross.

Gail: *sighs* "Don't let my son become a bad influence on you."

Darren: "That's it, Cross, grab your things. We're leaving."

Gail: "Okay, I love you too. Be safe."

Cross: "It was nice seeing you again, Ms. Gier."

Gail: "Likewise. Don't be a stranger, Cross. You're always welcome here."

Cross: "Thank you. I'll see you around."

The teens leave, enter the car, and drive off.

Cross: "Your mom seems like she's doing well. Does she know about this mysterious job you can't speak of?"

Darren: "No, she thinks I'm working a normal job... I hate lying to her."

Cross: "But you still do anyway."

Darren lashes out at him.

Darren: "My dipshit of a father is dead, Cross. He left us nothing. My mother has to work three jobs just so we can keep the house, and that's still not enough. If I have to work a job that puts me at risk, so be it... Sorry, Cross, I didn't mean to lash out like that. It's been tough for us."

Cross: "I understand that, Darren, but I don't want you getting hurt or, even worse, going to jail."

Darren: "...Right. Do you still live in the same house?" *changing the subject*

Cross: "...Yeah."

The two sit in an awkward silence for the rest of the ride to Cross's house.

Rayne: "Cross, where the fuck have you been?!!! ...Wait... Is that Darren?"

Rayne runs up to Darren, giving him a tight hug, then a powerful slap.

Rayne: "Never mind, Cross, where have you been?" *staring at Darren*

Darren: *chuckles* "Hey, beautiful, I missed you too."

Rayne: "Yeah, I can't tell. Shut up!" *turns around, blushing* "It's... It's been years."

Darren: "Too long, I'm sorry, Rayne."

Darren and Rayne glance at each other for what seems like an eternity, longing and desiring to see one another again. Cross smiles at them as he pulls Darren away.

Cross: "Alright, come on Romeo. You still need to say hello to our mom."

Darren walks into the house and greets Nora.

Darren: "Knock, knock, guess who's home?"

Nora: "Darren!" *she hugs him* "Wow, you've gotten taller. I remember when you and my kids were inseparable. How are you and Gail doing?"

Darren: "We're doing good. As good as we can be. Funny, my mom said something similar when she saw Cross." *smiles*

Nora: "Ah, so that's why he's late. I'm happy you two are reconnecting. Can you stay for dinner?"

Darren: "I'd love to, but I have to be at work in a few hours. May I have a rain check though?"

Nora: "We'll hold you to it."

Darren: *smiles* "That's fair. I'll be hanging out with Cross. If you need us for anything, just call. It was nice seeing you again, Mrs. Cross."

Darren walks into Cross's room. He notices that the room is in the same condition as his.

Darren: "Yeah... you're doing well for yourself."

He looks at the wall. Claw marks can be seen. As Darren touches the marks, he searches for any animals that could've possibly made them. Then, he takes a closer look, passing his finger over them when Cross catches him.

Darren: "Cross... These are human-size marks..."

Cross puts his head down, a worried look on his face.

Darren: "Cross... What's going on here? What aren't you telling me?"

Cross sits on his bed, speechless.

Darren: "Fine, tell me when you're ready."

It's so quiet in the room that Nora can be heard in the distance talking to Rayne. To break the awkwardness, Darren changes the subject.

Darren: "Tell you what, why don't you come to work with me tonight?"

Cross: "I can't, I have to take care of my mom."

Darren: "Late at night, when everyone is asleep."

Cross: "How late are we talking about?"

Darren: "Eleven p.m."

Cross: "I can try to work with that, but I don't know..."

Darren: "Come on, man. You've got to get out of this house, it's killing you. We won't be out long, I don't like being out late."

Cross: "Fine, eleven p.m."

Darren: "Great. Are you still good at running and climbing things?"

Cross: "You mean parkour?" *sarcastic tone* "Yeah, I could still move fast."

Darren: "Cool. Oh, almost forgot; wear clothes you don't really care about. After tonight, you can't wear them again. So make sure your face and skin are covered. I have to get going, need to pick up a few things for the job later. Meet me on the bridge over the train tracks at exactly eleven. No later."

Cross: "This all sounds really shady." *worried tone*

Darren: "I know it does, but I'll split the pay with you fifty/fifty. Trust me... and bring gloves."

Darren starts walking to the front door, when...

Cross: "Darren, wait..."

Darren: "Eleven, Cross, no later. See you then."

As Darren pulls off, Cross sits on the stoop, wondering: *What could his job actually be?*

Later that night, at 10:50 p.m., Cross, dressed in all black, is waiting for Darren on the bridge top.

Darren: "Well, look at you, all in black, looking like the Grim Reaper."

Cross sees Darren's wardrobe—he's also dressed in black, with a white hoodie.

Cross: "What are you supposed to be, a white knight like in fairy tales?"

Darren: "Ha! A white knight on a horse's back with Rayne in my arms, maybe."

Cross: *smirks* "What am I going to do with you?"

Darren: *laughs* "Same thing you always do."

Cross: "What are we doing out here? I thought we were working."

Darren: "We are getting ready." *looking at his wristwatch* "We have five seconds until eleven. Fifty-six, fifty-seven, fifty-eight..."

Cross: "What happens at eleven?"

Darren: *smirks* "Our rides show up."

The train speeds under the bridge, and Darren jumps onto the top of it. He turns around and yells.

Darren: "Better jump or you're going to miss our ride!"

Cross: "Jesus, Darren. What are you getting me into?!"

Cross jumps onto the train before it finishes passing the bridge. He carefully catches up to Darren.

Cross: "Y.. You know we could've just bought tickets."

Darren: "Nah, where's the fun in that? Besides, if we bought tickets, our faces would be caught by the cameras."

Cross: "Are you going to tell me what this job is before we go any further?"

Darren: "The job will kinda speak for itself when we do it. I'll tell you what it's all about when we get to where we need to be. Speaking of names, we have to go by different names just in case. I figured I would go by Mike, and you as Mikey."

Cross: "My real name is Michael. Not much of a difference."

Darren: "Ah shit, that's right, my bad. I'm so used to everyone calling you Cross that it completely slipped my mind. Any ideas for code names, smartass?"

Cross: "Hmm... How about Romulus and Remus?"

Darren: "Wait... The story from Roman mythology? About the brother who kills the other brother?"

Cross: "Yup, that's the one."

Darren: "I call Romulus."

Cross: *chuckles* "Fine, just don't kill me."

Sometime later at the Cross house, Rayne goes to check on her brother. She gets to his room, only to find an empty bed late at night.

Rayne: *angry* "Damn it, Cross!"

Rayne pulls out a cell phone from her pocket to call Cross. She waits and waits as it goes to voicemail.

Rayne: *thinking* "Sneaking out at night without me... Asshole."

Many hours later, Darren and Cross are in Manhattan, running, jumping, and climbing from building to building to reach their destination.

Cross: *thinking* "Darren's changed a lot since the death of his father... I guess we both have, from that time till now... Everything has changed... Like I feared it would when I was a kid, playing with Rayne and my best friend... Better times..."

Darren: "Hey, daydreamer, we're here."

The teens survey the perimeter around the construction site from the top of the building across the street.

Cross: "A construction job, this late? Oh well, at least it's legal. I thought we were going to do something that would be considered a crime."

Darren: "We are, we're going to break into the office and steal the blueprint of the building. That's the job. This is what I do. I guess like Robin Hood, without the tights." *laughs*

Cross: "What?! This is what you do? We could end up in federal jail for this."

Darren: "Remember that time when we were younger, you told me to steal that candy bar? I felt like shit after that." *smiles*

Cross: "Stealing corporate blueprints is a huge leap from a fucking candy bar."

Darren: "Well, if we get caught, at least we'll be cellmates." *smirks* "Look, if you don't want to do this, I'll understand. We can meet at the train station after I'm done here."

Cross: "...If we get caught, I'm calling top bunk. Someone's got to make sure your dumbass doesn't get caught."

Darren: *laughs* "Fine, you get top bunk. Don't worry so much, I've done this a hundred times. What could possibly go wrong?"

They infiltrate the site undetected, passing next to the guards and dodging cameras using shadows. The teens climb to the highest place to get a better vantage point.

Cross: "Do you know where the blueprints are?"

Darren: "Yeah, they're in a safe located in the manager's office. Let's go."

The two proceed stealthily to the office. Darren pulls out a stethoscope from his backpack. He uses it to listen to the clicking sound the safe makes when the correct combination is found, while Cross looks out for any guards.

Cross: "We're clear for now. What do you need me to do?"

Darren: "Do what you're doing now, hold on. I almost have it open."

Cross: "So who exactly hires you to steal the building plans, and how'd you get into something like this?"

Darren: "Some asshole politician, Davis?... Donovan?... No, something Deville. He wants these prints to blackmail some politician. They're all corrupt. No loyalty to anyone but their own wallets. How did I get into something like this? Search long enough on the dark web and you'll find jobs like this, or worse. So it isn't as bad, but it pays well, and I get to keep my humanity."

Cross: "So you're a mercenary for hire?"

Darren: "You can sort of call it that."

Cross: "You're not afraid of all this biting you in the ass later? What if those politicians need someone to take the fall once they get screwed over?"

Darren: "Luckily for me, they only know me by the name Silence. I hide my face and skin whenever I have to meet up with them. I made sure when I first started all this that they wouldn't be able to track anything back to me."

Cross: "You should still quit while you're ahead."

Darren: "Look, I know you're just worried about me, but you know I'm only doing this for a good reason."

Cross: "I know, but you're all your mother has left. There's plenty of other ways to make good money legally. From what I noticed tonight, all this is also for the rush of possibly getting arrested or dying. Since the train, you haven't calmed down, showing me that the adrenaline has risen extremely."

Darren: "...Just keep a lookout and shut up." *in a frustrated tone*

Just as he opens the safe, an alarm goes off at the site.

In the present, in West Virginia, Blak Hart gets out of the hotel to begin his hunt for Collin. He searches high and low, finding anything paranormal but no sign of Collin. He walks into an abandoned house. As he slowly heads to the second floor, he hears a commotion in the basement. He runs downstairs toward the basement door, when the door flies open. Out comes Mezun, covered in blood, tackling Blak Hart.

Mezun: "A few minutes earlier, and it would've been you."

Blak Hart: "What have you done?"

Mezun: "What everyone wanted to do, but thought he would be useful later."

Blak Hart pushes Mezun against the wall and rushes downstairs. He sees blood painting the walls, an exploded torso, limbs all over, and what appears to be the head of Collin. He goes back upstairs.

Blak Hart: "What happened? Did he outlive his usefulness?"

Mezun: "What concern is it of yours? He couldn't do anything right. He kept attracting your attention, so I took care of a necessary nuisance."

Blak Hart: "So, am I next? If I am, then let's go."

Mezun: "That's the problem with you kids in this era—always in a rush to die."

Blak Hart: "That's the problem with the old generation, always procrastinating when things should get done right at the moment."

Blak Hart charges at Mezun, but Mezun dodges and punches him with all his strength. Blak Hart is sent through the wall into the next room. He gets up with a smile on his face.

Blak Hart: "I hope that wasn't the best you have to offer. I expected more from an ex-warlock. I thought you became a powerful demon." *mocking him* "What a joke."

Mezun runs at him with his nails fully extended and chanting in an angry tone. Suddenly, Zima appears.

Zima: "Stop, you fool. I have bigger plans for this mortal."

Just like that, he grabs Mezun and vanishes.

Blak Hart: "I don't like this, they're up to something. I gotta let Ark know that these two are plotting again."

He calls Ark and lets him know what happened.

Ark: "In other words, Collin was already taken care of before you got there?"

Blak Hart: "Yup. What are you thinking?"

Ark: "They're definitely up to something, but what? I'll send your payment to the account. Then, if you want, go join the rest in Florida. Who knows, you might find out that having fun is actually a real thing." *joking around* "Enjoy yourself, life doesn't have to be all about hunting."

Blak Hart: "I'll try, but if you get anything..."

Ark: "I know, I know, I'll send it to you. Remember, keep the group safe. There were reports that they were being followed. Protect them."

Blak Hart: "They're my family. That goes without saying. Come on Ark, you know me better than that."

Ark: "I know, I just wanted to hear you say it."

Blak Hart: "Say what?"

Ark: "That you finally acknowledge that you have a family, like you always longed for."

Blak Hart: "I always had a family. I never felt like I had to say it. I consider you one, I thought you knew."

In Florida, Cross continues to tell Rayne about Darren as they head to the airport to pick up Luna.

Darren grabs the plans, putting them in his bag as a security guard runs into the room.

Security Guard: "You boys are in a lot of trouble when I get my hands on you."

The guard cracks his knuckles as he walks toward them.

Darren: "Shit!! Remus, run!"

Cross: "Please, after you, Romulus. Not like we're about to get jumped by a juiced-up security guard."

Darren: "Just shut up and run!" *annoyed*

Darren and Cross team up to give the guard a few hits, just enough to get him off balance. While he's off balance, the teens run past him, avoiding the other guards or knocking them out as they escape. They spot an exit, running and climbing to areas that appear unreachable. Once they reach the top of the site, a security guard is already there waiting for them.

Security Guard: "You two are pretty good, but as you can see, I'm not your typical security guard. This is the end of the road for both of you. You have two options: you can hand over the blueprints and all you get is a lecture, or I can haul your asses to jail. The choice is yours."

Cross: "Nice job, Romulus." *mocking Darren* "I've done this a million times, what could possibly go wrong? Who the fuck jinxes themselves before doing something illegal?"

Darren: "Are you done bitching, Remus?"

Cross: "Kiss my ass, I warned you, and now we're going to jail!"

Security Guard: "Seriously? Romulus and Remus?" *laughs at them* "Is that the best cover names you two could come up with? Wow, we have some Roman mythologists here. Cute."

Darren and Cross: *in unison* "Fuck off!"

Security Guard: *half a smile* "I swear, you two are very entertaining. Too bad you're going to be cellmates, so you can comfort each other from being someone's bitch."

Darren looks down as he sees an open dumpster filled with trash. Then, he gets an idea.

Darren: "Hey, Remus!" *points down* "Jump!"

Cross and the guard watch as Darren free-falls and lands in the dumpster.

Cross: "Crazy bastard." *he jumps down*

As the teens climb out, the security guard jumps down after them. When he finally gets out, the teens have gotten a good distance away from him. He looks at them and laughs at the whole ordeal.

Security Guard: "Luckily, Ark always has missions for me, 'cause I'm about to get fired for this bullshit."

The teens hear and observe from an abandoned building as police cars speed toward the construction site. As Darren enjoys the view, watching the police haul away the last security guard, he looks behind him and sees Cross's back against the wall, shaking.

Darren: "Um... Are you okay? The cops arrested that stupid guard."

Cross: "Remus doesn’t want to go splat." *nervous tone* "That guard was only trying to do his job."

Darren: *teasing* "Come on, not like your brother Romulus is going to push you off the edge... Hey, I bet you'd make a good looking pizza on those taxis down there. We don’t owe that guard anything, not like we’ll ever see him again."

Cross: "Fuck you!!!!"

Darren: *laughs* "Hang on, I need to make a call."

Darren pulls out a pre-paid flip phone from his bag and proceeds to dial.

Corrupted Politician: "Hello?"

Darren: "It’s me... This is a surprise. Normally your secretary answers first. You know, the cute one with the short skirt..."

Corrupted Politician: "What do you want? This better not be a normal late-night call."

Darren: "Anyway, I guess you're right. This isn’t just a normal late-night call..."

Corrupted Politician: "Did you retrieve the construction site's blueprints?"

Darren: "Yes, of course I have the plans. Come on, it’s me, remember. By the way, you failed to mention the safe had an alarm system attached to it. Plus, extra security. I’m sure you understand why I’ll be increasing my pay because of it..."

Corrupted Politician: "You don’t decide how much you get paid. You’ll get what we agreed on."

Darren: "Oh really? So I should just burn these prints or maybe leak information leading this robbery back to you..."

Corrupted Politician: "Are you threatening me? You don't want to make an enemy out of me, boy. I don't have time for this. You'll get five hundred more and be happy you get that!"

Darren: "See, now was that so hard? I knew you would see it my way. I'll leave the plans at the usual spot. Have your men drop off my pay."

Corrupted Politician: "You have no idea who you're dealing with."

Darren hangs up the phone and then crushes it.

Cross: "Can't they track that back to you?"

Darren: "Nah, I was paying for it using a stolen credit card from a drug dealer that was selling that crap to elementary school kids." *laughs* "He's in for a rude awakening tomorrow."

Cross: "Well, look at you trying to be a hero. You should still quit. This was stupid."

Darren: "Not this again, Cross. I mean, come on, you know my reason."

Cross: "I know, but you won't be if you end up behind bars. We almost didn't make it out tonight."

The two stand there, paused in silence, staring awkwardly at one another.

Darren: "Maaaann, I don't like doing these jobs. It's not the thrill or the chases that get me. It's knowing my mom is going to be okay. I'm saving up enough so..." *sighs*

Darren looks out toward the distance at the city with an expression of anger. Cross walks over to Darren, next to the edge of the building.

Cross: "Yeah... I know the feeling. You're all your mom has left. I'm sure you wouldn't abandon her with a bank account filled with money she hasn't got a clue where it came from."

Darren: "Yeah, you're right, I'm stuck here for a few more years. Until I'm alone again."

Cross: "You'll never be alone. You have Rayne and me. The only way you'll be alone is if you choose to."

Darren: *breaks the moment by joking* "Yeah, well, unless I do go to prison for a vacation or something. You know, just in case, I want to be the top dog in the pound."

Cross: "Yeah, right, you top dog? Keep dreaming. More like how low, sir." *busting his balls*

Darren: *chuckles* "Screw you too. I bend over for no one. My face is too pretty." *caressing his face*

Cross: "Exactly, no need for bending when they can use your face." *laughing and making a squirting sound changing the subject* "So, now what?"

Darren: "Did you bring that second pair of clothes like I told you to bring?"

Cross: "Yeah, it's in my bag."

Darren: "Good. We're going to change clothes and burn the ones we used for the job, until there's nothing left. Then, we'll take the train home like normal people this time."

Cross: "I thought you were going to drop off the plans."

Darren: "I’m going to after we get you home. I’m going to be making a second trip out here."

Cross: "Why are you going to go without me?!"

Darren: "Don’t take it the wrong way, some of these employers like to send trackers after me once I collect my pay and drop off what they need. I have ways of losing them, but they’re good at what they do. I can’t risk them following you back to your house."

Cross: "Wow! I guess you have no faith in my skills. I can be your backup in sketchy situations."

Darren: "I don’t doubt your skills, I don’t want to be the reason for anything to happen to you. Yeah, just trust me on this one, it’s riskier than the job."

Cross: "Alright, but just contact me when you're in the clear."

Darren: "Deal."

A week passes, and Darren walks into a church late at night carrying a bag. He looks forward to an old blind priest next to the altar, sneaking a shot of whiskey from his flask. Darren smirks and walks to the priest.

Old Priest: "Who's there?" *mumbles* "Fucking teenagers."

Darren: "It's me, old man. Nice flask. Not sure God would approve though."

Old Priest: "Oh, the Gier brat. What the hell do you want?"

The priest takes another swig from the flask as Darren throws the bag by the priest's feet.

Darren: "You're too funny, old man. I have no idea how anyone would let a miserable old drunk become a priest."

Darren begins to head out the door.

Darren: "Make sure it goes to the food drives."

Old Priest: "Wait."

Old Priest stops to hear him out.

Old Priest: "You must remember, God doesn't discriminate as long as you're His follower. Also, you're here every week right before we close, always with a bag filled with hundreds. I don't want to know how, but I want to know why, son. I could tell just from hearing you, your faith ain't strong. So why do this every week?"

Darren: "Well, all my bills are paid and I make a little more every week. Besides, do I need a reason to help those less fortunate?"

Old Priest: "Aawww, a true healer of the people. I take it this will be another anonymous donation?"

Darren: *as he walks away* "Yeah, it's better that way. See you next week, old man."

The following afternoon, Cross sits at his desk. He stares out the window, thinking to himself.

Cross: *thinking* "It's been a week since I did that job with Darren, he never called to let me know he's alright. Hope he didn't get caught. Like always, he's never in class. Wonder if I'm ever going to hear from him again."

Rayne walks over to Cross's desk as the bell rings.

Rayne: "Cross, mommy got an anonymous letter in the mail with no return address. It had a couple of hundreds in it. Do you know anything about that?"

Cross: "W-what? No... no clue." *thinking* "Darren?"

Rayne: "Well, either way, mommy said that's going to help her catch up with the bills. She said we could go out to dinner tonight, our pick. I'll see you after school, think of where you want to go later."

Cross: "Y-yeah..." *Thinking* "Thank you, Darren. I guess things will start to get better, I can only hope. Only time will tell."

Then Rayne's phone rings, bringing them back to the present. It's Luna.

Rayne: "Before I answer Luna's call, I would've understood. We'll talk later." *answering the call* "Hi."

Luna: "Hi, I'll be waiting for you in the restaurant right across from the package claim."

Rayne: "No problem, we should be there in a few."

Luna: "Are you coming alone or with Cross?" *in a giddy voice*

Rayne: "Yeah, who else did you think I was coming with?"

Luna: "Darren, your mom, it could've been anyone."

Cross: "Did she ask about me? What did she say? Does she sound excited?"

Rayne: "Why? Are you going to do your heroic voice again?" *puts her hands on her hips, mocking Cross and laughing at him*

Cross: "No." *puts his head down and blushes* "Punk."

Suddenly, Cross's phone begins to ring. It’s Louie.

Cross: "Hey bro, what’s up?"

Louie: "I’m on my way to chill with you guys. I have important news, and you’ll be pretty shocked when you hear."

Cross: "What’s the news? Is it about the female you are talking to?"

Louie: "I’ll tell you all when I get there."

CHAPTER 8.
WHAT'S GOING ON?

Cross: "Alright, we'll see you when you get here. By the way, do you remember the girl I was talking to you about, Luna? She is joining us for the rest of the vacation."

Louie: "Yeah, I remember. Guess you're excited, are you going to let her know how you feel?"

Cross: "I'm going to try, hopefully I don't get shot down."

Louie: "From the way Rayne talks about her, I have a feeling that you both are going to be a couple of shy romantics nervously looking at each other worried about seeming like an idiot in front of the other." *laughing* "Either way, see you all in a few hours. Later."

A couple of hours later, Louie finally arrives in Florida. As he gets off the plane, he gets a call from Ark.

Ark: "Louie; I know you're on vacation, but there are a couple of bounties in Florida with a hefty price upon their capture."

Louie: *sighs* "How many and how much? How much time is this going to take from my vacation?"

Ark: "Depends on how fast you get them done, all the info will be sent to your email. Good luck, talk to you then."

Louie calls Cross to notify him of his arrival.

Louie: "I'm in Florida, are you guys picking me up or are we going to meet up somewhere?"

Cross: "We're with Luna at the restaurant across from the package claim, so just go to the restaurant. We're sitting in the back next to the jazz players."

Louie: "Ok no problem, I'll see you all in a few."

In the restaurant, the sound of the slow saxophone solo almost soothing Cross as the teens order their meals. The sounds of the clinking of wine glasses, the scent of sizzling meat, and the warm breeze drifting in through the windows. Cross orders a bacon cheeseburger and a side order of fries with cola, and Rayne orders spaghetti with sausage and meatballs along with a cola. *She looks at Cross and chuckles* Luna orders a salad with a large diet lemon-lime drink. Cross fiddles with his straw and nervously taps his foot under the table while Luna brushes her hair behind her ear when she sees him smile.

Rayne: "What the fuck are you two ordering? Who are you two trying to fool? Cross, that burger won't fill you up and Luna, since when do you eat salad?"

Cross: "I'm not really that hungry, this is just something to hold me off until later."

Rayne: "Yeah right, I guess the love bug is working overtime." *teasing as she leans back smugly, arms crossed, watching these two nervously interact. Cross punches Rayne on her arm as he blushes*

Luna: "I eat a salad every once in a while." *said with a sassy attitude and blushing and thinking* "He doesn't have to try so hard."

Rayne: "Since when, was it when we left? If so, you don't have to keep in shape for him." *in a sarcastic tone while pointing at Cross*

Luna: *pouting her face changes from embarrassed to annoyed*

Cross: “Rayne stop being a bitch to Luna, shut up and eat your... Meal.” *looks at Luna* “You see what I have to put up with.”

Just as they begin to eat, Louie walks in. He greets everyone and as he goes to introduce himself to Luna, Cross jumps in front of him. *Blabbing and stuttering, he begins to speak. Louie looks at him and cuts him off while chuckling a bit.*

Louie: “Cross, are you alright?”

Rayne: “He's going through a love anxiety attack.”

Cross: “No I'm not, just... Louie, let's go outside for a moment.”

Louie: “Ooook.”

Cross quickly walks with Louie outside.

Louie: “What are you so worried about? She seems really into you so stop wrecking it for yourself.”

Cross: “I... I... can't breathe.”

Louie: *laughing* “After all the things we've seen, this is what scares you?” *laughs so hard begins to tear*

Inside, the girls are talking about the guys.

Rayne: *giggles* “He's so nervous, poor guy.”

Luna: *looks at Cross from the window* “Is it bad? I find it cute that he is trying so hard when he doesn't have to.” *smiles and blushes*

Rayne: “He can be a bit thick-headed, but you should see when he talks about you; there's a light in his eyes that I haven't seen in a while.”

Luna: "Really!?" *looks at Cross lighting a cigarette* "He smokes now?"

Rayne turns around, she looks at Louie and Cross smoking. Rayne turns back.

Rayne: "Yeah, this new habit he thinks it makes him look..."

Luna: "Like a bad boy, sexy, daring?"

Rayne: "What?"

Luna: "Huh... nothing. hehe" *drinks some of her soda* "So how's everything going between you and Darren?"

Rayne: "Don't try to change the subject, we can talk about him later. Right now it's about you and Cross."

Cross looks back at Luna and smiles.

Cross: "She's so beautiful, smart, and has a sweet heart." *saying all that with a smile from ear to ear* "I don't want to mess it up."

Louie: "So, why are you freaking out? Just talk to her."

Cross: "I'm freaking out because of everything I've seen and been through, I know things are just getting started and I don't want anything to happen to her because of me... She is already in danger just for knowing us... if something were to happen to her I wouldn't forgive myself."

Louie puts his hand on Cross's shoulder.

Louie: "I understand how you feel but, don't be afraid. You will be there to protect her and remember I got your back." *smirks*

Suddenly, Louie's phone vibrates. He looks at it and sees it's the information for the missions sent from Ark.

Louie: "If you want more time to think about the whole scenario, help me finish these missions. Depending on the outcome then decide whether your fears are valid or just being overthought."

Cross: "Alright, just let me say goodbye to the girls."

Louie: "Ok, let's say our goodbyes then prepare for whatever Ark has got for us."

Louie and Cross walk back inside. They walk up to the girls, Cross looks at Rayne and tells her that he has to take care of some business with Louie. Rayne nods and Cross looks at Luna blushing.

Cross: "I'm sorry, I have to go," *rubs the back of his head nervously* "I owe you ok? When I get back, just the two of us. We'll go out and get some real food." *smiles*

Luna: *blushes* "You don't have to, I understand how busy you can be."

Cross: *holds her hand* "I promise when I get back, I will take you out; I'm not taking no for an answer." *kisses Luna on the cheek close to her lips*

Luna: *cheeks are a bright red and tries to hide her smile* "...ok see you then." *blushing and flustered*

Louie: "Ladies, nice seeing you again. Rayne, in case of anything I'll call. As for Luna, it was a pleasure to meet you."

Cross and Louie leave as Rayne sits in disbelief with what just happened, Luna is smiling and can't stop blushing.

Rayne: "Did he just ask you out!?"

Luna: "Mmmm" *in her own world*

Rayne: "Earth to Luna, hello? I lost her to my brother's charm. Who knew he actually had game?" *smiles proudly*

In the car, Cross and Louie look over all the bounties. They carefully try to plan a strategy, all according to how powerful and dangerous the opponents are. As they read the files, they notice that each one has a higher bounty than what they're used to.

Cross: "I guess these are the top bounties for Ark."

Louie: *gears up for battle* "I've seen some of his other bounties that put these to shame. With these bounties, we can get your mom a new house and I can get some new toys." *he says with a smirk*

Cross: "So, who's going to catch an ass-whoopin' first?"

Blak Hart: "I guess the Skunk Ape. It has been hiding in various parts of Florida, North Carolina, and Arkansas. They are masters of blending into their environment even though they can be tracked by the fowl stench that is left behind."

Cross: "How is it that something with a horrible smell can't be found?"

Blak Hart: "Simple, they're not us. The rest of the missions are to put the spirits to rest. We can probably split up and send these spirits off to their destinations. Then leave Saint Augustine for last."

Cross: "Sounds like a good plan."

Back at the hotel, Nora and Gail plan an eventful day for the group. Darren gets a phone call from Rayne.

Darren: "Hey, how is everything going with Luna and Cross?"

Rayne: "Yeah, Louie came and picked up Cross; then they went to do a job."

Darren: “Oh, do you two need a ride back then?”

Rayne: “If you don't mind.”

Darren: “I will be right there, see you in a bit.”

Darren asks his mom for the keys to the rental car and goes to get the girls. On the way there, Darren spots Suma walking from the beach. Darren pulls up to him.

Darren: “Hey, Suma right?”

Suma: “Hey yeah, that's me. Your D... D... Darren right?”

Darren: “Yeah, what are you doing? Do you need a ride?”

Suma: “It's ok, you don't have to.”

Darren: “I'm heading that way anyway, come on, it's the least I can do.”

Darren opens the door.

Suma: “You sure?” *feeling uneasy*

Darren: “Yeah, come on.”

Suma jumps in the car and then starts looking around, Darren notices.

Darren: “Are you ok, Suma?”

Suma: "Yeah, I haven't been in a car, I spend most of the time on the beach, and if not there I'm in the water."

Darren: "If you don't mind me asking, where are you from?"

Suma: "Well, that's a long story but, I'm from Atlantis."

Darren: "Wow, Atlantis is real."

Suma laughs, showing his sharp teeth.

Suma: "Yeah, but my father wanted me to be this great warrior and fight by his side. I did all the training but nothing was enough for him, so my mother requested for me to take a break, so here I am."

Darren: "Wow, sounds like your dad is an asshole."

Suma: *laughs* "Yeah he can be, I'm just worried for my little brother. I don't want him to deal with my dad's BS."

Darren pulls up to the beach by the airport and drops Suma off. Then he picks up the girls who were waiting at the beach. Suma starts walking to the water, Darren calls out to him.

Darren: "HEY SUMA, IF YOU EVER WANT TO HANG OUT; STOP BY."

Suma: "ALRIGHT, I MIGHT STOP BY THEN." *waves*

Sometime later, deep in the woods of the southern region of Florida, Blak Hart and Cross are getting close to their destination.

Blak Hart: "Once you get out, be on guard. We don't know how strong this thing is."

Cross: "You act like my guard is ever down."

Blak Hart: "Good point, if you smell something before I do please let me know. Don't try to track it down without me."

Cross: "I jumped the gun once, you still keep bringing it up."

In South America, Zima and several demons are having a minor ceremony for their head witch. Shortly after, the ritual begins. The torches flicker, the voices of the people chanting, and the smell of the burning herbs fill the air.

The head witch: "Zima, do you see? These people worship me."

Zima: "Relax, bitch, these people were scared into doing so by being shown a couple of parlor tricks. This would impress the stupid, why are you making a big deal over this shit? You called me here for this? Kill an archangel or any chosen warrior or heaven's army, then I might be impressed. A key word might."

The wendigo king: "What are we going to do with my son and his annoying friends?"

Mezun: "We're working on something, isn't that right, Zima?"

Zima: "Always trying to take credit for something that you had no part of, I have a plan, but it's on a need-to-know basis, and right now it's none of your damn business. Is that understood? Now what are you doing to try and eliminate these pain in the asses? Hecate, you should be strong enough now for the first phase of our plan."

Hecate: "Yes I am, emperor of darkness."

Zima: "Good, flattery will get you nowhere! If I wanted an ass kissing, I would've gotten Amon to do it." *laughs*

At the hotel room in Florida, Darren and the girls are waiting for Suma to arrive. Nora and Gail are cooking for the guests, Cross and Louie.

Gail: "Darren, where did you say you met up with Suma again?"

Darren: "On the road several blocks away from the airport. I was on my way to get the girls when I spotted him."

Nora: "I wanted to thank him for finding Cross."

In the woods, Cross and Blak Hart are on the trail of the beast known as the Skunk Ape. The heat and humidity are making it an uncomfortable hunt, the buzzing of various insects just being a nuisance while, in the distance, the sounds of leaves rustling are heard.

Cross: "Damn, this thing lives up to its name."

Blak Hart: "I know, at least we can't lose him." *said jokingly*

As they continue searching, Cross senses all is not what it appears to be.

Cross: "I have a bad feeling, seems like our bounty has been watching us this whole time."

Blak Hart: "There's no way that it could've been expecting us. Fuck it, means it's ready to catch an ass whooping or brought its friend to the party."

Cross: "I feel there are a handful of creatures."

Blak Hart: "Luckily we have two hands." *laughs* "Two questions, what do you consider a handful of creatures? Can you hold that many and how closely have you been to our enemies that you can tell the species?" *laughing*

Cross: "First off, that was three questions, wise-ass. Second; doesn't matter as long as we get the job done." *annoyed*

Right as they were about to fight the creatures, Rayne decided to call Cross. Cross quickly looks at his phone, sighs, then sends it to voicemail.

Blak Hart: "Who called you?"

Cross: "Who else?"

Blak Hart: "Who, your girl?" *mocking*

Cross: "No, shut up. It was Rayne."

Suddenly, they're attacked by several Tik Tiks and several low-level wendigos.

Blak Hart: "Where the fuck is the Skunk Ape?"

Cross: "It's still close by just waiting for us to lose to these things." *grins as his fangs begin to grow*

The guys transform into their battle forms. Blak Hart's eyes start glowing an orange/red with yellow. His muscles expand, making his body capable of withstanding more attacks. Cross's fangs are fully exposed, his hair grows, and his nails extend, turning into claws, now his appearance is of a partial werewolf.

Blak Hart: “Since when have you been able to do that?”

Cross: “Back at the hotel in West Virginia when I fought a vampire.”

Blak Hart: “Ok, we’ll talk about that after we're done with this mission.”

The Skunk Ape charges at Blak Hart. He commands all the other creatures to attack Cross. He uppercuts Blak Hart, then throws him deep into the nearby swamp.

Cross looks over at Blak Hart, he wants to help him but is swarmed by the Tik Tiks and the Wendigos. A wendigo grabs him from behind while the Tik Tiks assault him simultaneously. Cross tears the arms off of his captor, then turns around and rips him to shreds. The onslaught continues, and more and more creatures repeatedly proceed to appear as quickly as Cross destroys them.

Cross: “What the fuck? The more I kill, double manifest, and attack. What's going on?”

In the swamp, Blak Hart is getting frustrated with the Skunk Ape. Every attempt to end the fight quickly resulted miserably. Blak Hart punches the beast as hard as he can, and just as he tries to escape and help Cross; he is pulled back by his opponent.

Blak Hart: *thinking to himself* “I hit him with my all, yet it comes back unfazed. From what I've read, they're not supposed to be this strong. I hear Cross fighting, but this looks more like a demon has had a hand in this situation."

The Skunk Ape started to get more aggressive with its attacks. It grabbed Blak Hart by his leg as he tried to jump away and began slamming Blak Hart into trees and the ground. Blak Hart surrounds himself with his flames, and the Skunk Ape jumps back and screams in fear.

Blak Hart: "Now I got you." *his anger makes the fire around him create a mist and smoke that makes it harder to see him, using it to his advantage he throws fireballs*

The Skunk Ape's eyes glow red and it punches the fireballs, then speaks.

Skunk Ape: "Rrraaahh I- I don't want this!" *grabs onto his head* "WHY AM I SO ANGRY?!!!!!!!"

Blak Hart tries to communicate with the Skunk Ape.

Blak Hart: "Hey Skunk Ape, where did you get the demon blood?"

Skunk Ape: "MY NAME IS NOT SKUNK APE!!!!!!!"

Blak Hart: "Easy there, what's your name then? I don't want to hurt you if I don't have to."

Skunk Ape: "Dakota, that's my name."

Blak Hart: "Well, Dakota, where did you get the demon blood?"

As Blak Hart tries to walk to Dakota, Cross is launched into the swamp by one of the wendigos, and it starts attacking Blak Hart as a figure appears behind Dakota. Cross stands and tries to help Blak

Hart as the wendigo bites his arm and swings around. Dakota tries to run but the figure stops him dead in his tracks. Dakota falls to the ground and begins shaking.

Dakota: "No, not you again. Leave me be, Zima!"

Zima: "Tsk tsk, Dakota, your job isn't done yet. Either you do this willingly or I will make you do it."

Dakota stands up and punches Zima but, Zima doesn't move an inch. Blak Hart burns the wendigo off while Cross slashes it to pieces, Blak Hart jumps to help Dakota as Cross tries to go after Zima.

Zima: "HAHAHA!!!!! Oh Dakota, my stupid friend, you should have done this the easy way." *Zima grabs Dakota's arm and breaks it, Zima starts pouring blood into the wound*

Blak Hart throws fireballs at Zima, while Cross tries to slash Zima but, Zima disappears and reappears on top of a tree.

Zima: "It's been awhile boys, how is everyone? Has Darren talked about me?"

Cross: "SHUT THE FUCK!!!!"

Zima: "Ooooh, so angry. Hahaha, what about you, Blak Hart?"

Blak Hart: "I'm gonna rip you apart!"

Zima: "Aww, you do care. Well... sorry boys, but you two seem busy, so goodbye for now." *fades into the shadows*

Behind Blak Hart, Dakota stands up; taller and bigger than before. His eyes are black and he's foaming from his mouth. He grabs a tree and then pulls it from the ground. He smacks Blak Hart with it, sending him airborne. Cross jumps on Dakota and claws at his face. It has no effect; Dakota grabs Cross and slams him to the ground. He tries to slam the tree on him but, Blak Hart draws his sword then, launches himself with his flames cutting through the tree in half and punching Dakota in the stomach. He lights his fists on fire and then begins hitting Dakota with a barrage of fire punches. Cross jumps up afterward and slashes out one of Dakota's eyes. While it's distracted from the pain, Blak Hart and Cross try to reason with it.

Blak Hart: "Fight the urge, don't let it take full control!"

Cross: "You don't have to listen to him, you can fight it!"

For all their efforts, Dakota pays them back with violent swings.

Blak Hart: "I have no choice but to end him, if we leave him like this he can hurt or kill someone."

Cross: "But he was talking to us, showing signs of intelligence and the ability to reason."

Blak Hart: "That was before Zima force-fed him the demon blood."

In a blink of an eye, Blak Hart darts toward Dakota with a sword in hand.

Blak Hart: "I'm sorry, but you leave me no alternative."

With a heavy-hearted swing, Dakota's remaining eye locks onto his. For a split second, the rage fades as a tear rolls down his cheek. He cuts off Dakota's head. Right after, they surveyed the area. Since the coast was clear, they proceeded to give him a warrior's funeral. Cross moves the body as Blak Hart kicks the head next to the torso.

Cross: "You didn't have to do that, you could've picked it up."

Blak Hart: "I didn't want to bend down."

They pray for the warrior, and then Blak Hart begins to torch the body. At that very moment, out of the woods emerges a female Skunk Ape with two small ones. Blak Hart looks at her, stops what he is about to do, and bows his head.

Blak Hart: "I am so sorry for what has happened to your husband." *in a regretful tone*

Female Skunk Ape: *tears in her eyes* "What did he do... to deserve this?"

Cross: "He did nothing, he wasn't himself, a demon force-fed him their blood."

Blak Hart: "I was trying to help him, I didn't want to hurt him... but please understand whatever he is now, it is not your husband."

Female Skunk Ape: "I told him not to trust that demon, but he kept promising it would be ok." *holds her husband's body in her arms as the children cry behind her* "Is it ok now?"

Blak Hart walks close to the female skunk ape with his head down as Cross follows behind.

Blak Hart: "I'm so sorry. If there's any way I can help you, please tell me."

Cross: "We will gladly help in any way we can."

Female Skunk Ape: "He was a gentle soul, he used to sing to our little ones when the storms came. He said no darkness could touch us if we stayed together." *she quietly weeps and says* "I want to take my husband back to our village, can you help me get him there?"

Blak Hart and Cross pick up the body.

Blak Hart and Cross: "Show us the way."

After hours of walking through the swamps and marshlands, they can see a small village. The female skunk ape takes the lead and approaches an older-looking skunk ape with paint on his face. They are speaking in another language. The children guide Blak Hart and Cross to their burial grounds. They place the body on a stone slab and the children weep as they place a cloth over the body. The elder skunk ape approaches Blak Hart and Cross as he introduces himself as the shaman and leader of the village.

Skunk Ape Elder: "My daughter has told me what has happened and how you helped the best you could. On her behalf, I thank you."

Blak Hart: "We just did what we could, no one deserves what Dakota got. His wife doesn't deserve to be put through this."

Cross: "We promise, they will pay for this."

Skunk Ape Elder: "Thank you, young warriors. I knew that demon was trouble when he showed up with his puppets."

Blak Hart: "May I ask, what happened when the demon came here?"

Cross: "We know it's a hard time right now, but any info can help us."

The skunk ape nods and tells them that Zima showed up with a little wendigo, a huge demon, and a female. He offered to give them power along with taking back their land. The elder turned him down and just wanted to be left alone, but Dakota said he would help the demon if he could take their land back. He wanted to help but once he drank the blood, he lost control. The little Wendigo summoned his helpers and the woman followed them out to the swamps as Zima stood in the village tormenting the rest of the tribe. The woman appeared and whispered something. They both left, and that's when his daughter left to check what happened to her husband.

Blak Hart: "Thank you for the information, does any of your people need medical attention?"

Skunk Ape Elder: "We are alright, he never left any wounded, we took care of the bodies as you came in."

Cross: "Sorry, wish we could've gotten here sooner."

Skunk Ape Elder: "Easy, young one, there was nothing you could have done."

Cross: *clenching his fists* "But…"

Skunk Ape Elder: "So many emotions in one person, learn to control that. There are some things in this world you can't stop from happening."

Blak Hart: "We should be going. Again, sorry for your loss and thank you for the information." *bows his head*

Blak Hart and Cross say their goodbyes; they start walking back through the swamps and marshes. They walk the same path as before, when Cross tells Blak Hart.

Cross: "We keep saying we protect people, but what about those that aren't human… but sometimes it feels like we're just cleaning up after monsters we couldn't stop." *Cross sniffs the air*

Blak Hart: "What do you smell?"

Cross: "We are being followed."

Blak Hart: "Do you think the elder sent someone to keep an eye on us?"

Cross: "No, it's not one of them. Whoever it is, has perfume on."

Blak Hart: "The elder said that Zima had a female with him."

Cross: "I bet this is her."

Blak Hart: "WE KNOW YOU'RE OUT THERE, COME OUT AND STOP PLAYING AROUND!!"

Footsteps can be heard from in front of them. Next to the tree, a curvy figure appears with glowing red eyes and begins to speak.

Mystery Woman: "Wow, you boys are no fun. I like to play with my prey."

Blak Hart: *laughs* "Hehe, maybe if you would have asked nicely, we could have played along."

The figure slowly steps into the moonlight. The mysterious woman is dressed in an all-black hooded robe that presses against her hourglass figure, glowing red eyes, and is wearing a black plague doctor mask.

Mysterious Woman: "Now that I'm in plain sight, what are you going to do to me? Hopefully, it involves pain." *in a seductive tone*

Blak Hart: "Well, if that's how you like it." *flirting voice*

Cross: "This isn't the time to try and get suave. Let's finish her off before she conjures up a surprise."

Blak Hart unsheathes his sword and walks toward her.

Mysterious Woman: "My, my, what a big sword." *teasing* "Is that for me?"

Blak Hart: "I can show you something bigger and it'll be all for you." *flirting perversely*

Cross: *thinking* "We're gonna get swarmed by demons again, just because he can't keep his junk in his pants. Oh, damn."

Mysterious Woman: "You talk a good game, can you back it up? You burned Dakota, I wonder… will you burn me too or will you beg me to stop?"

Blak Hart: "Why don't you come this way and find out? Things can get as hot as you want, how about, I'll go to you."

Blak Hart slowly walks in her path; she starts walking backward. Every step he takes, she does the same in the opposite direction.

Blak Hart: "What's wrong, can't handle all this?" *thinking* "If I keep her talking and distracted, Cross could flank her."

A slight breeze blows toward him; he gets a whiff of her perfume. Blak Hart calls Catherine.

Blak Hart: "Hello?"

Catherine: "Are you on your way back?"

Blak Hart: "No, just making sure you're ok."

Catherine: "Don't forget to call me when you get here."

Blak Hart: "I'll call the minute my plane lands."

As the phone hangs up, Zima laughs.

Zima: "Humans are so predictable, so trusting. What would happen if he found out something were to happen to his lovely Catherine? I guess I'll find out at a later date."

In Florida, the woman stays playing mind games with her foes.

Blak Hart: *thinking* "The perfume smells almost like the woman in Shepherdstown, only with a mild hint of evergreen."

He walks a little quicker to the woman, but she creates a thick dark fog, and Cross pulls him back. Out of the fog, several beings with glowing red eyes glare at them.

Mysterious Woman: "Until next time; Blak Hart, you won't be so lucky. I'll see what you're made of."

Blak Hart: "Why wait till then, you tease? Aww, you brought playmates. You are too kind." *sarcastically said*

The woman disappears in the fog, while the warriors prepare to fight the beings. Speeding out of the fog, several hellhounds attack them.

Back at the hotel, Suma is getting to know everyone. Nora shows her gratitude by serving him first.

Nora: "I would like to thank you for saving Cross back at the beach."

Suma: "Your smile is thanks enough."

Nora: *blushing* "I'm sorry my son isn't here to meet you."

Suma: "It's ok, he's probably busy."

Rayne: "Where are you from?"

Suma: "An island off the coast of Florida, close to the beach where your son was."

Back at the Heart property, Nite and Ark are following a lead.

Ark: "Nite, I think your lead might have a few screws loose."

Nite: "My source is rarely ever wrong."

Ark: "He stated that the proof would be where the ground opened up. I see nothing unless you see something that I don't."

Nite: "Don't be hard-headed, patience, my brother. We must search carefully, maybe then, we'll be able to find clues that can prove what was told to us."

Ark: "If he is right, what then?"

Nite: "We save them before it's too late."

Ark: "For Louie's sake, I hope it's not."

The sun is setting, Blak Hart and Cross must defeat their enemies quickly.

Cross: "We must beat these things before nightfall, they're strongest at that time."

Blak Hart: "Easier said than done. But the fun part is trying."

They fight the creatures, one after another, killing them by decapitation and then tearing their limbs off. The rest of the hellhounds flee after seeing the burnt corpses and bloody body parts that were left from the battle.

Blak Hart: "Well, at least they know better than to try to go against us."

Cross: "It was easier than I remember."

Blak Hart: "Of course, you're stronger now than you were before you got out of the hospital."

Cross: "Now we can finally head back to the hotel."

The guys get in the car, eagerly expecting Nora's great meal that should be ready by the time they get there. Cross calls Luna to let her know that they are on their way.

Luna: "Does your mom know?"

Cross: "I didn't tell her yet, could you let her know, please?"

Luna: "Sure."

Blak Hart: "Tell her I said hi." *teasing*

In New Jersey, Ark and Nite follow the hole in the house that leads to an underground tunnel. They look around, noticing blood stains on the walls, ground, and pieces of bone all over. As they keep searching, Nite sees a gold watch. He picks it up and cleans the dirt off the glass along with the back of it. He finds an engraving saying "Our Hearts are Sealed Forever."

Nite: "Ark, look at this. I told you we would find something."

Ark: "What did you find?"

Nite: "That's why I told you to come here."

Ark: "Ok, ok, let's see your incredible find."

Nite: "I believe this was Albert's watch."

Ark: "You're right, he received this on his birthday from Aggy. Great job bro, let's see if we can find anything else."

Nite: "Oh, now all of a sudden you have an urgency to find things."

Ark: "Alright, I admit it. This time you might have a solid lead."

Nite: "My intel is never wrong, unlike your untrustworthy informers ranging from demons to our fallen brothers."

Ark: "Can't you ever take a compliment without putting my accomplishments down?"

Ark walks away from Nite and focuses on finding anything that may prove who killed Louie's parents and who might have sent them. It's been almost thirteen years since their demise, and he's been trying all he can to give Louie any information he can to put that part of his life behind him.

Nite: "I must commend you on your persistence and keeping the promise you made Albert's son. Why must you blame yourself for what happened? You did your best considering the circumstances."

Ark: "I knew what was coming, they trusted me. I should've come by and helped."

Nite: "You were on a mission already; nothing could be done."

Ark: "I was on a mission, but they were my responsibility. I was supposed to protect them." *gets emotional*

Nite puts his arm around his brother's shoulder to console him.

Ark: "I appreciate your words, but the only help I need is to find the truth, whatever it may be. You should know how I feel better than anyone else, you lost a charge as well."

Back at the hotel, the guys finally arrive. They are greeted by Nora, Rayne, Darren, Luna, and Suma. Louie looks at the kitchen and sees a meal fit for an army. Nora goes to the kitchen to serve dinner.

Three hours later, Suma thanks everyone for an enjoyable time, and then leaves. As everyone is winding down, the ladies go to one room and the guys go into the other. The girls are going to make it a slumber party. The fellas are going to make it a night of horror movies and comedy specials.

Later that night, Louie goes for a walk on the beach. It was thirteen years ago today that his parents were killed. Every year on this day, he takes time out of his day to remember them and remind himself of the reason why he became a warrior.

As he walks on the beach, his phone goes off. He answers it.

Blak Hart: "Hello?"

Ark: "Hey buddy, how are you holding up?"

Blak Hart: "Ok for now, why?"

Ark: "Because of what today symbolizes. If you need me there, I'll go."

Blak Hart: "I appreciate it, but you don't have to. I just want to look at the waves and meditate on the sounds of the water. Are you alright? Do you need me to go to you?"

Ark: "No, I'm just checking in; making sure you're ok. By the way, your car needed repairs, so I went ahead and fixed it."

Blak Hart: "Really? Thank you, you didn't have to do that, I would've gotten to it when I got back."

Ark: "Just wanted to have everything ready for your return."

Blak Hart: "Aawww, you miss me." *mocking him*

Ark: "Shut up, I'll see you when you get back." *voice lower than normal as it cracks when speaking* "I found something, a watch, it was your dad's, with an engraving on it saying 'Our Hearts are Sealed Forever.' I thought you should know." *he composes himself* "Don't forget to do the mission in Saint Augustine; the other ones I'll take care of, but Saint Augustine must get done by any means necessary."

Blak Hart: "By the way; thinking back, I am reminded of the time I was working as a security guard several years back and got fired. Could you believe that Cross and Darren were the cause for me losing that job?" *laughing* "Never in a million years would I have thought that I would be saving, helping, and working with them. That's the Ark I know, consider it done. Thank you, where did you find it?" *Ark tells him then tells him to call him in case of anything then hangs up* "That means we're getting closer, I'm ready for whatever." *the waves crash softly, like whispers from the past, he takes off his boots and stands barefoot in the sand, the moon casting a silver halo around him. He closes his eyes, remembering the warmth of his mother's hug, the strength in his father's voice, then he says* "I'm still fighting, I haven't forgotten. I never will."

In the caves, Nite and Ark are restlessly searching for anything that could help answer the questions: who, what, when, where, why.

Nite: "Checking up on Louie?"

Ark: "Yeah, how did you know?"

Nite: "That's what a concerned father would do and what I used to do with Will." *looks down*

Ark: "A concerned father? I guess, never would I have imagined that a job would have me taking a human as a son. A stubborn one at that." *smiles then looks at his brother* "I'm sorry, I didn't mean to bring up old memories."

Nite: "Remember, you're not the only one who lost someone who was like family."

Ark: "I know, believe me. You were the one who helped me through it, just as I wanted to be there for you but you never allowed me." *looking away*

Nite: "It's an older sibling's job to protect and help through any situation without letting your own feelings or emotions cloud judgment."

Ark: *listening to Nite as he continues searching* "I think I got something."

Nite: "What is it?"

Ark: "It was Aggy's bracelet, it has the symbol that Albert told me to carve into it." *gazes at the bracelet* "We're on the right path, we should keep moving forward."

Ark moves further into the caves, searching everywhere, making sure that nothing has been overlooked. Nite begins to realize that the tunnel keeps leading downward. The atmosphere in the cave has changed from typical temperature to a sudden density in the air as they get deeper.

Nite: "Ark, do you smell that?"

Ark: "The smell of rotting flesh, and a slight hint of sulfur."

Nite: "We better keep our guard up, from this point on, this is uncharted territory."

They march on; after walking about a mile, a husky figure walks toward them. The brothers brace themselves for the worst.

The Figure: "Nite! May I have a word with you?"

Ark: "Mercury." *clenching his fist* "What do you want to say?"

Nite: "Ark, stand down. Mercury, what are you doing here?"

Mercury: "I need to speak to you, Nite."

Ark: "You have nothing important to tell him, you're a fallen! Your words are lies!"

Nite: "Ark, shut up! There's an area I didn't check, could you please search in there? It might have something of importance, but it doesn't lead anywhere."

Ark: "Fine, I still don't trust him."

Nite: "Advice taken." *waits for Ark to leave* "Ok, so what do you want to inform me about?"

Mercury: "I know that you have heard many stories about my disappearance. Please don't let that cloud your judgment on what I'm about to tell you."

Nite: "Go ahead, you are procrastinating."

Mercury: "Don't proceed down the tunnel, what you both seek isn't exactly what you want to find."

Nite: "Why not, you have no clue why we must continue."

Mercury: "I know what happened to Louie's parents was horrible but, if you keep going, Ark isn't going to deal well with what you find."

Nite: "What is over there that I must worry about Ark?"

Mercury: "I cannot tell you."

Nite: "Why not? Who is controlling you?"

Mercury: "I am not under anyone's control, please brother, believe me when I tell you, there are many things that I would like to reveal. But if I reveal the truth, I will get in serious trouble."

Nite: "How much worse can you get in trouble if you're a fallen?"

Mercury: "The story you've heard is partially right, I was told to fall. I do what I have to for a reason. If we ever must battle, remember this, I will fight but will never kill either one of you. I'll make it seem like I did finish you off but in reality, you will be unconscious."

Nite: "Why do you feel like this is relevant to going in the tunnel?"

Mercury: "To let you know I'm still me, this time I must do things that no one is allowed to know of at this time."

Ark: "I found stones with symbols that were before my time. Can you help me put these together?"

Nite: "I'll be right there."

Mercury: "Take care, see you soon, please heed my warning. Don't venture any further into the cave."

In hell, Zima is torturing Amon just for fun. Mezun is laughing and the Wendigo King guards the entrance.

Zima: "Time to have some fun, you have been a pain in my ass for a while now. Let's see how strong your will is."

Amon: "Our lord will have your head for this, you'll pay for this, Zima."

Zima: "HAHAH! You pathetic waste of space, you think I care about what your precious lord thinks? I was here before he fell, he doesn't scare me!" *starts punching Amon in the stomach and ribs*

Amon: "ZIMA, YOU PIECE OF SHIT! HOW DARE YOU SPEAK OF OUR LORD IN THAT MANNER!!!"

Zima: "HAHAHA, HE'S YOUR LORD, NOT MINE. YOU SORRY EXCUSE FOR AN ANGEL!!!!!"

Zima takes out his blades and starts stabbing Amon in the stomach. He pins Amon's wings down with his blades and commences to twist the blades in his stomach along with the wings. Amon spits up blood as he screams in pain.

Zima: "Aahhh, music to my ears."

Amon: "What about your little bitch?" *coughs up blood* "Where is she now, she didn't do her job."

Zima: "HAHA! *snaps his fingers and the curvy figure shows up* I keep track of my toys unlike your lord."

Zima walks up to her, grabs her face, then licks her face as she tries to pull away a little. Her eyes flash not with fear but with fury.

Zima: "Don't get soft on me, bitch, remember who you belong to."

The figure pulls away from Zima, he tells Mezun to have fun with Amon as he grabs the girl.

Zima: "Who do you belong to!?"

The Female Figure: "Yours, I will obey every word you give me. I'm not getting soft; I just think it will be fun to break his mind then his body."

Zima: "That's my girl. Hehe, sometimes, I think you're more twisted than I am, if that's even possible? HAHAHAHAHAH!"

The following morning, Nora and Gail let the group know that they'll be shopping on the main strip.

Nora: "In case of anything, call us. Behave, I don't want to receive a call from the hotel management giving any complaints."

Louie: *he arrives from his late-night walk, once they leave; he asks* "I have one more mission before I must head back, who wants to join me?"

Cross: *Everyone looks at each other, then Cross responds* "We wish we could, this is the only time we have had alone since we started the vacation."

Louie: "I understand, just making sure. After I finish the mission, I'll call you to let you know when I'm heading back. Enjoy the rest of the vacation, nice meeting you Luna and I'll see you all in New York."

Louie: *searches for a taxi, then travels to the Saint Augustine haunted locations. He travels from the cemetery to the lighthouse, any place that had a slight rumor or story involving paranormal occurrences; in each place, Blak Hart either helped the spirits move on or he exorcised the evil spirits.*

At the hotel; the couples headed to the rooms, things begin to get hot and heavy, while enjoying their alone time. Hours later, Nora calls Cross. The phone rings and rings, no response. Cross calls Nora.

Cross: "Hey mom, did you just call?"

Nora: "Yes, what are you guys doing?"

Cross: "Nothing, we're watching movies then planned to go to the pool."

Nora: "Either way, we'll be there in ten minutes. I love you, I'll see you then, and next time I call pick up." *annoyed*

As Nora hangs up, Cross goes to the room door.

Cross: "They're coming back, hurry up and finish whatever you're doing. They're ten minutes away."

Rayne: "Oh shit, keep an eye outside."

Cross: "I have shit I have to do before they get here. You watch or have Darren guard the window."

Rayne: "Fine. Darren, could you be the lookout for our parents? They're on the way."

Darren: "Ok babe."

In the tunnels, the deeper they walk into the cave, the smell of sulfur keeps getting stronger.

Ark: "Seems like this leads directly to hell."

Nite: "This might be a part of it, but not the main part of hell."

An incredible distance from them, sounds of agony, screams, and screeching from something that isn't human. As they closed in, the noises got louder and louder.

Nite: "Ark, do you think we'll find anything in there?"

Ark: "Why else would we keep going if there was no hope of finding anything?"

Nite: "Point taken."

They venture forward; at the end of the tunnel, they see people having their skin ripped off by demons and other beings. Nite looks around.

Nite: *thinking* "I hope we don't find Albert and Aggy, if Ark sees them he'll lose it."

Gail and Nora arrive at the hotel, while the teens scramble to make the suite look normal. The door opens, and Nora looks around, as Gail goes directly to Darren to talk to him.

Gail: "You were behaving while we went out, RIGHT?"

Darren: "W... w... we were watching movies."

Nora: "Cross, why are you sweating? The movie was that intense, huh?"

Cross: "After the movies, we were cleaning so the suite would look just the way you two left it. Jeez, always think we're up to something." *looks and smirks at Rayne*

Nora looks at Rayne and Cross with a look of uncertainty.

Nora: "Usually when you two are in agreement and you don't try to outdo the other, something is up."

Gail: "Darren starts stuttering when he's nervous, always been that way. Yet, he still thinks that he can get away with it."

Rayne: "It's so messed up how you have such little faith in us. How else can we prove that we're maturing if you don't give us the benefit of the doubt?" *hoping this would work*

Nora: "That was a nice speech, but I trust my gut and my senses tell me you all were up to something. When I find out what it is, I'm gonna whoop that ass."

Rayne: "Well, when you find nothing; I'll be waiting for an apology."

Nora: "Girl, you better be happy I'm not kicking your ass now and ask questions later. Watch how you talk to me!"

Cross: "How about we talk about how you and Ms. Grier enjoyed your time shopping? What did you two see? I know there must have been plenty of interesting things, please tell us about it." *trying to divert the situation and the tension in the room*

Gail: "We saw many things you all like much cheaper than at the malls where we live."

Cross: "Really?" *thinking* "What else can I get them talking about so they can stop trying to figure out what happened here?"

Luna: "Are you going back to the stores? If you are, may I accompany you?"

Nora: "Sure, is there anything specific that you're looking for?"

Luna: "I'm not sure, I've never shopped out of state before. Do either one of you have any ideas or seen anything that I might like?"

Gail: "I saw a few bathing suits that would look cute on you."

Nora: "I saw a few art shops that might have some of the supplies you need at lower prices than the store next to us."

At the airport, Louie has just arrived in New York. He calls Catherine to notify her of his arrival.

Louie: "I just landed in Queens, do you have any plans for later on? If not, what time should I pick you up?"

Catherine: "About time; I have a few things I need to take care of, what time is good for you?"

Louie: "It's up to you, I have plenty of time."

Catherine: "Pick me up at eight tonight."

Louie: "See you then." *thinking* "I should call Ark in case he has more work for me." *he picks up his phone and calls Ark*

Ark: "Can I call you later on, I'm sort of occupied at the moment."

Louie: "Uh, sure no problem. Is there anything I can help you with?"

Ark: "No, no, I appreciate it. I'll call you in a few hours." *abruptly hangs up*

Louie: "Damn, I wonder what he's going after? Must be something major for him to act unlike himself. Unless the boss is getting a little something, something." *laughs* "Hopefully this will mellow him out a little."

He catches a train and heads home. Once he's at his stop, a workout seems like a perfect thing to do to kill time. Louie begins his routine when he realizes that he's being watched. A child appears to be very interested in him, whatever direction he goes; the young spy follows without missing a beat.

Louie: "Is the kid lost? What could he want, children aren't usually allowed in here."

Louie tries to lose him, but no matter what he does, the child is right there. He plans to trap him, by asking him some questions. Louie walks to the locker room, the kid follows him. He turns on the shower and then hides as the kid walks in. Then the boy starts walking to the shower as Louie follows behind quietly. The kid turns around and bumps into him.

The Kid: "Ouch! Hey, what's your problem!... Oh, crap.... alright you got me."

Louie: "Why are you following me?"

The Kid: "I got dared to follow you by some of my friends. They said no matter what, just keep following you and here we are. It's stupid, I know."

The kid's eyes light up a dark red and sink in, it glows just for a second. Louie grabs him and slams him against the wall.

Louie: "WHO DO YOU WORK FOR!!!?"

The Kid: "Geez, is this how you treat kids where you're from?"

Louie: "ANSWER THE FUCKING QUESTION!!!"

The Kid: "Yeah, can't do that."

Louie: "You say that like you have an option."

The Kid: *horns grow from his forehead* "Let's just say, I know more than you know."

Louie goes to punch the demon; it turns to smoke and disappears.

Louie: "What the fuck just happened? Who was that and what was the purpose of this whole thing? Oh well, whatever. Catherine is late, what's keeping her?"

While Louie is trying to figure out what the small demon is up to, Catherine sneaks up behind him and then puts her arms around him.

Louie: *jumping and almost swinging at Catherine* "Shit, you startled me."

Catherine: "Sorry, I just couldn't wait to grab you."

He takes her hand and kisses it as they walk off together to his car. Once in the car, they gaze into each other's eyes. They get closer, he leans in and kisses her passionately with more intensity and love than she has ever felt before. Catherine gasps at the intensity and passion behind the kiss as she melts into it, wrapping her arms around his neck and responding eagerly. Then Louie gets a message from Ark to go to the next mission.

Catherine: "Who is it?" *annoyed but breathing heavy, still hot and bothered from the kiss as she caresses her lips, still feeling the warmth of his lips*

Louie: "It's my boss, I was supposed to have the week off. Fuck it, one week won't make a difference. I thought him getting laid would've kept him busy."

Catherine: "He'll be fine, obviously, he'll just have someone cover your jobs while you're out."

Louie: "I guess you're right."

In the next state over, an angry Ark is impatiently waiting for Louie's reply.

Ark: "Why isn't he answering back? I have important information regarding his parents."

Nite: "Maybe he's busy, tell him whatever you're going to ask later. Let's focus on the matter at hand. We're heading into this opening and we don't know what to expect."

Ark: "We'll handle this like everything else, kick some tail, and ask questions while their bones break." *chuckles*

Nite: "This is no time to take things lightly. You know better than anyone that demons and ghouls are very unpredictable, so get your head out of your rear and prepare for the unexpected."

Ark: *sees a rolled-up paper right before the entrance* "Hey, I think I found something." *unrolls the paper*

Nite: "What is it?"

Ark: "It looks like a note from Albert." *he begins to read*

"To my boy: It seems like an eternity since your mother and I last saw you. I hope Will found you, he'll be able to help you in ways that I can't..." *the dried blood damaged the rest of the note*

Ark: "This is dated April, can't see the day, but the year is 1998. This might have been on the day when it all happened." *puzzled yet hopeful* "This proves that they might have escaped before the blast."

Nite: "The coroner took out two bodies, if they escaped; who was wheeled out in their place?"

Ark: "An even better question is who planned this out?"

Back at the stores in Florida, Gail, Nora, Rayne, Darren, Luna, and Cross are enjoying themselves. No sign of any supernatural beings or the possibility of anything going wrong as long as they're together.

Rayne: “Luna, are you and Cross finally a thing?”

Cross: “Rayne, shut up. That is none of your business, worry about your relationship and I'll worry about mine!” *blushing*

Rayne: “Thank you, idiot.”

Cross: “Shit, sorry Luna.” *looks down blushing* “Luna, can I talk to you privately?”

Luna: “It's ok, I was wondering when we were going to be asked.” *blushing and giggling* “Sure babe.” *they walk away from Rayne* “We’ll be back.”

Cross: “I’m scared sometimes. That this world will take you from me.”

Luna: “Then hold me tighter. I’m not going anywhere.”

Nora: “It's about time, so who asked who? Did Cross ask you to be his girlfriend or did you ask him?”

Gail: “I think she asked him, Cross was too shy around her.”

Luna: “Not to sound disrespectful, but, it was a mutual decision. In a relationship, there must be an agreement in decisions. It starts from the beginning and then progresses from there.”

Gail: “Wow, how old are you again? You have wisdom beyond your years.”

Luna: "Thank you, I watch a lot of romance movies and talk shows." *smiles*

Nora: "Let me know which shows they are so Rayne can learn from it. The soap operas I watch still haven't taught her a thing."

Catherine and Louie are walking around Manhattan, as they watch the boats pass by; Catherine looks deep into Louie's eyes, still thinking about the kiss.

Catherine: "I think I'm falling for you, I know I shouldn't. My job isn't going to be happy about this." *her heart flutters the closer she gets, but the fear of her higher-ups linger in her mind*

Louie: "I am too, why will your job be upset with you living your life? Do you want me to have a nice talk with them? I have a way of convincing people to change their minds."

Catherine: "No please, they'll hurt you. They're not just my employers. They're watchers. Judges, if they think I've compromised the mission… I know you can handle yourself. I beg of you, please don't try to speak on my behalf. Promise me!"

Louie: "Fine, I promise. But if you ever need me to put them in their place, let me know." *sternly saying it as he makes a fist*

Catherine: "I think I love you. All the time we spent together showed me how different you are from other guys."

Louie: *speaking very quickly* "I love you too." *speaking normally* "I just treat you the way a man is supposed to treat his girl; with respect, and loyalty, and show all the affection constantly."

Catherine: "I know you do. It's just that no one ever treated me this way, it's a little hard to get used to. Wait a minute, did you just say that you love me too? Did you think I wasn't going to hear you?" *blushing*

Louie: "Yeah, I thought you didn't hear it and of course, I said it, why?" *blushed so much his ears turned red*

Catherine: "Making sure I hear you correctly." *she was a bright red while fanning herself*

Several feet away, a male with dark brown hair and hazel-blue eyes, in a business casual outfit, one of Zima's lackeys, is overhearing the entire conversation.

Zima's Lackey: *thinking* "With this information, Zima will promote me. I think it's time for him to teach this bitch a lesson for betraying him and let me have the honor of destroying Blak Hart to prove my total allegiance to him."

He contacts Zima in a semi-blood sacrifice.

Zima: "Yeah Daemon, what is it? Better have some good information or you'll wish I had killed you instead."

Daemon: "I think you'll like the juicy info I have for you."

Zima: "So out with it, I don't have time to waste."

Daemon: "I overheard a conversation between Blak Hart and Catherine, they told each other the whole mortal 'I love you' exchange. She is getting soft-hearted for the mortal."

Zima: "Show me proof, why should I believe you? Where are they currently?"

Daemon: "In Manhattan at the pier. Is there anything you would like me to do? I could rip Blak Hart to shreds for you or kidnap Catherine, then he'll come to her rescue. We can get him there."

Zima: "If I want you to do something, I'll tell you. I have another use for you, meet me at the four corners. As for them, my wonderful mind has plans." *maniacal laughter*

At the four corners, Zima arrives and waits for Daemon.

Zima: *sighs* "What the fuck is taking him so long, I gave him a simple task quite some time ago."

Daemon appears with a black bag, his body steaming, and he's out of breath.

Zima: "About fucking time, what took so long?"

Daemon: "You didn't tell me there were hellhounds guarding it."

Zima: "Well, I thought you could handle it. Was I wrong?"

Daemon: *gives Zima the bag* "No, I got it."

Zima: "Good boy, it's been a while since I was able to use this." *smiles as he sticks his hand in the bag and pulls out a suit made of skin*

Daemon: "Uh, Zima? What is that exactly and why was it guarded? ..."

Zima: "It's something that was taken from me a long time ago."

Zima starts to put the skin suit on, it starts to steam and form to Zima's body as he slides it over his head. His hair starts to grow. Daemon steps back in shock.

Daemon: "What the fuck is going on?"

Zima walks out of the steam smiling, he looks like a slim, well-toned human with long jet-black hair.

Zima: "It's been so long." *stretches* "Time to update my look. Hehe."

Daemon, still in shock and shaking.

Daemon: "Z-Zima, what the fuck was that?"

Zima turns around as his eyes turn a light blue and he smiles.

Zima: "Let's just say, it's part of my past. Now let's go and see if what you say is true."

Gail and the gang are leaving the Sunshine State.

Nora: "When we get back home, the twins need to prepare for the prom and graduation."

Gail: "Darren too, we have more shopping to do."

Nora: "Luna, what about you?"

Luna: "I don't know if I'm going to the prom, I don't have enough money to get a dress."

Rayne: "You told me that some money was saved. How much do you have?"

Luna: "Not enough."

Nora: "You will accompany us when we go shopping; whatever you need to complete for the dress, I'll complete it."

Luna: "No, it's too much. You don't have to do it."

Rayne: "She knows that she doesn't have to, she wants to. Now shut up, say thank you, and let's plan what idea you have for the dress and what color your nails are going to be and what style your hair is going to be."

Luna: "I appreciate it, I don't want her spending so much."

Gail: "Next exit, I'm getting something to eat. Those who are hungry, join me inside the restaurant."

Darren: "I'll go with you; I might be able to buy something they might like."

They arrive at a restaurant off the main highway. While they walk in, Rayne runs to the gift shop. Cross and Luna go around looking for a lobby shop. Gail and Nora head to the counter to order food. Darren tries to go after Rayne but, loses her in the crowd. He squeezes through the people, and in the horde, a possessed man charges at Darren. He grabs him by his arm.

Possessed Man: "You're coming with me. Please give me trouble."

Darren: *while throwing a punch* "I'm not going anywhere with you."

Back at the tunnel, Ark and Nite walk into the opening. Nite gets tackled by two demons. Ark runs to Nite's aid when he looks on the other side of the cave. He sees cells that could have possibly held Albert and Aggy at one point or another. He goes to free Nite from the demon's grasp then, a figure engulfed in flames slashes Ark on the neck; not enough for him to bleed out, just enough to distract him as Mercury punches Ark back to the entrance. Nite draws his sword, he slashes the demons one by one. He stabs the last one as Mercury sprints back, grabs Nite by the throat, and throws him directly at Ark, rendering both brothers unconscious. With one punch, Mercury causes the cave to crumble; various-sized rocks along with boulders block the entrance of what could've been purgatory.

CHAPTER 9.
PROM PREPARATIONS

Nite gains consciousness. He looks at Ark and notices that his brother's neck is bleeding. Nite rips part of his shirt and puts it on Ark's neck to stop the bleeding.

Nite: "What could affect Ark? Unless... No, it can't be."

Back at the rest stop, Darren is using his training to fight the being. It hits and grabs, pulling him closer to the rear of the place.

Possessed man: "You're coming with me. You can't defeat me."

Darren: "That's what you think."

Darren takes out a knife that Louie gave him when he finished his first training. He stabs the man, and thick black blood pours from his neck. He still tries to attack Darren right before Darren's killing blow. Rayne speeds through the crowd and tackles the man.

Darren: "Rayne, back up! I got him."

Rayne: "You're taking too long."

Darren: "Stop being impatient and let me finish him off."

Rayne: "I can do it! Damn it, why does everyone have such little faith in me?!" *in an angry tone*

Darren: "I know you can, I just want to do it myself. Please, just make sure no one else steps in."

Rayne looks at the guy, realizing that Darren wasn't doubting her. He had to prove to himself that he could do it. So, Rayne throws him to Darren, with a swift slash from his blade. The man drops to the ground, and a dark grey mist seeps out of the mouth and nose of the man, then disappears into the air.

Darren: "What the hell was that?"

Rayne: "That's what comes out when a person is possessed."

Darren: "How do you know that?"

Rayne: "If you paid attention when Louie was explaining, he showed us how to save the person and exorcise the demon."

Luna and Cross catch up with them. They see Darren's shirt ripped.

Cross: "What did we miss?"

Rayne: "Darren killed a guy and released the demon."

Luna: "How did he do that?"

Cross observes Darren's hand and sees the knife.

Cross: "At least he used the blade that Louie gave him, so technically he did exorcise it."

Rayne: "How?"

Cross: "The blade is blessed for the situation. The person may be dead, but the soul is free to go to the rightful place."

Rayne looks down in embarrassment, then glances at Darren with an unspoken apology. Darren thanks Cross for backing him up, while Gail and Nora call the teens to start heading back to the vehicle. The young adults continue their conversation about what has transpired.

Cross: "We should call Louie, to find out if this has anything to do with one of his missions."

Darren: "Why? I already took care of the threat. Also, why bother him?"

Rayne: "If Cross thinks we should call Louie, then that's what we should do."

Luna: "Why not tell Nora and Gail? They might know something."

Cross: "I think Luna's right. Let's tell our moms. They probably know exactly what you dealt with and if they've had any previous encounters."

Darren: "I did something awesome, yet no one acknowledges it." *pouting*

Cross: "We're not saying that it wasn't cool. We need more info in case any of us have to deal with a similar scenario."

In the tunnel, Ark is waking up with Nite attending to his wound.

Ark: "What are you doing?"

Nite: "Cleaning your injury. For some reason, it's not healing."

Ark: "That's proof that they're in there."

Nite: "It could've been one of the high-level demons that might have done this to you."

Ark: "This is probably Albert's way of warning me to stay back."

Nite: "I still wish there was a way I could prove that Will is alive, but you and I both know that he's not. I live with that every day, I know how you feel. Don't dwell on the what-ifs. Just worry about protecting Albert and Agatha's son. He's still here."

Ark: "Brother, you'll never understand. You have proof of Will's... Well, you know. I don't have that. You lost one. I lost two. I refuse to believe they died on my watch. I examined the bodies. Neither of them had the sacred symbols. The corpses were deceased way before the explosion. This is why I have been searching for all these years."

Nite: "Why didn't you share this information with me?"

Ark: "Because you would tell me that I'm in denial and to put it out of my mind. That's it, all part of the job. You constantly kept making me feel like a failure..." *starts breaking down*

Nite: *puts his head down, knowing that Ark is right* "Let's check on the Crosses and Louie. I'll help you search, after we know that they're safe. Please don't tell Louie anything until you have concrete evidence about his parents. We don't want to get his hopes up in case we can't find them. I only tried to protect you, but instead, I kept pushing you away because I didn't know how to handle my feelings. I didn't want to accept that 'the great warriors' failed in protecting their own charge and couldn't help their brother in his time of need." *tears slide down his face* "I never meant to make you feel that way. Please forgive me for not being there for you and having it seem as if you aren't good at what you do, because you are one of the best—aside from me." *joking to relieve the tension* "At what you do."

Ark: "Sounds good. And don't put this all on yourself, brother. You had plenty to deal with. I was disappointed in you because I was too blinded by grief to see you were hurting also. Please forgive me for not being the sibling you needed without you asking." *They both look at each other in acknowledgment and hug it out, then proceed with the proper phone calls.*

Back in New York, Nora, Gail, Darren, Rayne, Luna, and Cross are preparing to go to the mall.

Rayne: "I have to go to the restroom. I'll be right out."

She goes in and opens up a box. It's a pregnancy test. Rayne does as it instructs her. Once the results are shown, she goes to Darren.

Rayne: "Look at this." *she starts thinking* "I'm not ready. Maybe... maybe I will be if Darren stays by my side."

He looks at the test. She tells him to keep it to himself. She'll inform everyone after the prom.

Moments later, they all arrive at the mall close to their house. Cross called Louie to notify him about Darren's encounter and how he dealt with it. Louie told him that Darren did pretty good against a possessed person, but he should've called for Cross to assist. Louie said that he would be by the house in a couple of days.

Cross: "My mom is dragging everyone all over the mall in search of the perfect clothes for prom."

Louie: "Enjoy it. I never went to one, and at least you have the chance to get annoyed by a parent..." *stops himself from getting emotional* "Sorry, just enjoy it." *He receives another call* "I'll call you right back. Ark is calling me."

Cross: "Alright."

Louie: *Switches the call* "Hello."

Ark: "I've been trying to call you. Why haven't you picked up my call?"

Louie: "I was busy at the moment and I was on vacation. What's up?"

Ark: "I'm sending you to the Marines, Black Operations Forces division."

Louie: "Why? Is this some kind of punishment for not answering your call? I finally found someone who loves me for me."

Ark: "I'm happy for you. If she truly loves you, she'll wait for you. I'm not punishing you, Louie. There are forces moving. Ones that know who you are and who your parents were."

Louie: "Oh wow, you want me to disappear into a black file while people I care about get hunted? No, not unless you tell me everything."

Ark: "I have several missions for you on a grander scale. You're the only one I trust getting these jobs done. The others will go after graduation."

Louie: "I feel like an ass. When am I getting shipped off?"

Ark: "In about three days. It'll give you enough time to tell everyone. Also, you'll be in the same base your father was from."

Louie: "Sounds reasonable enough. Will I need any kind of new training?"

Ark: "I have a hand in picking a specific person to train you for military and supernatural battles."

Louie: "You act like I'm just learning how to fight."

Ark: "The battles you are about to face are unlike anything you have ever experienced or dealt with. Your opponents are ten times smarter, stronger, and more experienced than what you're used to."

Louie: "Will I have my own squad, and when can everyone else join my squad?"

Ark: "You will have several folders of the best candidates to choose from. As for Darren, Cross, and Rayne... Each will be training in the other branches of the military until it's time for them to join you. I'll call you in three days to take you to the base." *Ark hangs up the phone.*

Ark feels a bit guilty for not telling Louie about his findings, then walks to Nite to talk to him. Nite looks at Ark and informs him that he'll talk after he gets off the phone.

Nite: "Nora, I'm just checking up on you and the kids."

Nora: "We're just getting ready for the prom. We got the outfits. Thanks for checking up. Is everything alright with you and Ark?"

Nite: "You know Ark. Always reaching for the impossible, but he does shock me from time to time. As for our brotherly bond, we had a breakthrough, just as you said. Thank you."

Ark: "You do know I'm like three feet away from you. I can hear everything you're saying about me." *he says with attitude and laughing*

Nite: "Like I was saying before I was rudely interrupted, we're doing fine."

Nora: "Good to know. Please try to keep your cool when dealing with your brother. He looks up to you, and you know in your heart he means well. The issues you have with him, he's experiencing them with Louie." *she chuckles*

Nite: *snickers* "Karma is a ruthless teacher."

Nora: "Gail says hello. I'll let you go so you can attend to Ark." *Nora hangs up.*

Nite glances at Ark and smiles.

Nite: "What is it?" *he sounds angry, but he says it with a smile*

Ark: "I told Louie that I'm sending him to the Marines, and I was wondering if I should tell him?"

Nite: "Tell him what exactly?"

Ark: "I want to let him know who I really am."

Nite: "Why haven't you? He should've been told years ago."

Ark: "He was young. I didn't want him to freak out or become angry with me."

Nite: "Why would he be upset with you?"

Ark: "Cause he'll say, if I am what I am, why didn't I save his parents?"

Nite: "You already know what to say. He's a smart man. Remember, he has the Heart bloodline. You'll be fine."

In New Jersey, Louie takes a walk with Catherine, then talks about being shipped off in two days.

Louie: "My boss is shipping me to the Marines in two days. I know it's a last-minute thing; we can make the best of it until I return."

Catherine: "Why does he have to do that to you? You don't have to go, just stay with me." *eyes get watery* "If you love me, you'll stay."

Louie: "I can't stay, I do love you. If you love me, you'll wait for me. This is all part of being a hunter, I go wherever I am needed. Let's make memories instead of being depressed. And when my tour or mission is done, I'll be coming back to you and only you."

Catherine: "Does your family know?"

Louie: "No, tomorrow I'm going to tell them. I've grown very close to them, they're the only family I have left." *gets a bit emotional*

Catherine: *in a seductive voice* "Ok, tonight you're all mine."

Louie walks Catherine to her house; he closes the door behind him. She puts on some romantic music, turns on the candles, and shuts off the lights. While Louie pours wine into the glasses, she looks in his direction, taking one piece of clothing after another. As Louie quickly takes his clothes off, he picks her up and carries her to the bedroom.

In Long Island, Nora and Gail have the young adults training.

Nora: "Tomorrow, Louie has important information for us. Please hear him out and realize that no matter what, we're family."

Rayne: "Is it something bad, something that we should be worried about?"

Cross: "We'll find out when he gets here, if it were bad news, he would've been here already."

Later that night, Catherine leaves Louie's side for a moment. She walks to the bathroom.

Catherine: *she looks at Louie sleeping, her fingers trembling on the phone and whispers to Louie* "I'm sorry, baby." *then heads to the bathroom and dials as the person on the other end picks up* "In two days, you'll have the opportunity to do what you want."

Louie: "Baby, you ok in there?"

Catherine: "I gotta go, we'll talk tomorrow." *ends the call* "Yeah, I'm fine. I'll be out in a sec." *she silently sobs as the waves of guilt hit her like a ton of bricks*

Louie: "Hope you're ready for round two."

In hell, a conversation is in progress.

Shadow person: "We have to get out of here."

Shadow person 2: "I have a way, but you must give me time."

Shadow person 3: "Can't take being here anymore. I'm sick and tired of being controlled, I know you are too."

Shadow person 2: "I understand it's a struggle, but believe me when I tell you... He will pay for his actions, THIS I CAN ASSURE YOU."

Shadow person 1 & 3: "I can help you, if you can guarantee our freedom."

Mercury walks by as they were talking.

Mercury: "Get back to your cells, speaking like this will continue to get you in trouble. I figured by now you would have gotten the point, no one gets out. So, quit your whining and complaining, this is your lives now."

Shadow person 2: "Mercury, our quarrel is not with you. Please don't involve yourself in matters that don't concern you."

Shadow person 1 & 3: "We'll go back, but don't get in our way when the time comes."

In the Pine Barrens in Suffolk County, Zima, Daemon, and the Wendigo King are having a meeting.

Zima: "Wendigo King, tomorrow, have your strongest wendigos ready for an interesting test. Daemon, have a few different demons with you, make sure they are S-class only. When I call for you both, I'll let you know where to play."

Daemon: "Will Hecate be there, or are you sending her back to the Crossroads?"

Zima: "I'll probably send her to Noneya."

Daemon: "Noneya, where is that? I never heard of it."

Zima: "None of your damn business; now go do as I ask, and Wendigo King, let's give you a better name. From now on, you will be known as Nuzem."

Nuzem: "As you wish."

The following day, Louie pays the Cross family a visit. Cross is outside waiting for him.

Cross: "So, what's the big news?"

Louie: "Not really that big, just that Ark is sending me to the Marines."

Cross: "Really? Mom made it seem more cryptic than that."

Louie: "How did she know before you?"

Cross: "Most likely, she spoke to Nite."

Louie: "Yeah, probably. How's it going with you and Luna?"

Cross: "Going great, my mom is happy for us. How's everything between you and Catherine? You look like you've seen heaven and hell in one night. What's the deal?"

Louie: "Pretty good, spent all day and night with her. I told her that I was leaving tomorrow. Well, heaven, yes; hell will be the deployment." *he smirks and smiles from ear to ear*

Cross smiles knowing what he meant and nudges him in a brotherly manner, then proceeds with the conversation.

Nora comes outside and gives Louie a bear hug. Rayne follows shortly after.

Rayne: "What's the mysterious information?"

Louie: "In two days, I'm going to the Marines base."

Darren arrives, and Louie continues to explain.

Louie: "I'll be there until the corruption in there has been taken care of."

Darren: "What about us?"

Louie: "You will be going to different branches of the military. Once you're fully trained, then you'll be joining me for some black ops work."

Cross: "When are we getting deployed?"

Louie: "Most likely right after graduation."

Darren: "I thought you were going to be here for our prom."

Louie: "I wish I could, but..."

Rayne: "What about our graduation?"

Louie: *looks down* "I'll try, but there's no guarantee. I'm sorry."

Cross: "Damn, that shit sucks."

Louie: "I know, but we'll meet up in the military. For today and tomorrow, let's make the best of it."

Nora: "If something were to happen to you, will they notify me?"

Louie: "I'll have you down as my emergency contact, and Ark will keep you posted when I'm allowed to." *Nora hugs him*

Nora: "I love you as if you were my own, please be very careful. We'll miss you." *a tear rolls down her face* "Please take care of my babies."

Louie: "I will, I'll be leaving the day after tomorrow. Well, at least I'll be following in my father's footsteps." *nervously smiles*

In heaven, everyone is preparing for the events that are about to unfold. The unknown angel walks into the meeting room and sees the head angel along with the second angel already in their seats.

Head angel: "We're still waiting for a few more, then we'll commence."

The unknown angel looks at the head angel, then tells him what's on his mind.

Unknown angel: "As head angel, why don't you start and fill in the rest later on. Brother Michael, you don't have to wait for them. They know you called for this meeting at the time specified."

Head angel/Archangel Michael: "You are right, brother Zechariah; I want to make sure our brothers know what is to come. Waiting a bit longer gives time for the rest to arrive before I begin giving punishments for not attending. Raphael, please close the door in exactly ten minutes."

Second angel/Archangel Raphael: "No problem, anyone after that will get what's coming to them."

In New York, Luna arrives at Cross's house. Cross fills her in on the conversation that his family had with Louie.

Luna: "Is everyone going into the military?"

Cross: "We're all trying to follow in our parents' footsteps, in the process, we'll be ridding the world of horrible individuals."

Luna: "How does Nora feel about it?"

Cross: "She's worried about us, but at least we still have time before going. Louie is the one that is leaving in a couple of days."

Back at the conference room in heaven, everyone that was called to the meeting showed up. The Archangel Michael begins the meeting.

Archangel Michael: "Louie is going to our section of the military, he doesn't know it yet, but he will be one of our greatest assets. Louie will face trials that would break most. But if he endures… he will be the blade that cuts through the veil."

Archangel Raphael: "His time in there will give him the training needed to help us in the war against evil."

Guardian Angel Zechariah: "When are we sending the others to train?"

Archangel Gabriel: "Who's going to be training them? May I train them, brother?"

Archangel Michael: "The other three will be sent after graduation, they all will be trained by someone that will be selected specifically by their guardian. Gabriel, you are needed here, but you will have the chance to choose the trainer for the teen in your care. We'll talk privately about a different matter."

In New York, Darren, Cross, and Louie are going to have a guys' night. They start off the day with bowling, then head to the shooting range to show off their skills.

Meanwhile, Gail, Nora, Rayne, and Luna are grocery shopping. Shortly after, they start weapon training and technique training. Several hours later, they begin preparing for the guys' return.

Luna: "This will be a meal to remember when he's in the Marines."

Nora: "I'll send him a few treats when he's out there."

Gail: "Is that possible?"

Nora: "Aggy and I used to send all types of food to Albert and Will..." *looks down and gets misty-eyed*

Gail: "I'm sorry, I didn't mean to make you cry."

Nora: "It's alright, it's a great memory. It's tears of joy and sorrow all together." *Rayne runs up to her and hugs her tightly*

The guys are planning to go play pool until it's time to go back to the house.

Louie: "Cross, do you have an idea what your mom is going to make for dinner?"

Darren: "Whatever it is, you know it's going to be awesome."

Cross: "I have no clue, but I have a feeling she's going to go all out. I can't wait to return home to see what we're getting surprised with."

Louie: "How good are you guys with pool?"

Cross: "I'm pretty good."

Darren: "I never played it."

Louie: "It's not a hard game to learn."

They walk into the pool hall and start playing. Cross and Louie are teaching Darren the basics of the game. Cross is showing off his talents in the game while enjoying himself.

Louie: *begins to think* "I'm ready to fight, but am I ready to lead, to be like my father was?"

Cross: "This is how it's done."

Darren: "It's not as hard as I thought it would be, but Cross, take it lightly on the new player."

Louie: "Damn Cross, that's cold-hearted." *chuckles*

Cross: "I don't give chances, either he tries harder or gets his ass handed on a silver platter." *smiles at Darren*

Darren looks at Cross and hits a combination of plays that ends the game, making him the winner.

Cross: "Beginner's luck, let's see how you do on the next game."

In heaven, the meeting has just ended. The trainers for Louie and the teens have been chosen. Archangel Michael pulls his younger brother, Archangel Gabriel, to the side.

Archangel Michael: "You must spend time with the one you're protecting before having him go to an even harder training. He still doesn't know who you are."

Archangel Gabriel: "What if he doesn't like me or if he won't listen to me? I don't want to fail."

Archangel Raphael joins in the conversation.

Archangel Raphael: "I haven't had the most perfect record in guarding either, but I still try my best to make sure that he's safe."

Archangel Gabriel: "Yeah, but you have a warrior forged by grief and a drive to do whatever it takes to do the mission."

Archangel Michael: "It takes time, which you don't have on your side. If you need advice, we will be here. Raphael and I have done this several times, we can give insight on how to do a good job depending on how he reacts."

Two hours later, Cross, Darren, and Louie are leaving the pool hall. The excitement of Nora's food fills their minds.

Louie: "The quicker we get there, the quicker we can stuff our faces with great food. Oh, and by the way, I'm really glad I didn't get you two arrested that night."

Cross: "Wait, what? You knew?"

Darren: "So me either. Would be pretty weird that you were the cause of our prison experience." *all three laughed about it and decided to never speak of it ever again and it was agreed*

Cross: "You have to beat me to it."

Darren: "I'll beat both of you there."

Darren starts to run; Louie pushes him into the bushes while Cross passes both of them up. They are laughing and making fun of each other all the way to the house.

Darren: "Told you I would get here before you."

Louie: "With your speed, you have no reason to lose."

Cross: "You cheated, powers aren't to be used."

Louie: "How do you think we got here so fast? We all have some type of speed in us. Not as fast as him." *punches Darren on the arm lightly*

They enter the house; Nora is hard at work in the kitchen. Gail is baking enough for an army. The girls are busy cleaning the house. Luna realizes that Cross walked in, and she jumps on him, hugging and kissing him.

Rayne: "Get a room, we have some stuff to do before dinner."

Luna: "Ok, ok, I'll be there in a sec."

Cross: "Babe, please help her before she turns into a bitch."

Louie: "Hey Luna and Rayne, need a hand?"

Rayne: "No, we got this. Look at this, the only one who offered. That's a real man."

Nora: "Well, if you really want to help, could you please go to the store and pick up some drinks? We were so preoccupied with getting the ingredients for the food, we completely forgot the beverages."

Louie: "Sure, no problem. Is that all?"

Gail: "Yup, that's everything. Thank you."

As Louie drives off, a vehicle stops in front of the house. Rayne and Darren were outside talking when they notice Ark and Nite with a stranger.

Rayne: "Hey Nite, hey Ark. Who's the stiff?"

Nite and Ark: "Hi Rayne, he's Spaz, our baby brother."

Rayne: "He's pretty old for being the baby."

Spaz: "Nice to meet you too, Rayne."

Ark: "Has Louie been by?"

Rayne: "He went to the store for my mom, he'll be back soon. Are you guys coming, or are you staying outside?"

Ark, Nite, and Spaz walk inside to be greeted by Nora along with Gail. Nite greets them back, then introduces himself to Luna.

Nite: "I wanted to introduce myself."

Luna: "I know who you are. To deal with Cross and Rayne, you are definitely not mortal. What are you?"

Louie returns only to see Ark waiting outside for him.

Ark: "Louie, I have some important news."

Louie: "What's the news, am I going somewhere worse than the Marines?"

Ark: "I should've told you this years ago."

Louie: "What? Are you dying?"

Ark: "No, I am not human. I'm an..."

Louie: "I know you're not human, that I knew. You're in too many places in such a short period of time."

Ark: "Shut up and let me finish. My real name is Raphael; I am an Archangel. I have been your family's guardian for centuries."

Louie: "I'm not trying to be rude, but where were you when my parents were being attacked?"

Archangel Raphael: "I called your mother that day warning them of the attack. I tried getting there after I fought off the legion that was heading to your house. Sadly, I wasn't able to fight off all of them. By the time I arrived to you, it was too late." *he looks down with tears rolling down his face* "I have sworn to prevent you from falling under the same fate. That's why I have been keeping a close eye."

Louie: "Why didn't you tell me that Nora and the kids were alive? I wouldn't have been alone for so many years."

Nite tries to interject but isn't allowed to.

Archangel Raphael: "Michael, not now."

Archangel Michael: "I can help if you need me."

Louie: "Wait a minute, you are both angels. Yet, where were you when I was running away from foster homes, getting odd jobs from cleaning police precincts and them paying me with food, along with some self-defense lessons to staying in various dojos learning from the sensei of that dojo to be able to go after the scum that killed my family!" *spoken with a shaky voice and crying*

Archangel Raphael: "We were there helping you every step of the way. Every time you ran away, I was the one who took you in. My appearance may have been different, but I was always with you."

Louie: "Why didn't you tell me the truth back then? Why now?"

Archangel Michael: "My brother has done so much for you; do you think he doesn't replay that night every day and pray that it would've been him instead of them? When you're not around him, he's been trying to find out who sent out the bounty and researching that day over and over since it happened." *puts his hands on Raphael's shoulder*

Louie: *sits on the steps outside of the house, looking down* "I appreciate all that you have done for me, but you should have made yourself known. The transition would've been a lot smoother for me knowing I had someone in my corner."

Archangel Raphael: "I thought you would hate me if you knew the truth."

Louie: "I'm upset with you, then again, I understand that you always tried the best you can. As far as I've known you, you have done nothing less than your best. I believe you did the same at that time. Like I said, I'm not mad, just hurt that you hid something like this from me."

Nora comes outside.

Nora: "I hope you're not mad at me."

Louie: "Why would I be angry with you? Nite... I mean Michael has been your guardian, so I put two and two together."

Nora: "I've been holding this for a while now, the kids don't even know."

Louie: "Are there any other surprises before we go eat Nora's incredible meal?"

Archangel Raphael: "That's everything."

Everyone goes into the dining room and eagerly waits for the food. Moments later, Nora, Gail, Luna, and Rayne come out. One after another, they keep coming out of the kitchen with tons of food. They made enough food for a Thanksgiving feast to feed twenty to thirty people. Before everyone partakes in the feast, Michael prays over the food and everyone's health. Everyone bows their heads in prayer. Then Cross clanks his glass with his spoon.

Cross: "Let's raise our glasses for my brother. To Louie, our brother, our fiery blade, our hope. Come back stronger and kick ass for all of us." *they all cheered*

Before Cross could take the first spoon, Michael rushes to get the first scoop.

Raphael: "Damn, Michael, what's the rush?"

Michael: "Great food like this doesn't come every day."

This is the first time Michael has been seen acting normal.

Spaz/Archangel Gabriel: "Raphael, is he normally like this?"

Raphael: "No, it's just creepy."

Once the meal was done, Nora decided to take pictures of everyone.

Nora: "Life is too short, I'm taking these pictures so we can always look back and remember these wonderful moments."

Luna: "I'll clean up while you two relax a bit."

Nora: "You don't have to, but thank you."

Gail pulls Nora to the living room.

Gail: "Why are you acting like this is the last time we're all going to be together?"

Nora: "Whenever things are going well, that means something horrible is coming our way."

Gail: "You know it doesn't always happen that way."

Nora: "You know me. When I get a feeling, I'm usually right. That's what I'm afraid of."

Gail: "What are you feeling?"

In the other room, the three archangels and the young adults are talking about the future.

Michael: "Louie, Raphael already has a game plan for you. Look, you've walked through fire, now you'll learn how to wield it." *he takes a deep breath* "Cross, we'll get a plan set for after graduation. Darren, Gabriel and you must get acquainted, he is your guardian. Rayne, you will meet your guardian during the week."

Gabriel walks with Darren to the dining room; they begin to talk. Meanwhile, Michael and Raphael prepare to leave. Louie thanks Nora, Gail, Rayne, and Luna for everything.

In hell, Zima gives his group a motivational speech.

Zima: "We all know what I'm gonna say. Do a great job and you'll get rewarded. Fuck this up, die over and over until I decide otherwise. Tonight, Blak Hart will be leaving to go to the military. Tomorrow, we attack. No worries of him getting in our way, this time we'll finish the Cross family once and for all. Does anyone have any questions?"

Daemon: "Should we attack in the morning, or do we go at night when we're strongest?"

Zima: "First group will go during the morning. At sunset, the second group will attack with whatever survivors from the first attack."

Daemon: "As you wish, my lord."

Zima: "This is how you all should be."

Back at Nora's house, Louie heads home as Gabriel chats with Gail and Darren. Nora talks to the twins.

Nora: "What are we going to do for Louie's last day here? Does anyone have ideas for tomorrow?"

Cross: "I have a few ideas, but I'm not sure it would be memorable."

Rayne: "Have you asked Darren? He might have more thoughts of what we can do."

Cross: "Not yet, I'll talk to him before he leaves."

Right at that moment, Darren walks out with Gabriel.

Darren: "I'll see you guys tomorrow, I have to start training with Archangel Gabriel."

Cross: "Hold on, I gotta tell you something."

Darren: "We'll talk tomorrow, see you then." *Darren darts off with Gabriel*

The following morning, everyone gets ready for Louie's arrival. Luna and Darren will come by later. For right now, Cross, Rayne, and Nora will take Louie to Extreme Paintballing.

Cross: "Teams are set, you and mom against Rayne and me."

Louie: "Nora, these two think you don't have what it takes." *joking around*

Nora: "I may be rusty, but we're going to kick their butt."

Rayne: "Yeah right, you can't move as fast as you used to." *mocking her mom*

Nora: "Here are the rules. Each of us gets ten bullets. Second, after the bullets are done, you must do whatever it takes to render the opponent incapable of doing any harm or capture the opponent. Third, these bullets are live. Last but not least, to win, both must be captured or unable to continue. So, don't hold back because I'm not. Everyone has ten seconds to pick an area, after that, the game begins."

The groups start the countdown. Nora and Louie go deep into the woods while Rayne and Cross hide in the self-made bunkers. By the time they stopped counting, the groups notice there are more players on the field.

Louie asks Nora to call Gail while he calls Cross.

Louie: "Cross, let Rayne know to get ready, this training has actually turned to the real thing. Your mom is calling Gail right now to stay put until we handle this situation."

Cross: "No problem, this isn't a trick to get our guard down, right?"

Louie: "You know me better than that. Let's save our weapons for later, we're going to start the battle with our God-given weapons."

Nora: "I already notified Gail, she'll be waiting for my call. What's the game plan, Hunt... Sorry, your parents and Will tried to play extreme paintball, but this always happened. Your father was the one always with a plan ready for any scenario, seems like that's something you also inherited."

Cross: "What should we do first?"

Louie: "I need you to change into a wolf and distract them while the rest of us attack. Once the attack starts, feel free to have fun."

Cross: "Can I try my new stuff?"

Louie: "Knock yourself out."

Cross runs out in mid-transformation, several wendigos chase him as Nora attacks them. Rayne jumps out and is swarmed by several low-level demons. Once Nora finished helping Cross, they both joined Rayne. Nora searches for Louie, but there is no sign of him. She calls for him, and after Rayne and Cross are done, they assist Nora in the search. They keep scouting the area when Nora looks up. She sees a fireball in the sky followed by several explosions. The twins look at her and they all have the same thought.

Cross: "Let's hurry up, that might be him."

Rayne: "Or him getting killed by demons."

Suddenly, a loud yell comes from that direction with an explosion that shakes the trees. Louie is busy chasing the demons that looked almost like the ones that killed his parents. The demons scratch and bite him, but Louie dismembers them with ease. The last one he was dealing with tries to run away. Blak Hart pulls out his pistol and shoots the beast in the head, killing it instantly.

Louie: "You wanted Blak Hart, you got 'em."

By the time Cross, Nora, and Rayne catch up to Blak Hart, the forest is engulfed in flames. There's flames and smoke everywhere as Cross, Nora, and Rayne get there.

Cross: "Aahhh! The smoke is burning my nose."

Rayne: "Turn back, you can control your senses better."

Cross turns back to human and starts running into the flames. As he jumps, a hand grabs him by the face and pushes him back, knocking down Rayne and Nora.

Cross: "What the hell, who did that?!"

Zima walks out from the flames, smiling and laughing.

Zima: "HAHAHAH! Hey there, pup, oh look, the whole Cross clan is here. Well, what's left of them anyway. HAHAHAH!"

Cross: "Zima, you bastard!"

Cross charges at him with Rayne right behind him. He gives him a devastating punch to the stomach as Rayne kicks him, sending him sliding.

Nora: *yelling* "This is for Albert, for Aggy, for every soul you've twisted."

Nora appears behind him and goes for a punch of her own. Zima smiles and grabs her punch.

Meanwhile, Blak Hart finishes off the demons, and through the smoke and flames, Blak Hart emerges, with eyes blazing.

Blak Hart: "Zima, you forgot one thing, I'm still here!" *he charges in and hits Zima back a few feet before Daemon appears and teleports him away several feet from the others.*

Daemon: "You!"

Blak Hart: "Who were you expecting, the Easter Bunny?"

Daemon: "You're not supposed to be here."

Blak Hart: "Surprise!"

CHAPTER 10.
WHERE'S RAPHAEL & URIEL

Zima, still dazed from Blak Hart's hit, recovers and sees him dealing with Daemon, then returns his focus on the Cross family.

Zima: "Well, well, well, what have we here? Nora or Wolf Queen, whatever, it's been a while. Hahahahaha!" *starts to crush her hand*

Nora/Wolf Queen: "Grrrr!" *her eyes start to glow*

Cross punches Zima in the face, but he doesn't move. Rayne starts hitting him with a flurry of punches and kicks.

Zima: "Your pups are a handful, huh?"

Nora/Wolf Queen: "Don't you dare lay a hand on them!"

Zima: "HAHAHA!" *breaks Wolf Queen's arm and throws her at Cross and Rayne*

Zima rushes after them and then grabs Cross. He starts punching and kicking him until he coughs up blood.

Zima: "Oh my, did I break something?" *sarcastically speaks to Cross as he kicks him through a tree*

Zima keeps up the assault, and Rayne charges after him. Zima slaps her to the ground and then kicks her. Cross gets up, grabs Zima by the head, and slams him down. He starts punching Zima repeatedly into the dirt, creating a crater.

Cross: "HOW DARE YOU TOUCH MY MOTHER AND SISTER!!!" *eyes glowing yellow, his claws extend*

Zima uppercuts Cross, knocking him out, then begins kicking, punching, and throwing Cross all over the place.

Zima: "Just like your father, you don't know when you've lost. Accept the inevitable and join me, or end up like Wolf—six feet under." *boasting and grinning*

Zima turns around and starts walking toward Nora. Rayne jumps in front of him. It starts to rain as she begins punching Zima, but he doesn't budge. Rayne's nails grow into claws; she stabs Zima in the stomach.

Zima: "Ouch! That actually hurt, you bitch!" *Zima punches Rayne, grabs her by the neck, and grabs Nora at the same time* "Now, it's time we finish this bullshit!"

Zima applies pressure. The air starts to get cold. Zima can see his breath as it starts flurrying.

Zima: "What the fuck is going on?"

Zima turns around. Cross is standing in a royal blue aura as snow swirls around him. His eyes glow as Cross starts walking toward Zima.

Zima: "Oh my, someone is angry. Oooooo, that's new." *drops Rayne and Nora*

Zima rushes toward Cross, but he teleports in front of him. Cross grabs Zima by the neck; Zima starts to freeze. Cross creates an ice blade and stabs him in the stomach with it.

In mid-conversation, Blak Hart draws his blade and slashes Daemon in half. Daemon looks at him and then throws a fireball that blasts Blak Hart. After the smoke clears, Blak Hart walks toward him with a devious look on his face.

Blak Hart: "So, this is what you thought would affect me. Tsk, tsk, you really should've done your homework."

Just as he raises his sword to give the final blow, Zima appears from the ground in a cloud of smoke, still with the ice blade in his gut. He drags Daemon, and they disappear as quickly as he appeared.

Blak Hart: "I wonder who that guy was? Seems like Zima is pretty fond of that one."

Nora, Cross, and Rayne show up at his location. Cross doesn't seem like himself.

Nora: "Now that we're all here, is everyone alright?"

Louie: "I don't think we are. Cross looks out of it. By the way, what's with the Christmas appearance surrounding him? Is that part of his evolution? If it is, that's awesome."

Nora: "I know my bloodline has water abilities, but as for ice, that's unknown."

Louie: "Cross, snap out of it! Don't let the power control you—get control over it! Listen to my voice; you are Michael Cross, and you kick ass and take names!"

Nora: "Let's give him time to get used to the power. I'll try to get his mind back. Just make sure no one tries to attack."

Louie: "You got it. Rayne, help me keep everyone back."

Rayne: "No problem. Do you think Mom will be able to fix this?"

Louie: "If anybody can do it, she can. Just give her space and let her work her magic."

Nora goes into a trance and begins talking in Latin. Cross's body begins to shake; his body turns into ice. Nora chants in another language, and his eyes slowly start changing from yellow back to their normal hazel color. His body thaws out, and he drops to the ground. Nora chants a little longer as she places her hand on his head. To end the ritual, she says several prayers and then hugs him.

Nora: "He should be back to normal. This new ability he has is going to take time for him to master, but he'll be fine."

Rayne: "Will he have any side effects from this evolution?"

Nora: "Right now, it's hard to say. We'll know as time progresses."

Louie: "Let's help him to the car and deal with this at the house."

Nora: "Sorry your last day with us isn't what I wanted you to remember out on the field."

Louie: "It's fine." *gets cut short*

Nora: "No, it's not. I want you to have great memories so you'd want to come back to us."

Louie: "All of us being together is a memory that makes me want to come back. This is home. Remember—home is where the heart is, and my heart is here with family." *he hugs Nora as she cries*

Rayne: "Mom, let's get Cross home so we can attend to him properly."

Nora: *wiping her tears* "You're right, we must hurry up before more things try to attack."

In Hell, Zima is trying to heal Daemon. While healing, he realizes that Daemon will have a permanent scar showing where Blak Hart nearly ended his life.

Daemon: "I thought Blak Hart wasn't going to be around. Why was he there?" *still clenching his scar in pain*

Zima: "That's a great question. Where the fuck is this bitch? She has some explaining to do! HECATE, GET YOUR ASS OVER HERE, NOW!"

Hecate: "Why? What happened? Didn't the plan go as expected?"

Zima: "No, bitch! If it would've gone well, you wouldn't be here! Blak Hart was there. All the ones I sent were killed by them. The only one that survived was Daemon, so why was he there?"

Hecate: "I don't know. I told you what I was told."

Zima: "Well, I guess your charm isn't as strong as you think." *smacking her around* "Maybe he doesn't love you as much as you thought." *continues smacking, then begins punching her face*

Hecate: "I'm sorry, please no more." *sobbing*

Zima: "Don't worry, I'm done for now. Daemon, take her to the torture cell. I'll deal with her later on; I hate liars. Daemon, come here." *punches him on the forehead* "Next time, prepare for the unexpected. I shouldn't have to heal you. Now look at you—you have a scar that will be a constant reminder of it."

Daemon: *returns from taking Hecate to the torture cell* "How's your wound? Seems like you weren't expecting that either." *chuckles* "How did you prepare for the unexpected?"

Zima: "Don't be a wise ass, Daemon. Cross, unlike Blak Hart, just got new powers. What's your excuse, huh?"

Daemon: "Like I said before, he wasn't supposed to be there. If I knew he was there, I would've—"

Zima: "Would've what... killed him? Hurt him?" *laughs maniacally* "What do you know about Blak Hart anyway? Study your foes before facing them. This is the only way to be able to mentally and physically torture them. Knowing their history will help assist you in ways to defeat them before attacking them."

Back at the Cross household, Nora attends to Cross as Gail, Darren, and Luna arrive.

Luna: "How's Cross?"

Nora: "He'll be fine; he just needs time to recover."

Louie: *in a different accent* "Remember, Cross is strong like bull." *chuckles, then in his regular voice* "He's a warrior. He'll pull through, and with all of our support, he'll be back stronger than ever."
Nora, Rayne, Luna, and Gail look at Blak Hart and laugh, easing the tension that was looming in the area.

Gail: "What exactly happened? Is there anything I can do to help?"

Nora: *after updating her on what happened* "I appreciate the help, but I just need to find out what's going on with him."

Darren: "Did you call the Archangel Michael yet?"

Nora: "Not yet, but I'm going to in a few. Louie, please tell Raphael to have Michael come by to help me figure this out."

Louie: "You got it." *he calls Raphael*

Raphael: "Hey Louie, what's going on?"

Louie: "Nora said to please get your brother Michael to come by the house. A situation happened, and she needs help figuring something out."

Raphael: "What happened?"

Louie: "She'll explain everything once you guys get here."

In the blink of an eye, the archangels Michael and Raphael show up. They are greeted by Louie and Rayne.

Michael & Raphael: "What happened? Who's hurt?"

Louie: "No one is hurt. Nora will explain the reason why she needs you here."

They head inside, and Nora explains to Michael the whole incident. She takes him to see Cross. He walks over to him and puts his hands above Cross. A bright yellow glow is emitted from his hands. He turns to her to explain.

Michael: "This new power he was blessed with is a family trait from your bloodline."

Nora: "None of my family members ever had it, that I know of."

Michael: "This is a dormant gift that's activated when a combination of events occur. He's meant for greatness—probably another reason why he was given it. When he wakes up, call me, and I'll train him how to use his gifts, along with any other manifestations that may arise."

Louie: "Do you think he'll be up before I gotta go?"

Michael: "I'm sorry to say I believe not, but we can relay the message if you like."

Louie: "Nora, if anything happens, could you notify Raphael so he can keep me informed?"

Nora: "Sure, but I can keep you up to date when you call."

Louie: "I mean in emergency cases."

Raphael: "Louie, it's time to go."

Louie says his goodbyes individually.

Louie: *hugs Nora* "Thank you for everything you've done for me. I'm blessed to have a godmother as awesome as you, with incredible god siblings to have as a family." *calls Rayne* "You're the sister I always wanted. You're a warrior I'd proudly want by my side. Trust your instincts—they're very important in the midst of battle." *hugs Rayne as she hugs him tighter, like it's a competition to see who can hug tighter, then he calls Luna over* "Please take care of Cross. He's the brother I always wished for, and you're my sister-in-law who will always have my backing. If any of you need me, I'm there." *goes to Gail* "You're a great mother—never second-guess yourself." *shakes Darren's hand and whispers in his ear* "I'm proud of your progress. Keep up the great work. Help them out as best as you can, and don't be afraid to ask your mom for help. She can help out in any situation—just ask. If she can't, she'll find a way."

Darren tears up and walks away. Louie waves while walking away with Raphael.

Darren: "Archangel Michael, will Louie be alright?"

Michael: "Call me Michael. If he does as he's trained, he'll be fine."

Darren watches as Louie and Raphael walk away until they can no longer be seen. With his head down, he walks back into the house and hears Cross.

Cross: "Where am I? What happened?"

Michael: "You have evolved into something incredible."

Cross: "Where's Louie?"

Nora: "He had to go. He wanted to be here when you got up, but someone told him that you wouldn't be up for a while."

Cross: "When's the next time we're going to see him?"

Michael: "You'll see him when you are fully trained and have passed army boot camp."

Nora: "Do they all have to go?"

Michael: "I'm sorry to inform you that they're all destined to be great. I'm not allowed to say more, but everything they're going through now is necessary."

Darren: "What branch of the military am I going to?"

Gabriel: "Navy SEALs. You must pass Navy boot camp before going into any black ops organization."

Darren: "Black ops? What about Rayne?"

Gabriel: "You'll know more when the time comes. As for Rayne, she'll be in Louie's unit."

Cross: "What about me? Where am I going?"

Michael: "Army—just like Wolf."

Darren: "Why do we all have to be in different branches? Why can't we all be in the same one?"

Michael: "To have a specialist in each form of the military. That will make one powerful team once reunited."

Cross: "I like the strategy. How long was this plan made?"

Michael: "Since your birth."

Cross tries to stand, but his legs begin to buckle, and he sits back down.

Cross: "What the hell is going on with me? I feel cold, but I'm not shivering."

Michael: "Some new powers have manifested. It's going to take time for your body to adjust."

Cross: "How long is it going to take?"

Michael: "It all depends on you—and how your body adapts to the changes needed to continue and advance."

Back at the Marines’ base, Raphael introduces Louie to the Marines’ head soldier, Guardian Angel Uriel. He tells him to call Uriel “Corporal” when in the presence of the unit.

Louie: "No disrespect, Corporal... Raphael, how is he going to prepare me for what’s to come?"

Raphael: "He has vast knowledge of fighting techniques, war strategies, and much more. He can help you in areas where I may lack. Do as you’re told, and you’ll be fine. Try to go against any authority figure, and you’ll be punished more severely than the rest."

Louie: "That sounds messed up. Why do I get punished worse than anyone else?"

Uriel: "You are not the only one with special talents here. This unit was hand-selected by several angels to train the elite of you mortals into Heaven's earthbound protectors. Not all privates have abilities. Those with gifts who are insubordinate will be made an example of for the normal soldiers. Is that understood, Grunt?!"

Louie: *salutes* "Yes, sir!"

Uriel: *looks at Raphael* "Relax. I have a feeling all that you've said about him might be right. But I'll be the judge of whether he's more hype or the real deal."

Raphael: *pulls Uriel to the side* "When you see him in battle, you'll realize you have an asset instead of a liability."

Uriel: "I know who this kid is, and his bloodline. That still doesn't mean I'm going to treat him any differently. Once he's proven himself to me, then we'll deal with it when it's time."

Months pass. Michael, Gabriel, and Raphael have been training the teens, preparing them for the military.

Cross: "How's Louie doing in the service?"

Raphael: "He's done with boot camp and now he's in a unit on the field. If he keeps doing well, he'll be leading a squad."

Nora: *smiles* "Sounds like a proud father talking about his son. I know he thinks of you as a father figure. You don't have to hide how you feel just because you're not supposed to get too attached."

Raphael: "Well… am I wrong?"

Nora: "How are you holding up?"

Raphael: "I'm fine. He's with someone I trust. Are you ready for the prom?"

Nora: "You mean the kids, right?"

Raphael: "No, I mean you, the whole part of them growing up so fast."

Nora: "Oh great, now I feel old." *joking*

Raphael: "You shouldn't. Compared to me, you're an infant." *laughs* "I gotta go. If you need me, just call. Thanks for everything."

Nora: "My pleasure. Take care, see you soon."

Michael: "Nora, since the prom is in a couple of days, training will be postponed until Sunday."

Gail: "What about us, the parents?"

Michael: "If you like, we can train in the areas you feel are needed."

Two days later, Darren accompanies Cross to get Luna her corsage.

Darren: "Can you help me get one for Rayne?"

Cross: "Sure. Do you have an idea of what kind you want to get her?"

Darren: "I have no clue."

Cross: "When we arrive at the florist, we'll ask for any corsages with Rayne's favorite colors."

Darren: "Sounds good."

Back at the house, Luna and Rayne help each other get ready for prom. Nora and Gail begin cooking before the teens have to go.

Rayne: "Mom, what are you making?"

Nora: "Baked ziti and garlic bread. Don't worry, it'll be done soon."

Gail: "Missing Louie, huh?"

Nora: "Yeah, but he's a strong man. I just worry."

Gail: "I'll set the table."

Nora: "Thank you, sweetie."

The girls are putting on makeup as the boys walk in wearing suits.

Luna: "Wow, I could just—"

Rayne: "Nasty, keep that to yourself. Don't look, boys. You get to see us when it's time, until then, go."

An hour later, Nora calls everyone to the dinner table. Darren stares at Rayne as Cross admires Luna and smiles.

Rayne and Luna: "Stop drooling, it's not attractive."

Both Darren and Cross blush. Nora stands and looks around the table.

Nora: "I look around this table and see my babies all grown up, great friends from the past and present here with us. I'm proud that this year, even though we've had our bad times and good times, we managed to persevere with Louie's help. Hopefully, God will grant us the gift of having more times like this in the future."

Gail: "I believe He will. Together, we're stronger than we'd be alone. Enough depressing thoughts. Let's enjoy this meal Nora and I made. Tonight is a special night, enjoy, be careful, and make it memorable."

Cross: "Amen. Let's eat. The prom starts in an hour and a half."

After dinner, Gail drops the teens off at the dance. Once inside, they survey the area and realize it might actually be a good night. They begin pairing off and hit the dance floor. Darren and Rayne slow dance beside Cross and Luna.

The first two hours are magical, the lights set a romantic mood, the songs are just right, and the feeling of love fills the air, until Cross notices someone staring through the window.

Cross: "Damn it, not now."

Luna: "What's going on?"

Cross: "Go get my sister and Darren."

Luna: "Why? Tell me, what's wrong?"

Cross: "I'll tell you when we're all together."

Luna gathers the couple, and Rayne looks at Cross. She sees how angry he is.

Darren: "What's wrong?"

Cross: "I saw Mezun and a few of his lackeys looking through the glass. I don't sense anyone in here with power, we have to get ready. Darren, call your mom and ask her to bring us a few items."

Darren: "Like, what items?"

Cross: "She knows what they are. Her and my mom know that whenever Louie called for 'a few items,' it meant the same thing."

Darren: "Done. My mom said she's on her way with Nora. So what do we do now?"

Cross: "Rayne and Luna, start getting everyone out. Darren, back them up in case trouble starts, help them if you need to. I'll keep my eyes on Mezun and his goons until the place is empty. Then we'll bring the fight to them."

Luna moves to one side of the dance hall while Rayne goes to the other. Cross and Darren stay in the center. The girls tell the students to form two lines and calmly head to the doors. As they're leaving, all the glass windows shatter. Mezun's goons pour in like a flood.

Cross: "I'm sorry to rush everyone, but let's go, speed it up!" *Cross starts putting up a barrier.*

They hear car brakes screeching outside. Cross looks at Darren.

Cross: "Darren, you know what to do."

Darren: "Got it." *runs and grabs the bag that Gail and Nora brought*

Gail and Nora jump out of the car and start purifying the area.

Gail: "Why does this stuff keep happening?"

Nora: "Are you really surprised?"

Gail: "No, not really. Just thought our kids wouldn't have to deal with this crap."

Nora: "It's in their blood."

Darren returns and hands the bags to Cross. He gives him his blessed blade. Cross starts taking out demons while Darren eliminates others, throwing holy water, oil, and salt as he fights toward the girls.

Rayne: "Luna, start making a circle!"

Luna: "Got it!"

The girls create a circle and seal the room, then grab their blessed blades.

Rayne: "Get my back and stay close."

Luna nods, and the teens attack the swarm until only a few demons remain. Cross spots Mezun trying to escape. Before he can reach him, Darren rushes forward and chases Mezun down.

Cross: "Darren, don't lose your head!"

Darren: "I know, I got this!"

Rayne follows Darren as Luna stays behind Cross, covering his back.

Cross: "Sorry the night got ruined."

Luna: "You can't control this, plus, you can make it up to me." *blushes and giggles*

Cross: *blushes and smirks* "Hehe, alright. You're right about that."

They finish off the remaining demons as Darren and Rayne catch up to Mezun. Darren sprints past him and slams him into the ground. Rayne freezes for a moment, shocked by Darren's fury, as he delivers a series of devastating punches before pinning Mezun down with lightning.

Darren: "Who the fuck sent you?"

Mezun: "*Hehe,* you're nothing like your father. Why won't you join me?"

Darren: "SHUT UP!" *punches him in the stomach*

Mezun: "*Hehe,* oh, I felt that."

Rayne: "Darren, don't let him get to you!"

Darren: "I know. I just need to let some anger out."

Rayne: "I understand, but don't get carried away!"

Darren: "I wo—"

Mezun teleports behind him and lands a roundhouse kick to Darren's stomach, sending him flying. He grabs Rayne by the neck.

Mezun: "My grandson has great taste in women, but sadly your relationship isn't going to last." *tightens his grip*

Darren stands up, then stabs his blade into Mezun's side. Out of reflex, Rayne stabs his arm. Mezun lets her go, and Darren kicks his blade deeper into Mezun's ribs. Mezun screeches in pain as his claws grow longer. He slashes Darren, but before he can land another strike, someone steps in front of him. Darren freezes when he sees Cross and Luna arrive. Cross sees Rayne impaled by Mezun's claws, stabbed straight through her stomach.

Cross: "NOOOOOO!!!" *breaks Mezun's claws and jumps away with Rayne in his arms*

Darren: "No... no, not her. Mezun, you BASTARD!!!"

Darren appears behind Mezun and punches him with black lightning. The blade still in Mezun's side acts as a conductor, electrocuting him from the inside out. Darren unleashes a storm of furious blows, hurling Mezun like a rag doll while Cross desperately tries to heal Rayne.

Cross: "Come on, sis, don't give up on me." *his hands glow with a deep blue aura as he pulls out Mezun's claws; Rayne's wounds begin to heal*

Rayne: *coughs* "My dress is ruined. Where's Mezun and Darren? What happened?"

Luna: "Darren's beating up that Mezun guy."

Rayne: "I gotta help!"

Cross: *grabs Rayne* "You need to wait here, you were hurt bad. I don't know if I healed you well enough yet."

Rayne touches her stomach, and tears stream down her face.

Rayne: "I can't feel it anymore. No... no, not my baby."

Cross: *grabs his phone and calls Nora* "Mom, I need you to meet us by the back of the school, Rayne needs to go to the hospital!"

Nora: "Oh no, I'm on my way!"

Darren continues pounding Mezun into the ground, but a sudden punch from another figure sends him flying. Mezun vanishes as Darren lunges at the new figure, but it disappears in a cloud of smoke.

Cross: "DARREN, GET YOUR ASS OVER HERE!!!"

Rayne, still in shock, holds her stomach. Darren rushes to her side and hugs her as she trembles.

Darren: "I'm so sorry."

Rayne: "I can't feel our baby, Darren... why can't I feel our baby?"

Nora and Gail arrive. Darren lifts Rayne into the car. As Cross tries to get in, Zima attacks him.

Cross: *grabs Zima and yells to the others* "Go!"

Luna stays back to help while Nora speeds off toward the hospital.

Zima: "Hey, kiddo, it's a shame what happened to your sister. I thought Mezun would finally kill her, but nope, just took the baby out. Oh well."

Cross: "I WILL KILL YOU!!!"

Cross charges at Zima. Zima meets him head-on, unleashing a barrage of punches that knocks Cross out.

Zima: "You caught me off guard last time. Besides, I'm not here for you." *smiles*

Luna rushes out and stabs Zima in the chest with her blade. Zima grabs her by the throat.

Zima: "Oh, *ouch,* is that how you greet people? How about I teach you some manners?"

Cross tries to stand but can barely move.

Cross: "Don't you dare touch her!"

Zima: "*HAHAHA!* Or what? I broke a rib and damaged your spine, you're all talk! *HAHAHA!*"

Cross suddenly leaps up so fast it looks like teleportation. Zima counters, creating a shockwave that sends Cross crashing through trees.

Zima: "I gotta say, you impress me. You just don't know when to quit. But oh, I digress, I must be going now. Say goodbye to your honey bunny. *Hehe.*" *holds Luna by her hair*

Luna: "MICHAEL, PLEASE HELP ME!!!"

Cross: *struggling to stand, reaching toward her* "LUNA!!!"

Zima disappears with Luna in a cloud of flames and smoke. Cross collapses to his knees, trembling and crying.

Cross: *yelling and sobbing* "I won't stop until I find you, Luna! And as for Zima, prepare yourself! When I find you, I'll kill you! That's a promise!" *his eyes glow yellow*

Overseas, Louie is on a top-secret mission. He listens to his orders, then follows his unit into a building.

Lieutenant: "We go in quietly, take out the threats, then save the civilians. Is that clear? No hero bullshit, just stick together and get it done."

The Squad: *in unison* "Yes, sir!"

The team stealthily enters the building. The first floor is clear. They move to the second floor, same thing. The whole facility seems abandoned. They continue to the top floor, where they find an enemy soldier holding an operative at gunpoint.

Enemy Soldier: "Drop your weapons or he dies! Don't try any heroics. If I die, you all come with me!"

Lieutenant: "On my mark, fire! Louie, rescue our operative. Fire at will!"

The squad opens fire. Louie grabs the hostage and runs for cover as explosions tear through the walls. The building begins to collapse around them.

Louie: *speaking through his headset* "He must've had a dead man's switch!"

Lieutenant: "Just worry about saving him, I'll think of a way to get us out!"

Louie: "Better make it quick, we're on the fifth floor!"

Beams crash down, trapping soldiers. Suddenly, the lieutenant is sucked into an air pocket that opens beneath him. Louie carries the hostage to a window. Seeing the building fall fast, he hurls the man out a few feet before impact. Louie tries to leap out after him but gets pinned by collapsing concrete.

Hours later, Louie regains consciousness, dazed. He feels the crushing weight of debris on his back and legs. Looking up, he sees the rescued man struggling to free him, but then a shadow moves behind the operative. A figure picks up an assault rifle and shoots the man in the head and chest. The gun clatters next to Louie. He tries to see the shooter, but blood loss clouds his vision. Louie passes out.

When he briefly regains consciousness, everything is blurry. A group of dark, soldier-like figures approaches. Their shapes are indistinct. They pull Louie from the rubble. One of them places a glowing hand above him, and he blacks out again.

Moments later, a recon unit arrives at the mission site. They find Louie next to the operative's body. They drag Louie out and carry both men back to base.

At the hospital, Nora, Gail, and Darren wait for the doctors to give an update on Rayne's condition. A doctor exits the room and gestures for them to follow.

Doctor: "We did the best we could. She has a small wound, her MRI and X-rays show little to no injuries. The mind-boggling part is, the only thing found in the womb were blood and minor pieces of a fetus. Whatever injured her destroyed the infant but barely harmed her… yet it completely destroyed her uterus."

Darren and Nora break down, crying uncontrollably. They enter the room and hug Rayne tightly.

Darren: "We'll get the one who did this to you."

Rayne: "We lost the baby..." *sobs, holding onto Nora and Darren*

In Heaven, the angels are tense. They've discovered that Archangels Uriel and Raphael are missing.

Archangel Michael brings Cross to the hospital, where Gabriel is already consoling Darren.

Cross: "How's Rayne?"

Nora: *explains what the doctor said and holds him as they cry together* "Where's Luna?"

Cross: "She was taken by Zima." *his eyes begin to glow*

Michael**:** "We'll find her. Don't worry."

Cross: *glares at him* "Where the fuck were you while all this was happening?"

Michael: "Watch your tone. I was trying to find where he's hiding."

At the base, Louie wakes to a harsh interrogation.

High-Ranking Official: "Why did you kill him? It's your fault the entire troop died on that mission!" *punches Louie in the face*

Louie: *spitting blood* "The whole mission was a setup. The building was empty except for the fifth floor. He was waiting for us..."

High-Ranking Official: "Lies! What was the reason for the death of an innocent?" *shocks him with a stun gun repeatedly* "Throw him back in the cell. I'll finish with him later on."

Two privates grab Louie and drag him to his cell. When they get there, they toss him in like a pile of trash.

Louie: "When I get out of here, you two will be the first people I visit."

Two Privates: *mocking* "What are you gonna do, kill us? This place has enough weapons to take down a country. You think others won't stop you?" *laughing*

Louie: "Laugh all you want. Keep your guns close, you're gonna need them to feel safe, 'cause they won't stop me anyway."

Private 1: "Remember, we caught you."

Louie: "Does that really count? I wasn't conscious, and I didn't fight back. That won't be the case next time we meet. Can you two idiots do me a small favor?"

Private 2: "Why?"

Private 1: *whispering to the other* "Just humor him." *looks at Louie* "Sure, what is it?"

Louie: "Call Corporal Uriel for me and tell him Louie's asking for him."

Private 2: "Haven't you heard?"

Private 1: *whispering* "Just play along." *looks back at Louie* "He's out on a top-secret mission. He can't be reached right now. We can get someone else for you."

Louie: "Fuck it, never mind."

The soldiers leave the room. Louie calls out for Raphael, but there's no sign of him. Frustration builds until he starts yelling.

Louie: "Raphael! Where the fuck are you?!"

A creaking door echoes. Blak Hart peers out of his cell, sees nothing, and sits back down. Suddenly, a shadowed figure approaches his cell.

Unknown Figure: "Here's some food. I know what really happened. Don't say anything, I'm part of Uriel's elite. He and Raphael have disappeared without a trace. Sit tight. I need you to try and find any info you can about them. I'll check in from time to time."

Louie: "Okay, before I do, who are you, and why are you helping me?"

Unknown Figure: "Oh yeah, my name's Revis, but you can call me Reeves. I'm part of Uriel's elite battalion to fight evil. I was told to check on you, to make sure you're ready to lead our troop. With Raphael's training, I wouldn't be surprised if you could escape this place."

Louie: "You think I'm gonna try to escape from a government prison? Are you insane? That'd cause more problems than help."

Reeves: "No problem. You're not as hot-headed as I thought. Just do your best to find what you can, I'll see who I can contact to get you out."

Louie: "Sounds good. Whenever you need me, just find me. You know where I'll be." *he sits on the ground and thinks to himself* "I saved a life, lost a squad, got framed… and now I wait. But they won't break me, not now, not ever."

Back at the hospital, Rayne is being released. She's accompanied by Gail, Nora, Darren, and Cross. Nora helps her walk out. Cross runs ahead to hold the door open, while Darren rushes to the car and opens the door for her. Rayne stops in front of him.

Darren: "What's wrong, love?"

Rayne lowers her head as tears roll down her cheeks. Darren hugs her tightly.

Rayne: "D-do you hate me?"

Darren: "No. I could never hate you. It wasn't your fault."

Rayne: "I could've just run… but I couldn't let you get hurt again."

Darren holds her as she cries. He fights back his own tears.

Darren: "You did what you had to do to protect me. I feel like shit because I lost control. I should've been the one to get hurt. I'm so sorry, it's my fault."

Cross: "None of us could've known what was going to happen, but we need to be here for each other. Stop beating yourselves up over something we couldn't control. I know it hurts, but we have to make sure we never feel this pain again."

Rayne: "Michael..." *runs to Cross and hugs him tightly*

Nora: "He's definitely his father's son."

Gail: "Felt like a flashback, huh, Nora?"

Darren runs over and hugs Cross. Rayne, Nora, and Gail join in. They wipe their tears, pile into the car, and head home.

At the White House, the President of the United States, Benjamin Garrison, holds a meeting with his senior staff and military heads.

President: "So, any questions about the upcoming visit? No? Alright. Next order of business, do we have any information from that soldier we're holding? Does he know what happened, General Devin Deville?"

General Deville: "He believes it was a setup, but there's no evidence proving he's wrong."

President: "His fingerprints were on the weapon, and the ballistics from the bullet match the one found in the corpse. The assault rifle was next to him."

General Deville: "The issue I'm having, Mr. President, is that he was buried under rubble. How could he have fired a round with all that debris on him?"

President: "Please excuse us. General Deville and I need to talk privately. We'll continue this meeting tomorrow at 0800 hours."

Everyone leaves. The President pours himself a drink and loosens his tie.

General Deville: "Ben, I don't see why we should keep him locked up. He's been a great asset for us, he could still make a difference if we let him work again."

Ben: "You really don't think he did it, Dee?"

Dee: "He was top of his class. His record's spotless. Why would he betray his government like that?"

Ben: "War does crazy things to people. You remember that soldier who took his own life, right? What was his name again?"

Dee: "Robert Pierce. We used to call him Scatter. Remember? He'd always yell *'scatter!'* when he threw grenades." *laughs softly* "Luis Heart isn't like him."

Ben: "Why are you so sure?"

Dee: "I just know it. I feel it in my bones, it wasn't him."

Ben: "Dee, I can't release him just because you have a feeling. He's being accused of terrorism and was found at the scene of an attack. You need to give me proof, then we can talk. I'm sorry, old friend."

Dee stands and lays several photos on the table, images showing a shadowy figure standing over Louie at the scene.

Ben: "What are these? Why weren't they shown to me before?"

Dee: "These came from one of our drones. Someone tried to hide them, but I got them before they were tampered with."

Ben: "You've been busy, huh? Do you know who the man is?"

Dee: "Not yet, but my best people are on it."

Ben: "Alright. I'll release Louie to you, but keep an eye on him. If this was a setup, people will come after him."

Dee: "Come on, Ben, it's me we're talking about. *Hehe.* Thank you, old friend."

Ben: "*Hehe,* no need for thanks. You're good at your job."

Dee: "Glad you think so. It just feels like second nature to me. Any luck finding a new Vice President?"

Ben: "Not really. Unless…?"

Dee: "Ben, I'm good where I am."

Ben: "Well, if you change your mind, let me know."

Dee: "Alright."

The two old friends enjoy a few drinks and reminisce.

A month later, at the military prison, Louie endures his daily cycle of beatings and interrogation. His wrists are bound tightly, his body suspended from a beam in the ceiling. And still, Louie remains unfazed.

CHAPTER 11.
GRADUATION

Sargent: "Aren't you tired of getting beat every day? Just tell us what you know and all this can stop."

Louie: "I've..." *spitting out blood, going in and out of consciousness* "told... you... everything... you piece... of shit..." *spits in the Sargent's face*

Sargent: "You'll never learn." *punches him in the ribs, then grabs a pipe and begins hitting him with it* "Talk, don't talk, it's all the same to me. I can keep this up for as long as you can take it."

Suddenly, the door bursts open just as Louie is about to break the chains. General Deville walks in.

General Deville: "What do you think you're doing?"

Sargent: *salutes the General* "Doing as I was told, sir!"

General Deville: "I doubt that very much. He's getting released today. That means you defied a direct order from me!"

Sargent: "I wasn't notified that he was being released, sir!"

General Deville: "Cut him down!"

The General walks up to Louie and smiles.

General Deville: "Here's a few dollars to get your gear. When you're up and running," *throws him a business card* "go to this address. You report to me from now on. See you tomorrow at 0500, don't be late."

Louie: "Yes, sir!" *the hairs all over his body stand on edge*

A month has passed. Nora and Gail are getting the teens ready for their graduation. This is an incredible day for them. Through all the challenges they've had, they were able to graduate. With tears in their eyes, they all express their feelings.

Nora: "I am so proud of all four of you... Sorry, I'm just so used to having four here." *a tear slides down her face*

Gail: "What she's trying to say is, we're proud of all of you, no matter where you are." *her eyes begin to water* "This is your day, enjoy it."

Gail and Nora drop off the teens at the high school stadium, then go to their seats. The ceremony commences with the principal of the school, Brian Kinsley, telling everyone to rise for the Pledge of Allegiance. Then he congratulates all the students for their hard work and determination. He starts to calls the names of the alumni to be. He first calls for Luna Light.

Principal Kinsley: "I call Luna Light, in commemoration of her work with the school, her academic achievements, and her way of touching people's lives just by knowing her. Let's bow our heads for a moment of silence in memory of this exceptional young lady."
everyone lowers their heads

Nora, Gail, Rayne, and Cross cover their faces as they cry. Darren wipes his tears away, trying to stay strong for them.

After a few minutes, Principal Kinsley continues calling the students for their diplomas. He calls Cross, Rayne, and Darren last.

Principal Kinsley: "I now call up here the last three graduates for their diplomas, and to say a few words in memory of Luna Light. They were the ones most affected by her loss. Please come up, Michael Cross, Rayne Cross, and Darren Grier."

The three walk up with tears in their eyes and take their diplomas. As they turn to face the crowd, they see Nora, Gail, and a dark figure standing behind them. At first glance, the figure is unrecognizable. They look again. It's a husky man dressed in military gear, his hat shadowing his face. The man raises his hat and salutes them.

Cross, Darren, and Rayne: "It's Louie!"

They take turns saying a few words for Luna, helping each other through the ceremony. At the end, the valedictorian gives her speech. The graduates toss their caps in the air, completing their rite of passage.

The teens rush down to hug Nora and Gail. As they scan the crowd, Louie is nowhere to be found.

Nora: "What are you guys looking for?"

Darren: "Louie was in the stands, right behind you."

Gail: "We didn't see him. Are you sure it was him?"

Cross: "It was him. He had his gear on... and he saluted us."

Nora: "Did all three of you see him?"

Rayne: "Yup. The only difference was... he was scarred up."

They head to the car. Rayne runs ahead toward it.

Nora: "Wow, she's eager to leave."

Cross: "No," *starts running too* "Louie's there!"

Rayne gives Louie a bear hug. He laughs and hugs her back, the scars on his face and knuckles catching everyone's attention.

Louie: "Nah, you didn't miss me." *joking*

Cross: "Alright, wise ass, we're happy to see you. I thought you wouldn't be able to show up for this."

Louie: "What, and miss the proudest day of my life? How could I miss that?" *sarcastic tone* "I'm proud of all of you."

Nora and Gail rush over, smothering him with hugs and kisses.

Nora: "What's with all the scars? What did they do to you?"

Louie: "It's a long story. Today's their day. We'll talk when they go to sleep."

They all pile into the car while Louie follows behind on his motorcycle. Gail leads them to a restaurant to celebrate the teens' special day. Once parked, Louie pulls Cross aside.

Louie: "I know what I'm about to ask is painful, but please bear with me."

Cross: "It's about Luna, right?"

Louie: "Yeah. What happened? Is there anything I can do to help?"

Cross: "Well, now that you asked… Luna is still alive. I can feel it. It's like a heartbeat in the dark, faint, but steady. She's still fighting. What I'm asking is for you to try to find her."

Louie: "If she's still alive, I'll find her. I'll use every resource I have."

Cross: "All I ask is that you try. I know when you say you'll do something, you do it."

Louie: "Brother, you have my word, I'll do everything I can to find her."

Cross: "What's with the scars? You usually heal. What did this much damage to you?"

Louie: "It's a long story. The short version, I'll tell you when Michael comes over."

Louie calls Michael and tells him to meet them at Nora's house. Michael says he'll be there in about two hours. After the call, Cross and Louie join everyone inside the restaurant.

Lasagna, cheesecake, chicken parmesan, chicken cordon bleu, all types of food cover the table, satisfying every craving. Once they finish eating, everyone agrees to meet back at the house.

When they arrive, Michael is already there waiting.

Michael: "Congratulations on a job well done. Louie, may I have a word with you?"

Louie: "I'll be right back."

They walk a short distance away from the house.

Louie: "Any word on Uriel and Raphael's whereabouts?"

Michael: "Nothing yet. I have the best angels on it, but they've found nothing so far. What have you heard?"

Louie: "All I know is that Uriel was abducted in his office, no visitors, no witnesses. Raphael was on his way to you when he disappeared in thick smoke, gone without a trace. Reeves told me he'd help, and he did."

Michael: "He's one of the angels searching for Uriel and Raphael."

Louie: "What? Is Devin Deville one of yours?"

Michael: "Who's that? Never heard of him."

Louie: "He's a five-star general who takes direct orders from the President. His presence made every follicle on my body stand. I was too injured to process it at the time, but thinking back… it felt wrong, uncomfortable."

Michael: "Keep working with him until I gather more information."

Louie: "No problem. Do you have any info on Luna?"

Michael: "No, but if I hear anything, I'll notify you."

Louie: "Thank you. If you need my help, just call."

Michael: "Who did these wounds to you? Your body heals, but these wounds… they're not just physical. You've been marked by something ancient. While I research, you're needed in the Black Operations Division of the government to find out more about this Devin person. I'll follow up on any leads that reach me. Enjoy today, tomorrow, your lieutenant will request you back."

Louie: "You got it. Please keep me posted."

In a flash of blinding light, the Archangel Michael vanishes. Louie turns back to see the group waiting at the door.

Cross: "How's the Marines?"

Louie: "Easy. Just the assholes get in the way of plans."

Rayne: "What happened to you over there? You came back with scars, I thought you healed instantly."

Louie: "I've been through a lot. That's all I'm allowed to say, it's classified. As for the scars, demons of S-Class or higher can injure me in ways that slow down my healing. Sometimes, I don't heal completely."

Nora: "I didn't think that was even possible. Will and Albert rarely got scars."

Louie: "It depends on how powerful the demon is, and the effort it takes to defeat them. Michael said it might be something ancient that marked me."

Gail: "I guess that's why Raphael wanted you in the service."

Louie: "Probably. That, and to get paid on a grander scale while fighting evil alongside people who trained or taught my father, maybe my mom, too."

Darren: "Was she in the military too?"

Louie: "As far as I know, she was a medic. She also fought when enemies tried to take over the field hospital."

Darren: "That's badass."

Nora and Gail: "She was one hell of a person." *both tear up*

Louie: "Yeah. She taught me how to fight while Dad taught me weapons."

The rest of the night is spent reminiscing. Gail and Darren say their goodbyes and head home, leaving Nora and Louie talking after the twins fall asleep.

Nora: "So what really happened over there? And don't give me that 'classified' bullshit."

Louie: "It really is, but… I'll tell you the full story. I was captured and tortured to the point that once I healed, they reopened my wounds. They tried to frame me, said I caused my squad's deaths."

Nora: "How did you get out? Who set you up?"

Louie: "A guy named Devin got me out. Said he was a five-star general working directly under the President. I don't know who set me up, but I have a feeling my superior was involved somehow."

Nora: "Why would they set you up instead of someone with seniority?"

Louie: "Jealousy, maybe. I rose through the ranks fast. Completed missions with minimal casualties or injuries. That rubs some people the wrong way."

Nora: "That general's name sounds familiar. I can't remember where I've heard it."

Louie: "We can check the files, maybe we'll find something."

They search for hours. Nora finally finds an empty folder labeled "Devin Deville." Just as she's about to show Louie, his phone rings.

Reeves: "Come outside, it's time to go."

Louie hugs Nora, then leaves with Reeves.

Louie: "What happened? I thought you were going to get me out."

Reeves: "I was busy searching for Raphael and Uriel. What are you complaining about? You got yourself out."

Louie: "Actually, someone named Devin helped me."

Louie explains everything about Devin to him.

Reeves: "He's not one of ours. I know all the high-ranking generals, and that name doesn't ring a bell. But I'll look into it.

There's chatter about a ghost general, no records, no history, just orders being handed down."

Minutes later in New York, the Archangel Michael speaks softly, almost as if praying.

Michael: "Raphael, wherever you are, hold on. I'm coming."

He appears at Nora's house moments later and is greeted by her.

Nora: "How are you, Michael?"

Michael: "Just here to remind you, Cross and Rayne ship out in two days."

Nora: "Is there anything they can do to prepare?"

Michael: "Louie and I have been working with them. They're ready for whatever's coming. After all, they're your kids, I know they'll prevail."

Nora: "The teens are sleeping right now, but I'll let them know you stopped by and relay the message. Is anything going to be done about what happened to Louie?"

Michael: "Thank you, Nora. With all that's going on with my missing brothers and now the abuse of Louie, my sources are stretching in all unorthodox ways."

Nora: "Any word on the whereabouts of your brother?"

Michael: *his face changes from cheerful to worried and grim* "Nothing yet. I just hope we find him before it's too late. I don't want another brother to..." *he stops abruptly and excuses himself* "I must go. I'll keep you posted."

The following day, Gail is at home cooking when she hears a knock at the door.

Gail: "Who is it?"

Person at the door: "It's Gabriel. May I come in?"

Gail opens the door and invites him in. Gabriel glances around, then asks about Darren.

Gail: "He's teaching himself Navy jargon."

Gabriel: "Could you call him out here for me, please?"

Gail: "If you like, you can go to his room. Before you go, have you gotten any news about Raphael?"

Gabriel: *looks at the floor, then clears his throat* "No. My brother has the best recon angels on the job. Anyway, I'll go to his room. Thanks for asking about him. I see Michael pretending like it doesn't phase him, but I know it's on his mind. I can't believe a warrior like Raphael got abducted by an entity stealthy enough to remain unidentified."

Nora: "I was shocked when I heard the news. Raphael is always cautious on missions. For someone to sneak up and capture him, that being must have serious training. Not even Wolf could ever catch Raphael off guard, and Wolf was the best in covert ops."

Gabriel walks down the hall toward Darren's room when a loud boom shakes the house. He bursts into the room and sees Nuzem trying to grab Darren, but Darren snatches a lamp and smashes it over his father's head. Darren prepares to attack again when Gabriel raises his hand.

Gabriel: "Gail! Come up here, your ex-husband is attacking Darren!" *Gabriel's wings flare with golden light; the room trembles as he raises his hand* "You will not harm him!"

Gabriel tackles Nuzem to the ground, clenching his neck as Gail rushes in. Her eyes lose their color as she starts chanting, screaming at her ex-husband. He thrashes violently, trying to fight off both the angel and his former wife.

Nuzem: "You're not always going to be around him, Gail. I'll be back to finish you off. Then… son, you'll stand by my side."

Darren: "You don't get to call yourself my father! You gave that up the day you chose darkness! Mom, Gabriel, are you two okay?"

Nuzem gives a dark, guttural laugh before disappearing.

Gail: "I should be asking you that. Are you alright?"

Gabriel: "Great job retaliating."

Darren: "I had him!"

Gabriel: "That's what he wanted. Nuzem was hoping you'd fight back, and when you least expected it, he'd transport you to his realm."

Gail: "He's right. The anger you feel toward him could've been your downfall. This could've gone his way."

Darren: "I knew what I was doing."

Gabriel: "You did, but he had a counter for every move. My sources have seen him spying on you. Tomorrow, you're shipping out to begin training. The next time he tries this, it'll be his last."

The following morning, Nora wakes the teens up with their favorite breakfast and snacks.

Cross: "Who died, or what did we do wrong?"

Nora: "No one died, and none of you did anything wrong. I just wanted to let you know Darren's heading out today, so I invited them for lunch. We'll say our goodbyes before he leaves. Then tomorrow, both of you are leaving me too." *sobbing* "I know you'll be in good hands, but it's hard for me to let my babies go."

Cross: "We'll be fine. Louie and the archangels trained us, plus, you threw in a few things Dad used to do. We're ready for anything."

Nora: "I know, I know. It's a mother's job to worry. Even if you were immortal, I'd still worry, that's just who I am."

Rayne: "You raised us to be independent and strong-willed. No one can say we weren't brought up right. You're the best mom we could've had." *tears in her eyes*

Nora: *blushing* "Thank you." *hugs them while crying*

At the base, Reeves meets with Louie.

Louie: "Are the teens out of danger yet?"

Reeves: "Today, Darren ships out to the Navy. As for the twins, tomorrow, Cross joins the SEALs and Rayne heads to the Army."

Louie: "How's the search for Ark and Uriel going?"

Reeves: "Michael's working overtime. He's also gathering intel on this Devin and trying to understand why they tortured you."

Louie: "I've been digging too, but I keep hitting the same wall. There's no record, no birth certificate, no past, nothing. I hacked every top-clearance military database, still nothing. There's only one person who seems to know him personally."

Reeves: "Who?"

Louie: "The President. They served together years ago."

Reeves: "Before you get yourself into trouble, let me tell you this, the President's only in office *because* of Devin. Somehow, this man has deep ties and serious power. Anyway, on a separate note, I've filed all the paperwork for you to run an organization higher than Black Ops. Your second-in-command will be Alala Bellum, she has abilities that've proven useful in almost every scenario. The squad will be known only to you. Some people might fear or reject them otherwise."

Louie: "Thank you. I'll make sure those with special abilities are handled the way they should be."

Reeves: "I'll return in a few days with the squad's dossier. If you hear anything, contact me immediately. If I can't be reached, go to Michael."

As Louie leaves the conference room, he accidentally bumps into a pale-faced woman with dirty-blonde hair in fatigues. Her uniform is decorated with medals and rank patches, a lieutenant.

Lieutenant: "Watch where you're walking, soldier!" *she turns and notices his medals, badges, and higher rank; she immediately salutes, blushing* "Oh shit, sorry, sir!"

Louie: "Lieutenant, what's your name?" *in an authoritative tone*

Lieutenant: "Lieutenant Bellum, sir!"

Louie: "It's fine, Lieutenant. I've heard great things about you. Sorry for my rudeness, I'm—"

Lieutenant Alala Bellum: *cuts him off nervously* "I know who you are, sir! Permission to speak freely?"

Louie: "Granted."

Lieutenant Bellum: "If you don't mind me saying so, you're a legend. There are stories about your battles, about how you were tortured by our own military and still survived. You're a hero in more ways than you know." *blushing*

Louie: "Thank you, but I don't see myself as a hero. I was just following orders, didn't know who was friend or foe, so I made sure that if they killed me, they'd at least remember I was that son of a bitch who didn't know how to die. Little did they know, I've got a high threshold for pain." *chuckles* "Nice to meet you. Hope to bump into you again sometime." *winks*

In the Oval Office, the President and Devin Deville meet with global leaders to discuss international issues.

President: "What are the main issues in your countries aside from poverty?"

Devin: "If I may, there's one thing we all have in common. We all want that unclaimed land off the Andaman Sea."

Leaders of the world: *murmuring among themselves* "That's true. But how do we claim it without going to war?"

Devin: "There are two or three unoccupied islands. We'll take the smallest one for our research facility. The other two can be divided among the rest of you."

Leaders: "Why do you get your own island while we share?"

Devin: "Because we're taking the smallest island, perfect for our laboratory to develop cures. The larger ones are better suited for you. Do we have a deal?"

Leaders: *talking amongst each other* "First one to claim them controls them!" *they continue arguing as they leave*

President: "Devin, now what? You know they'll race each other for those islands. If they don't get what they want, they'll start a war over it."

Devin: "Don't worry. We're taking the one I told them about. The others? They'll handle things their own way. If a world war breaks out, I'll make sure *you* become the hero who ends it."

President: "And how exactly do you plan to do that?"

Devin: "There are things about my life you don't know, and that ignorance is what's been helping your career, including your presidency." *thinking to himself* "This fool really believes he became President on his own. War is exactly what I need."

At the Wolf household, preparations are underway for a farewell lunch for the teens.

Rayne: "You guys do know this luncheon is mainly for the parents, right?"

Darren: "Really? Why?"

Cross: "They want to say how they feel and make sure we remember it, that way we'll want to come back."

Gabriel arrives at the Wolf home. He knocks, and Cross opens the door, letting him in.

Nora: "Hello, Gabriel. Would you like to eat before taking Darren?"

Gabriel: "Sure. I'd be a fool to say no to such great food."

For several hours, they laugh and enjoy each other's company. Then the time comes for Darren to head out. Gail hugs him tightly, as if it were the last time she'd ever see him. Rayne hugs and kisses him, and Cross pulls him in for a firm embrace.

Cross: "Remember, whenever you can, call your mom. And if possible, try to write to us. We'll let my mom know where we are so you can send letters."

Darren: "Alright, brother. I hate writing, but I'll make an exception."

The following day, the twins prepare to be shipped out. Nora hugs and kisses them both, crying as she watches her babies leave. Michael takes Cross while Reeves escorts Rayne to her destination.

Several days later, Reeves meets with Louie, handing him a folder.

Reeves: "The following are under your watch. For some reason, I don't have access to the complete list, here are the ones I can give you at the moment." *he begins to brief Louie on the personnel files in detail.*

Sargent Alala Bellum

Canadian-born, 21 years old, with several exceptional abilities, a rare combination of speed, stealth, and heightened senses. She can instantly determine how many enemies are present and what they're equipped with, whether weapons or machinery. She graduated at the top of her class as a sharpshooter. Sgt. Bellum doesn't know the meaning of missing a target.

Corporal Frank Chapmen

A 27-year-old better known as Kage, a name he earned for his unmatched ability in stealth. Cpl. Chapmen single-handedly cleared an entire enemy base, and when asked how he did it, he simply replied, “Kage,” which means *shadow* in Japanese. Cpl. Chapmen’s mother is from Japan, and his father is from Brooklyn, New York. The rest of his information is classified.

Corporal Elizabeth Miller

Corporal Elizabeth Miller, A Native American woman, 34 years old, and head of the recovery team. She's located hundreds of soldiers who were listed as MIA during wartime. Some were wounded, others dead, but regardless of the outcome, she can track anyone, no matter how long they've been missing. Cpl. Miller's sense of smell is animal-like. She wears a gas mask in the field to protect her nose, removing it only when she's tracking allies or enemies.

Lance Corporal Robert Wilson

From Bozeman, Montana, a 20-year-old ex-mercenary discovered by Miller a month ago, shot and bleeding out. He remained hospitalized until yesterday. LCpl. Wilson's wounds have fully healed, and he claims to feel better than ever. The moment he woke up, LCpl. Wilson requested to join Cpl. Miller's squad. He's been rapidly rising through the ranks to serve by the Corporal's side.

Sargent Dorian Theodore Gray

He claims to be 300 years old but looks no older than 19. He's the head of the data recovery team, and his chosen specialty is interrogation, specifically, torture. None of his subjects have ever survived, but he always gets the information he needs. There are rumors that he takes pleasure in it. He's had multiple sexual harassment complaints from both female and male members of his squad, yet they all willingly choose to remain under his command.

Reeves: "The rest of your group will be sent to you on the day of deployment. Sorry I couldn't give you more info, but as soon as

I have it, you'll be the first to know. Did you find anything about Devin?"

Louie: "I found something that might be useful. I heard a voice that sounded like Zima talking to a high-ranking officer. I didn't catch the whole conversation, but what I did hear was about who would be leading each troop for the upcoming battle and where each one would be stationed."

Reeves: "That's interesting. Don't be surprised if Michael himself comes to you. We'll meet again the day before deployment."

At Nora's home, Gail and Nora are sitting together, sharing letters.

Gail: "I've gotten two letters from Darren. How about you?"

Nora: "Cross and Rayne each sent one. Louie writes to me weekly, if he can't write, he sneaks a phone call just to let me know he's okay."

Gail: "I hope they keep us updated once things get hectic."

Nora: "Louie did whenever he could. If he thought it would get too crazy, he'd notify me. He can always sense it before missions. If I'm not mistaken, Rayne and Cross wrote that they'd both call today."

Gail: "All Darren said was that he's doing fine and not to worry. He'll send another letter when he gets the chance, which probably means I shouldn't hold my breath."

Nora: "Don't be so hard on him. Being a seaman" *chuckles* "might be a tough life." *starts laughing harder*

Gail: *laughing through tears* "I just hope he can handle everything thrown his way."

The phone rings. Nora and Gail race to pick it up.

Gail and Nora: *in unison* "Hello?"

They listen, it's a telemarketer. Gail hangs up, irritated.

Nora: "Damn telemarketers. Can't stand them. Oh, did I tell you I found my mom's journal while I was cleaning last night?"

Gail: "No, you didn't. Did you get a chance to read it?"

The phone rings again, this time, it's Darren.

Gail: "Hi, sweetie. How's it been for you?"

Darren: "Training was a breeze. It feels like everything I did before this was preparing me for it."

Gail: "Well, Louie and Gabriel did warn you. If those two tell you something, they have good reason."

Darren: "Have you heard from anyone else?"

Gail: "Nora's been getting calls from Louie, and today Cross and Rayne are supposed to call. She's beyond excited."

Darren: "I gotta go, but if you talk to Rayne… please tell her I miss her."

Gail: "I will. Be careful, and Nora says God bless. I have everyone's addresses here." *she reads them to him*

Once Gail hangs up, the phone rings again. Nora picks up, it's Cross.

Cross: "Hi Mom, how's it going over there?"

Nora: "It's fine so far. I should be asking you that. Gail's been keeping me company."

Cross: "That's good. Training's about what I expected. Everything we learned before helped a lot. Have you heard from Rayne and the others?"

Nora: "Yeah, they're doing fine. By the way, Gail sends her blessings."

Cross: "Please tell her it's appreciated, I'll take all the blessings I can get for whatever's ahead."

Nora: "You'll be fine. You're Wolf's son, after all." *chuckles softly*

Cross: "Gotta go, there's a long line for the phones. I love you, and I'll call when I can."

He hangs up. Nora lowers her head, eyes glistening.

Gail: "Don't worry. Like you said, he's Wolf's son."

Nora: "I know… but look how Wolf ended up. All his strength and knowledge, for what?" *she starts crying*

Gail: "He'll be fine. Michael's watching over him." *trying to distract her* "You didn't tell me if you've read your mom's journal yet."

Nora: "Oh yeah, I started. Turns out, I do a lot of things like she did."

Nora shows her the cover of the journal, *Denise Frost Memoir*. She opens it and begins reading aloud, showing Gail the old black-and-white photos tucked inside.

Denise Frost: "My father has ordered everyone to stay inside. There's news of an attack. I don't know who would dare fight us, we don't bother anyone. While everyone's distracted, I'll go to the next tribe to see if Beth knows what's going on."

Denise glances around before sneaking away from camp. She slips into the neighboring tribe where Bethany is waiting.

Bethany Constanza: "Hurry! The warriors are making their rounds."

Denise: "I'm going as fast as I can. Remember, I wasn't gifted like you."

Beth: "Were you able to ask Rosalie what's happening?"

Denise: "Your camp is the closest one to mine. I came straight here."

Beth: "Fine. I'll get her. If you hear anyone, hide."

Beth pounds the ground softly, using the vibrations to locate everyone in her tribe. When the moment's right, she speeds off, so fast, the guards think it's just the wind.

In a blink, Beth reaches Rosalie's camp.

Rosalie Inferno: "Are you crazy? What are you doing here?"

Beth: "I'm here to take you to my camp. It's safer there."

Rosalie: "And why do you think it's safer there than here?"

Back at the Army base, the platoon has a few hours to themselves. Most gather in groups, laughing and talking, while Cross sits alone by the fence, holding a picture. A friend approaches, a tall, lean man wrapped in medical bandages, a gas mask covering his face.

Platoon Member: "Hey, Cross, you okay?"

Cross looks up and sees him.

Cross: "Oh, hey, Justin." *wipes a tear from his cheek*

Justin: "Is that her, the girl you told me about? She's beautiful."

Cross: *smiles faintly* "Yeah… that's my Luna. I miss her so much."

Justin places a hand on his shoulder.

Justin: "I know you do. But I also know you'll find her. Love like that, it always finds a way."

Cross: *smiling through tears* "That's so corny, Justin… but you're right. Still corny, though." *chuckles*

Justin: *laughing* "Maybe I should stop reading so many romance novels."

Cross: "You couldn't if you tried. You love them too much."

Justin: "This is true." *pulls out a flask and takes a swig*

Cross: "Hey, are you gonna share?"

Justin: "Oh, did you want some?" *pretends to drink faster*

Cross: "Greedy ass." *grabs the flask and finishes it* "When did you start drinking whiskey?"

Justin: "Eh, figured you could use a pick-me-up. I know you like whiskey. I'm not picky, liquor's liquor."

Cross: "Yeah. Thanks, Justin." *looks at the moon and smiles faintly* "I'll find you, Luna. I promise. And once I do… I'll never let you go again."

He stares into the distance, memories of Luna flooding his mind, every smile, every touch, every moment they shared.

Back in New York, Nora and Gail continue reading and imagining the scenes as if they're there.

Beth: "Hold on tight!" *she lifts Rosalie onto her back and speeds back to the tribe, but when they arrive, Denise is nowhere to be found.*

Beth: "Denise, where are you? I'm back with Rosalie!"

Rosalie: "I think she left… or she's hiding. Where would she go?"

Beth: "If she did, the only place I can think of would be the trunk." *they both rush to check the trunk, but Denise isn't there*

Outside Beth's dormitory, the sounds of battle erupt, shouting, clashing metal, and distant screams. The girls stay silent, listening closely to the warriors' voices.

First Warrior: "The next tribe over got destroyed. Chief Frost was tortured and killed. Seems his daughter escaped. We've eliminated the trespassers, but I fear more will come."

Second Warrior: "Chief Frost's tribe has no survivors. The attackers were searching for his daughter. Reports say Denise Frost was heading this way several hours ago. Our scouts will alert us when reinforcements approach."

Tribe Chief: "Get the archers ready! Have them take positions in the trees, the least guarded areas first. Half of the warriors will assist them; the rest will defend the village. Send all the women and children to the furthest camp until it's safe."

CHAPTER 12.
TURMOIL

At the Navy base, Darren is still trying to find his place. There's been controversy about his accomplishments, a few respect him, but many envy his rapid rise through the ranks.

First Sailor: "Grier acts like his shit don't stink."

Second Sailor: "In every field exercise, simulation, and real battle, he's not the best, but he gets the job done."

First Sailor: "That may be true, but he's still an asshole. I'm done with him. He'll learn he can catch an ass-whooping like anyone else."

Second Sailor: "When are you gonna fight him?"

First Sailor: "I'm not the only one who wants to kick his ass. He's gonna get what's coming during the next drill."

Second Sailor: "Is our drill sergeant in on it?"

First Sailor: "Hell no. The best part is, he left someone in charge tomorrow who can't stand Grier either. The medics better be ready."

The first sailor gathers everyone who's against Darren for a meeting in the cafeteria.

At 0400 hours, the drill begins as usual, light jogging, followed by a three-mile run, then fifty diamond push-ups and five-minute sets. As soon as they finish, the drill sergeant shouts:

Drill Sergeant: "NOW!"

The majority of the troop rushes Darren, striking him with punches, kicks, and blunt objects. Enraged, Darren explodes out of the group, throwing his attackers several feet back. He stands firm, eyes blazing.

Darren: "Come at me now! Stop with the cheap shots, you bunch of cowards! Let's see how tough you are when I'm ready to fight!"

The troop charges again, but one by one, Darren tosses them aside like rag dolls. Then the drill sergeant steps forward.

Drill Sergeant: "Grier! You did pretty well against the boys, now get ready for the ass-whooping of your life."

Darren: "With all due respect, sir, I still can't figure out how the fuck you got your rank. I know I can do your job, and better. Starting with giving you the first beating you'll never forget. Slow, painful, and with a smile on my face while I do it." *he grins*

The sergeant throws knockout punches, but Darren dodges every one with ease before retaliating with a brutal barrage. He gets so caught up that he starts flinging the sergeant around.

Out of nowhere, one of the few sailors Darren gets along with rushes to him.

Sailor: "Grier, stop! If you keep going, you'll kill him! I get it, he's a dick, but going to jail will only make him happy. Don't give him the satisfaction!"

Darren: "George, why didn't you tell me they were planning this?"

George: "They tied me up and drugged me."

Darren: "How did you escape?"

George: *laughs* "The idiots forgot we were trained to escape restraints two weeks ago!"

Darren slams the sergeant down next to the rest of the unconscious sailors.

Darren: "You think they'll try to kick me out?"

George: "I doubt it. Pretty sure they won't mess with you again."

Somewhere in a dark pit of hell, a voice echoes through the void.

Voice: "WHERE AM I? IS ANYONE HERE? SOMEONE SAY SOMETHING!"

Far off in the distance, another voice responds.

Other Voice: "HELLO! WHO IS THAT?"

Footsteps draw closer to the first voice, slow, deliberate. The unseen figure snickers.

Voice: "WHO'S THERE? SAY SOMETHING! WHY AM I HERE? RELEASE ME, IF YOU KNOW WHAT'S GOOD FOR YOU!"

Person Walking in the Dark: "Funny how you're still acting high and mighty when no one's coming to save you. You really think they care about you?" *laughs maniacally*

Voice: "Louie?"

Person Walking in the Dark: *laughs again* "Why would you expect him here? He doesn't think about you. If Louie cared about you, he'd have found you already."

Voice: "Then who are you? What are you planning to do with me?"

Person Walking in the Dark: "What fun would it be if I told you?"

Other Voice: "Raphael, is that you?"

Voice: "Yes. Who are you?"

Other Voice: "Uriel. Do you know who our captor is?"

Person Walking in the Dark: "You both cast me aside and left me to die. That's all I'll say for now."

Uriel: "Still no clue. Brother, any ideas who this is?"

Raphael: "None at all."

Person Walking in the Dark: "It'll all be clear soon enough." *vanishes into the darkness*

Moments later, the mysterious figure appears in another part of hell, equally dark and silent. A woman's scream cuts through the air.

Woman: "LET ME OUT OF HERE!"

Person in the Dark: "Scream all you want. No one can hear you."

Woman: "What do you want with me!?"

Person in the Dark: "You don't need to worry about that. We're just going to have a little fun together, that's all."

Woman: "You son of a bitch! If you touch me, my—"

Person in the Dark: "HOW ARE YOU GOING TO STOP ME!? You forget you're *my* prisoner. I'll do whatever I damn well please."

He rushes forward, grabbing her by the chin.

Woman: "LET ME GO!"

Person in the Dark: "Hehe… now for the real fun."

Woman: "Whatever you do to me, I'll make sure it happens double to you!"

Person in the Dark: "I welcome the challenge, my dear. But remember, you're stuck here with me." *leans in, face inches from hers* "No one can hear you scream. We can have all the fun I want." *taunts her* "And don't forget, I know your secret. This is going to be fun."

Woman: "NO, PLEASE! LEAVE MY.."

The captor puts the woman to sleep, smiles, and disappears into the shadows.

Back at Nora's house, she and Gail keep reading from the journal.

Beth and Rosalie return inside.

Rosalie: "Wow… all the villages are being attacked."

Beth: "Yes. We were supposed to help."

As Beth sits on the bed, Rosalie notices a piece of paper on the floor and hands it to her. Beth unfolds it and reads.

Beth: "It's from Denise, it says…"

I ran off looking for help. Soldiers are dying all around me. I remembered that my father used to ask members of the Heart family for help during crises. Since you two were taking too long, the only way I can help now is to find those hunters again. I just pray I make it back in time to save these villages. By the time you return, if everything's destroyed, go past the villages. Over the bridge, on the left, you'll find two farms. On the right, the Heart family home. I'll be there.

Sincerely,
Denise Frost

Rosalie: "Should we go? Your warriors need help, and if the Hearts are as good as the chief said, they might save what's left of these lands."

Beth: "You're right. Hold on tight again, we have to get there as fast as we can."

Rosalie climbs onto her back, and Beth speeds off. They pass through village after village, all destroyed, even Rosalie's.

Rosalie: "It doesn't look like anyone survived. How will your people make it if they're still heading this way?"

Beth: "I can leave a message here, let them know to move to another area."

She quickly writes a note for any survivors and posts it at the town's entrance. Then they continue on, following the path Denise described.

Death. Destruction. Blood. That's all they see until they finally reach the bridge. The girls follow Denise's directions precisely.

When they arrive, Denise greets them with a smile and hands each of them a drink.

Denise: "About time you two got here." *She turns around and looks at Sparta and Azrael.* "Girls, I would like you to meet my father's very powerful friends."

Everyone gets introduced, and then they begin to talk about what's happening in the villages nearby.

Azrael: "Do any of you know who started the battle?"

Rosalie, Beth, and Denise: *in unison* "We don't know."

Azrael: "What have the villagers done to save themselves from being massacred?"

Rosalie: "The chiefs that survived have sent all the women and children to the next tribes, but we kept leaving messages at the entrances of the destroyed villages to keep those looking for refuge to go elsewhere."

Azrael: "This is the plan, Sparta and I will find out the full information. Then, we'll deal with whoever is harming everyone. What I need from you three is—"

Sparta jumps into the conversation.

Sparta: "Go to those towns, try to find survivors, and bring them back here. When all the survivors are here, assist them the best you can. When we return, we'll handle it from there."

They proceed with the plan; the girls go town to town in search of survivors. Each village they walk through is a horrific sight, bodies cut in half, homes in shambles, some still in flames. Hours go by, and the hope for finding survivors slowly fades. Friends and family of the girls seem to have perished in the onslaught.

Denise: "Hopefully, Sparta and Azrael are having better luck than we are."

Beth: "If I'm not mistaken, they're going out there to destroy whatever or whoever did this mess."

Denise: "If there are survivors in their area, they'll do their job and rescue them also."

Rosalie: "You have a lot of faith in them. Have you ever seen them in action?"

Denise: "I snuck out one night to see what my father did after dark. He went with his two friends to protect areas bordering our villages and assisted those that couldn't fend for themselves."

Beth: “They look pretty mean. Are you sure they can kill the ones that destroyed our homes and still be kind enough to help the injured?”

Rosalie: “If she trusts them and her father trusts them, then that’s all I need to know that they are trustworthy individuals.”

In a remote location, Catherine is trying to hide. She has been moving from place to place awaiting Louie’s return. As she walks to the bed, Catherine hears a voice.

A haunting voice: “Why are you trying to run from me? No matter where you go, I will find you, bitch. Louie isn’t going to save you from the punishment I have for you.”

Catherine: “Just leave me alone!”

A haunting voice: “I am just coming for what’s mine. You owe me. That which you try to keep secret is not, give in or pay the consequences.”

Suddenly, a mist of dark gray smoke appears. She attempts to flee; a hand comes through following her. When Catherine tries to catch her breath, it grabs her by the arm. She fights it off, shooting magical blasts and throwing her most powerful hits, but it’s too strong and pulls her into the smoke. Shortly after, Catherine vanishes into the mist.

At a classified location, a storm has begun. A few Black Ops members split up and infiltrate several areas on the base. Once everyone is at their post, the wind begins to pick up and the clouds darken the sky. One of the members in the east sector is surrounded by enemy forces; he whispers into his communicator. Before the message is sent, the downpour begins.

East Sector Black Ops member: “Take ’em out at my coordinates.”

One by one, the enemy is eliminated. In the distance, a shadow figure drenched from the rain smiles at the fellow soldier. Immediately after, the sniper gets back into position.

The Black Ops member from the northern sector asks the sniper in a low tone.

Northern Sector Black Ops member: “How many tangos do you see in my area?”

The sniper points his laser directly in front of the soldier and flashes it several times, indicating how many are there. Shortly after, the north sector lights up from the shots fired. Luckily, the soldier uses a silencer, killing all in that area. The western and southern sectors are cleared by the sharpshooter, making it much easier for the rest of the squad to finish the mission.

Sniper: *notifying the rest of the soldiers* “All areas are cleansed of tangos. I’m trying to find out where we’re getting picked up.”

After the area is secured and all the important information is taken, the group departs and rushes to the extraction point. The rain slows down, the sky begins to clear up, the trip home looks decent. The chopper arrives, picks up the squad, and lifts off.

In Nora’s house, the ladies are still reading the diary. It continues from when Rosalie, Beth, and Denise search for survivors.

Beth runs over to rubble next to a stable and sees movement. A person is barely alive, trying to get her attention.

Beth: “Try to keep calm, I’m going to help you out of this. Rosalie, Denise, I found a survivor!”

The other two help her get the survivor out. The person has both legs broken, bone piercing through the skin, blood everywhere. His ribs are badly bruised, and he's horribly dehydrated.

Denise: "I'll take him to safety. Bring the survivors to the house."

Rosalie: "Will do. Beth will drop them off to you."

Denise carries the injured man while Beth and Rosalie continue searching.

On the other side of the woods, Sparta and Alpha look for the cause of all this destruction.

Alpha: "Let's split up to cover more ground. You take the east and I'll head west."

Sparta: "No problem."

They proceed through the destroyed villages, no sign of life. Suddenly, Alpha picks up the scent of several perpetrators. He lets out a loud howl to notify Sparta. Sparta rushes over to his location, but on the way he gets ambushed by figures with a grayish-black appearance. Smoke surrounds them; the closer they get to Sparta, the stronger the sulfur smell becomes.

Sparta: "Who ordered you to do this?" *His eyes turn fiery as he gets into his battle stance.*

The figures: "You will find out soon enough, mortal."

Sparta: "Tell me now or face my wrath!" *He draws his blade, flames dancing along its edge.*

The figures: "My, how impatient this one is. Never seen anyone so eager to meet their death."

Sparta attacks, swinging his sword. The figures vanish, then reappear several feet away.

Sparta: "Why are you backing up? Come and fight me, cowards!"

He charges again with the same result. Then the thought hits him, they're keeping him from reaching Alpha. He runs toward his friend, but they teleport into his path again. Frustration builds, and instead of losing control, he fires a massive blast of flame into the air. The blaze shoots high enough to alert Alpha.

Alpha: "Fuck, Sparta must be in some shit. I better get to him instead." *He shifts into a wolf and sprints toward him.*

He dodges trees, sniffing the air to catch Sparta's scent. Several feet away, he transforms back and ambushes the creatures.

Sparta: "Oh, sure, he can hit you." *Annoyed.*

Alpha: "Are you alright?"

Sparta: "These things are stalling. For what reason, I don't have the slightest idea."

The figures: "We're just waiting until the two of you are together. Now we can finish you off in one shot." *The beings merge into one massive figure.*

The figure: "We are legion. Grovel before our power."

Sparta: "No one has answered my question, who sent you?"

The figure: "We go where we are sent. We don't ask questions. Now bow before us and worship us."

Sparta: "Do you believe this motherfucker? What a fucking joke."

Alpha: "We won't grovel or kneel before the rejects of hell or any of its inhabitants."

The figure: "You have chosen your path, die for your stupidity!"

The figure dashes at the two warriors, attacking them at once. The creature moves like a blur, almost too fast for them to keep up. Alpha closes his eyes and relies on his other senses, hearing the faint shift in movement, then pivots in the opposite direction.

Alpha: "Sparta, left! Sparta, swing your sword right!"

While Alpha directs him, suddenly Beth sprints in and joins the fight. The men stand back, stunned, watching two blurs clash. Out of nowhere, a burst of ice shoots toward the figure, freezing it in place.

Rosalie: "Hit it now before it breaks out!"

Beth circles the frozen creature to keep it trapped as Alpha slashes and Sparta scorches it with fire. More figures manifest, swarming the two warriors and Rosalie. She burns several before retreating, and Beth rushes after her.

Beth: "We need you."

Rosalie: "I am helping. I need to get help from nature. My conjuring will assist—ask them to guide us from those things."

Before Beth could say another word, *Rosalie begins chanting. She sprinkles white powder onto the ground; Beth lights four areas of the powder as Rosalie continues chanting. The beings start climbing out of the fire.*

Rosalie: "Spirits of nature, rise and help us."

The figures approach Rosalie and Beth.

The nature spirits: "How can we assist you, child? We are at your command."

Rosalie: "I need you to protect us from those creatures."

The nature spirits: "As you wish. In return, we will tell you what we want."

The spirits rush the figures, attacking them. As Sparta and Alpha battle on one side, the spirits slay the rest of the adversaries on the other.

Once the enemies are defeated, the spirits turn to Rosalie.

The nature spirits: "As gratitude for our assistance, we only ask for your firstborn son—born from your bloodline."

Rosalie: *on one knee* "Consider it done."

The nature spirits disappear into the mist.

Beth: "Are you sure this is what you want to do? You must give them your son when the time comes. Do you really want to do that?"

Rosalie: "They helped us. For now, I'll tell them whatever they want to hear. If I do end up having a son, they'll have to pry him from my dead body before they can lay a hand on him."

Beth: "They're pretty powerful. Do you think you can fight them and win?"

Rosalie: “Honestly, I don’t know. But hopefully, in the future, I’ll be strong enough to protect my family—and I pray that my husband, when that time comes, will be a warrior. Together, I believe we’ll be able to protect our son.”

Beth: “Sounds like a fairy tale, Rosalie. I just hope for your sake, and for the safety of your future son, that it happens the way you wish.”

Rosalie: *changing the subject* “I wonder if Denise is okay. She’s alone with the injured villagers.”

Beth: “She’ll be fine. You know her—doing her famous miracles.”

They find a few more survivors and head toward the hut. When they arrive, they’re greeted by Denise, who immediately tends to the injured that were brought in or managed to walk in.

Denise: “Anything good happen?”

Beth and Rosalie: *They look at each other.* “Nothing out of the ordinary.”

Denise: “Are the two warriors on their way back?”

Rosalie: “No clue. We had our own little battles while searching for your patients.”

Hours later, Azrael stops by to relay a message.

Azrael: “Thank you for all your hard work. But, Rosalie, that deal you agreed to cannot be taken lightly. Agreements with spirits can be held even after our generation has long passed. Do you understand this?”

Rosalie: “I understand. I did it to save the surviving villagers.”

Azrael: "I know you did what you thought was for the greater good at the moment. I just want you to realize that this is a serious matter."

Gail and Nora put the journal down and realize it's 11:27 p.m.

Gail: "It's late. I didn't notice the time."

Nora: "We can continue reading next time you have time."

At the Vatican, bishops and priests are closing off a section of the church for their visitors.

Head Bishop Joseph: "We must make sure this area is in tip-top shape for the archangels to feel welcomed and not be interrupted. Also, our job is to keep them hidden from anyone."

Father Delaney: "Guarding our Lord's angels is an honor all its own. Fellow brothers, we all have the tasks given to us by the Pope. Let's make our Father proud to know he can count on us."

Hours later, the archangels Michael and Gabriel arrive at the cathedral. The heavenly brothers walk into the hall to meet with Joseph, Delaney, and the Pope.

Pope: *kisses the hands of Michael and Gabriel* "We are honored that we were chosen to orchestrate and help you with anything you need."

Michael: "We are here for information. Any whispers from the demonologists on the whereabouts of our brothers Raphael, Uriel, and Luna?"

Joseph: "Who's Luna?"

Gabriel: "That doesn't concern you. We would like to know if you've heard anything involving the two we mentioned."

Pope: "Please excuse Joseph. He's a very curious individual. It's his greatest attribute, yet his greatest weakness."

Michael: "I would like to speak to the head demonologist. His last name is Kennedy."

Father Delaney: "I'll get him over here at once."

A week later, at Gail's house, she is busy helping Michael find Raphael, Uriel, and Luna with all the information that's been given to her. Gail creates a map to mark all the locations where Michael has found clues. Suddenly, the phone rings, it's Nora.

Gail: "Hello, Nora. Is everything okay?"

Nora: "I have great news."

Gail: "What is it? Are the kids coming back?"

Nora: "No, no, I met someone."

Gail: "Really? Who? Do I know this person? Give me all the juicy details."

Nora: "Not much to tell. We bumped into each other and had some small talk about produce—"

Gail: *interrupts* "Produce? Where were you when you met this mysterious man?"

Nora: "I was grocery shopping when we met. Moving on, then he asked if I'd like to go to dinner with him."

Gail: "Did you tell him yes?"

Nora: "If you let me finish, I said yes," *giddiness in her voice,* "and we've been going to dinner every day since then. He's a great person."

Gail: "Wow, I'm happy for you! When am I going to meet this mystery man who has you acting like a schoolgirl again?"

Nora: "Soon. He's going out of town for the week, but when he returns, I told him he has to meet you."

Gail: "What does he do for a living?"

Nora: "He told me he's a bounty hunter. That's why we're getting together next week, there's a bounty that crossed several states and has been spotted."

Gail: *giggles* "How do you always find a hunter?"

Nora: "It's a talent, I guess." *laughs*

Gail: "Well, I can't wait to meet him."

Nora: "I'll call you later to tell you where and when to meet us."

Gail: "Sounds good. I've got to let you go, I'm in the middle of something. We'll talk more tomorrow if you're not busy."

Nora: "That's perfect. I'll cook a little something, then we can continue with the journal."

Gail: "We can make a night out of it."

Somewhere in the depths of hell, a prisoner gets an unusual visitor. Someone walks into Luna's cell.

Luna: "Look, I'm sick of you guys talking shit to me, I'M NOT AFRAID OF YOU!!!"

The person in the cell speaks from the shadows, it's a woman. She walks out, though her face remains hidden.

Woman: *in a soft, sweet tone* "You poor thing. Have you been eating?"

Luna: "Is that a joke!?"

Woman: "No, I really want to know."

Luna: "Sure you do." *sarcastic tone*

The woman gives Luna some food from a bag. Luna looks at it and begins to cry.

Luna: "Now what, you're going to eat the food in front of me?"

Woman: "Look, I know it's hard to believe, but I want to help you. Do you want to eat or not?"

Luna shows the woman that her hands are tied and she's shackled to the wall.

Woman: "I will feed you, then."

Luna: "How do I know you're not trying to poison me?"

Woman: "You don't."

The woman kneels down in front of Luna and eats some food to show it's not poisoned.

Woman: "Do you trust me now? Please, let me feed you."

Luna: "A-alright… t-thank you."

The woman feeds Luna and gives her some water to drink, then cleans her up a bit.

Luna: "Why are you helping me?"

Woman: "Let's just say I'm a friend. But, just like you, I'm kind of trapped here myself."

Luna: "Who's keeping us here?"

Woman: "Silence is key."

Luna: "What does that mean?"

The woman shows Luna a tattoo around her neck.

Woman: "This is the price for helping you. They put a spell on me so I can't tell you anything."

Luna: "That looks painful."

Woman: "Yeah, but whatever they put you through must have been worse. I couldn't take hearing them hurt you, so I asked if I could take care of you. They need you alive, so they let me. I don't know how long it's going to last, though."

Luna: "Thank you. What's your name?"

At the Marine base, Louie is being sent out on a mission. The hand-selected squad is going with him, this will prove why Reeves chose them. While on the chopper, he gives them a summary of what to expect.

Louie: "This mission is basically to clean up what the Seals screwed up. Be ready for anything. I was only told to cover their tracks, since we're being called, that can only mean it falls along the lines of the unknown."

They arrive at the location, the town is destroyed. The soldiers exit the helicopter; body parts and blood paint the streets and buildings. The smell of iron mixed with rotting flesh lingers in the air. Some of the facilities are still smoldering from flames that recently died down.

Lieutenant Alala Bellum: "What happened here? I thought they were here for a rescue mission, not to eliminate everyone in the town!"

Sergeant Dorian Theodore Gray: "Looks like they screwed up. Upon arrival, they killed everyone to make sure no one would get in their way."

Corporal Frank Chapmen: "I'll scout the area for survivors. Would anyone like to accompany me?"

Corporal Elizabeth Miller: "If it's alright with Blak Hart, I'll go with you. Never know if you might need an extra set of eyes."

Blak Hart: "Miller, go ahead and get Chapmen's back. Wilson, Bellum, check around to see if we have any tangos to worry about. Gray, you and I are going to find out what happened here, and why they felt the need to destroy the village. Any of you see anything, radio us."

Lance Corporal Robert Wilson, Bellum, Chapmen, Miller, and Gray: "Yes, sir!"

The group departs to follow Blak Hart's orders. Miller and Chapmen search high and low, all they find are carcasses. Wilson and Bellum scan all vantage points and possible entrances to the village. Blak Hart, with Gray's assistance, searches for evidence explaining the destruction of the town. They find a small bunker next to the hospital with a hidden compartment.

Blak Hart: *on the radio* "Everyone, meet us at the front of the hospital. We got something."

While waiting for the rest of the squad, Gray notices the guards of the bunker had their hearts ripped out of their chests. Blak Hart looks at the entrance and sees a few corpses that appear drained of all bodily fluids, skin wrinkled and dried up, almost mummified.

Gray: "What could be so important that even the guards had to be killed?"

Blak Hart: "Whatever it was, we're going to find out."

The group reunites and begins to report their findings to the leader.

Miller and Chapmen: "No survivors, just bodies of people that look like they were used in dark rituals. They all looked like bones with flesh. Very odd."

Wilson and Bellum: "No tangos. The only thing we found, all but one entrance was blocked off so no one could interfere or rescue. Who could be so cold-hearted?"

Gray: "We found this bunker. The entrance needs a certain chant to get in." *points to the door*

Blak Hart: "The way to get in without the chant would be to use TNT, but chanting with dynamite could be a little more efficient."

Bellum: "Why use the TNT?" *scratches her head*

Blak Hart: "This would push back any enemies guarding the entrance, giving us the element of surprise with a kick." *chuckles* "Not to mention, if they had any strategy for their enemy, the blast will have them rethinking it."

At a secluded location, Rayne begins to write a letter for Nora when she's told to get ready for the next mission. She and her group are being sent to find an artifact.

Moments later, the squad is picked up by helicopter and taken to the location. At the drop-off, Rayne leads her troop. They see blood and gore everywhere; the smell of death and smoke fills their lungs.

Rayne: "This place was destroyed with the civilians still in it. Keep your eyes peeled, I don't have a good feeling about this mission."

The group searches for clues that might lead to where the artifact is being held. They march on until Rayne spots footprints, several hours old. She signals the group to prepare, pointing to the tracks to let them know they're not alone. The group continues when they hear an explosion and notice smoke coming from behind the hospital.

Rayne climbs to the top of the building to get a better view. Looking down, she sees the area still smoking from the explosion. The sniper quickly joins her to keep the squad informed. As soon as they reach the spot, Rayne calls headquarters about possible thieves still in the vicinity.

Everyone creeps down into the bunker, the door inside has been blown open. They turn on their flashlights and proceed into what appears to be a tunnel. The smell of rotting flesh and skin residue coats the walls; bodies of what could have been civilians lay on the ground.

The rookie of the group, Frank Ramsey, slowly walks forward when one of the bodies grabs his leg. He yells for Rayne.

Frank: "Rayne! It has my leg!"

Rayne: "Relax, probie. I'll handle this."

She pulls the hand off Frank when she hears a faint voice speaking to her.

Faint Voice: "Get out while you still have a chance…" wheezes "The demons destroyed the whole village…" gasps "The team that arrived before you was attacked…" gagging on blood "I was the last survivor…" dies from open wounds all over the body

Rayne: "Everyone, lock and load 'em. We're in for some shit." wishing her brother, Darren, and Louie were here to see her battle evil without their help

The battalion walks for about a mile into the cave. Just as they let their guard down, shots echo through the cavern.

Out on the Navy ship, Darren has been transferred to the Navy SEALs. On his first mission as a SEAL, he's being sent to a destroyed town. During the briefing, he and his fellow soldiers are told to destroy all tangos in the area and retrieve all valuable intel before it falls into the wrong hands. A smile crosses Darren's face.

Darren: "When do we leave?"

The SEALs depart on a small ship. The trip takes less than thirty minutes. The group finds a cavern blocked by tree bark and leaves. They move the debris aside, and Darren peeks in. Seeing it's safe to proceed, he notifies the rest to follow him.

The squad leader instructs everyone to switch to night vision before continuing through the cave. He tells Darren to guard their blind side as they move deeper inside. While walking, they come across a statue resembling a warrior from the island's past. For a closer look, Darren darts toward it. He realizes the statue is holding an antique bow and arrow, and at its feet lie the remaining arrows.

Squad Leader: "We don't know if any of these things are traps. Do not, I repeat, do not, touch anything."

Darren reaches for the bow and touches it. The bow and arrow begin to glow; the rust and stone fall away. He adds it to his arsenal.

Squad Leader: "Damn it, Gier! What did you do?"

Darren: "It compelled me to grab it. It's not a trap. I think this is meant for me."

Squad Leader: "What the hell are you smoking? This is for us to bring back, not for your toy collection!"

Suddenly, several ghouls charge at the SEALs.

Squad Leader: "What the fuck are those?"

Darren: "Those are low-level ghouls! Just shoot them in the head before they get too close!"

The group: "What do we do, Chief?"

Squad Leader: "Let's see if what Gier says is true. Put the bow back."

Darren: "I'm taking it with me. It feels like it belongs to me."

The group, in unison, shoots down the beasts as Darren uses his new weapon. With one pull of the bow, he fires an arrow that takes down a row of enemies.

People see my demeanor and think I'm harmless, don't let my appearance deceive you. I can do anything the guys can, I can be by your side or against you. The choice is yours.

Rayne Cross

I may not be the person you'd expect to save your ass, but just be happy I'm saving it instead of ending your life. Pray that you'll never have to find out what I'm capable of.

Darren Gier

CHAPTER 13.
CONFUSION

Reeves: "This is an unnecessary battle. The leaders that sent them are clueless on the reason why the order was given."

Cross: "Who gave the order?"

Reeves: "I'm still trying to figure that out."

Cross: "If they are all on the same island, eventually whoever comes into Louie's path will fall. And if we're all here, who's watching to make sure this all goes according to their plan?"

Reeves: "This is why I'm coming to you. I need you and your group to try to get to his squad before they all kill each other."

Cross: "Tell our leader that you are giving us a top-secret mission, then he'll send us out."

Reeves: "Get your squad ready. You'll be sent in about an hour."

Sixty minutes later, Cross and his battalion get their orders for a rescue mission. Cross briefs everyone on what the mission is truly for and why it is important that it gets done precisely as stated in the manifest. As they get ready, Cross gets a specially made gas mask that appears to look like a skull with the helmet to compliment the mask.

Hours later, they arrive at the location. The sounds of gunfire are echoing. Cross looks at the map that was given to him by Reeves. It shows the last known location where Louie's squadron were. The group rushes over to the point where both groups would eventually meet up. Carcasses of non-human things were all over the ground. The closer they got to the cave, the more gruesome it appeared.

Suddenly, a few soldiers run past the group, screaming and yelling something about supernatural entities.

Cross: "What is wrong with them? Those soldiers, from the looks of it, are from Black Ops, Army, and Marine."

Sergeant: "What the hell could've scared off the military? If they are hauling ass, why aren't we?"

Cross: "Because we're here to do a mission; that's exactly what we're going to do."

Sergeant: "But at what cost? I don't think we should continue on…"

Cross (interrupting): "If you want to head back, go ahead and desert us at the very moment you're needed, but I'm going to complete the mission. I'm not getting court-martialed."

More people begin running toward them. This time, they are attacking anything in their path. Already, a couple of Cross' group were killed by the mob.

Cross: "We're being attacked by the Seals! Brace yourselves, this battle is going to be bloody!"

The Seals rush in. They are being led by a man wearing an unusual white helmet. Not too far behind that group, a sniper is clearing the way for its squadron. This group is attacking the Seals. On the other side, a soldier wearing a black mask and his platoon are sprinting toward Cross' squad. None of them knowing what to expect on the other areas. The masked man leads his members full steam ahead, the only thing standing in their way is Cross and his men. A female soldier begins the assault, shooting her way through several of Cross' men.

Cross: "This is bullshit! We're trained for this shit! Stand your ground, we're going to take these assholes out!"

The masked warrior's squad runs at Cross. As he prepares for the impact, they pass right by him.

Cross: "What the fuck?"

As Cross looks at the soldiers passing by, he gets punched by a soldier who only has a bulletproof vest, a couple of pistols, and is wearing a black mask. The two go toe to toe, punch for punch. Cross pulls his gun out and begins to unload his clip onto the soldier's head. The warrior shakes it off and continues to pound his adversary. Out of nowhere, the masked soldier gets shot in the shoulder by a sniper. Another fighter wearing the white helmet charges at Cross while shooting the unhumanly warrior. The three battle each other as the sniper keeps trying to choose the next targets. It tries to shoot the helmeted man but continuously misses.

Helmeted Soldier: "Who the hell has the guts to try and shoot me?!"

While he tries to figure out who's shooting at him, the masked fighter grabs him by his neck and picks him up off the ground, squeezing his grip. The helmeted man shoots the masked soldier over and over again in the face to see if he would get released; instead, it makes the individual angrier. From the side, Cross attempts to tackle him but ends up getting kicked several feet away. The sniper's focus now is to take down the one who seems very difficult to defeat. Cross gets up and sprints to the black-masked man, then stabs him in his chest.

Masked Man: “Fuck this!”

The masked warrior throws the helmeted man about a yard away, then keeps his eyes on Cross. He takes off his vest, throws down his guns, and unmasks himself. Cross tries to see who this superpowered enemy is, but it’s too dim to make out the facial features. When the masked man is finally ready to battle again, his eyes glow; it appears as if they were filled with fire. Cross goes to throw another punch. The man dodges and retaliates with a few punches of his own. It is then that Cross realizes who his opponent is.

Cross: “Louie, is that you?”

Masked Man: “Maybe I am, maybe I’m not; who’s asking?” He begins to dart toward Cross.

Cross takes off his mask and helmet. The masked man abruptly stops.

Masked Man: “Oh shit, what the fuck are you doing here?”

Cross: “I knew it had to be you. You’re the only person I know that can survive a shot to the head, dodge my punches, then send me flying.” *Chuckles.*

Blak Hart: “Sorry about trying to kill you, bro. Just following orders.” He explains what his mission was.

The helmeted man speeds in, then begins shooting both Blak Hart and Cross. While he tries to run past them, Blak Hart once again grabs him by the neck.

Blak Hart: “Who the fuck are you?”

The man looks at him and recognizes him.

Helmeted Man: "Louie?"

Blak Hart: "Take the helmet off or die!"

Cross: "Hold on, it sounds like Darren. Well, if you are Darren or not, I suggest you do as he says before he snaps your neck like a twig."

The helmet gets thrown to the ground. Blak Hart puts him down, seeing that he almost killed Darren.

Blak Hart: "Why are you here?"

Darren: "We were sent on a mission. Why are both of you here?"

Cross: "Same reason, duh."

Blak Hart: "So Darren, you're a Seal. Nice, must be fun. Cross, how's the Army treating you?"

Darren: "Sort of. How's the Marines?"

Cross: "I'm getting promoted to an organization that's higher than any military as soon as the paperwork goes through."

Blak Hart: "Congrats on your future promotion. I have your paperwork at the base. Being a Marine is incredible."

The sniper zoomed the scope to see who the targets were. Shortly after, they got down from the vantage point to go join the three.

Blak Hart: "Where are your soldiers, guys?"

Darren: "I have no clue." *A little shocked and disappointed.*

Cross: "I hope they went to a safe area. Where are yours?"

Blak Hart: "I told them, once they find their way out, to wait for me at the drop-off."

Cross gets a call on his radio.

Cross: "Hello, Reeves. Just on time. I got an update."

Reeves: "Go ahead."

Louie takes the radio.

Louie: "Hey Reeves, Cross found Darren and me. Also, whoever sent us on this bullshit mission wanted me to kill all the other squads, then try to set me up for the murder of the people before we landed here and the other squad deaths. In other words, this would've been the best way to kill off the charges of Nite, Spaz, and permanently have Ark's charge out of the way for those in high positions to cause havoc against the king. Keep in mind, this is just a working theory."

Reeves: "Good to hear. Look around. Rayne was sent out there also. As for your theory, you might be on to something."

They walk in the direction Louie's squad went. They hear the steps of someone running in their direction. In the distance, they see a woman with a sniper rifle coming toward them. Due to the sudden action, he accidentally hung up on Reeves.

Louie: "Halt! State your purpose or die where you stand."

Woman: "Shut up, Louie."

Cross: "Rayne, cut your shit and get over here."

Rayne: "You take the fun out of everything."

She runs up to Darren and hugs him, tackles Louie, then punches his ribs. Finally, she runs up to her brother and jumps on him. He pushes her off, then hugs her.

Louie: “Contact Reeves and let him know you found us all.”

Cross: “I’m on it.”

Cross contacts Reeves and informs him on Rayne joining them.

Reeves: “Check for any injured soldiers and casualties. Let Louie take care of the fallen warriors, then have all others on the side so I can heal them.”

Cross: “You got it. After the clean-up, we’re gonna compare notes. Over and out.”

Cross checks out the wounded and sends them to go with Reeves. Then he sees several fatalities. Upon searching, he finds two from his squad, Steven Wrigley and his best friend, John Peterson. He looks at John.

Cross: “I’m sorry I wasn’t there when you needed me the most. Please forgive me.” *Sobs.*

Louie: “Cross, is everything okay?”

Cross: “I’m fine. I just lost my right hand in command. Give me a moment.”

Louie walks away to give him time to mourn. He goes to Rayne and Darren to try to catch up. Reeves has his group finish up as he interrupts the reunion.

Reeves: “Louie, may I have a word with you?”

Louie: “Sure, how can I help?”

Reeves: "How's Cross holding up?"

Louie: "He's mourning."

Reeves: "As for catching up, you're all going to be working together very soon. Darren needs a little more experience under his belt, Cross needs to move on from this loss, and Rayne must try to push past her inner demons."

Louie: "How long before we're doing our own missions?"

Reeves: "Like I stated earlier, soon. I need help looking for Uriel and Raphael along with Luna."

Louie: "What about Catherine? I haven't gotten any letters or calls from her for several months now. I'm worried that something might have happened to her."

Reeves: "Don't take this the wrong way, but I'll put it on my list of finding people. Remember, I'm already looking for three. If I hear of anything, I'll notify you immediately."

Louie: "I really appreciate it. How about assist in the mission searching for all four? What did you really want to talk to me about?"

Reeves: "Devin has disappeared without a trace. Seems he has gone into hiding. He's probably plotting something. If you get wind of any news,"

Louie: "I'll contact you immediately."

Cross: "Hey Reeves, what's the miracle you're visiting?"

Reeves: "Cross, sorry for your loss. The four of you, take some time off for a few days, then prepare. The other forces are going to try to weaken you all. This was their feeble attempt to do what Louie's theory suggests."

Cross and Louie: "Who are they, and when are they going to attack?"

Reeves: "We don't know until the time comes. Rest up and train hard. I need you all to be ready."

Reeves disappears. The four reminisce as their respected squads await orders.

Louie: "As the highest-ranking officer at the moment, everyone get some R and R until your squad leaders tell you otherwise. As for my squad, take a little leave for now. In about two weeks, get ready for our next mission."

The troops head to their rendezvous as the four squad leaders walk together to get a few drinks and enjoy each other's company.

In New York, Nora and her significant other are walking hand in hand around Central Park when they bump into Gail.

Nora: "Hey Gail, what a surprise!"

Gail: "Hi Nora, what are you doing here? Wait a minute, is he the guy you were talking about?"

Nora: "Yeah, why?"

Gail: Gail eyeballs him from head to toe to see if she knows him or not. "Should I tell her, or are you going to?"

Nora's male friend: "Damn, why can't you just be happy for us?"

Gail: “She’s like a sister to me, so if you hurt her, I’ll make you regret it, Shawn.”

Nora: “What’s going on? How do you two know each other?”

Nora’s male friend: “You really don’t remember me? We all grew up together. I used to call her poopie head.” *Laughs.*

Nora: “Oh shit, you’re pain-in-the-ass Shay! Why didn’t you tell me?”

Shawn: “I thought you figured it out when I said my full name, Shawn Swift. Emphasis on Swift.” *Smirks.*

Nora: “I didn’t put two and two together. Wow, I feel so stupid.”

Gail: “Don’t feel bad. I didn’t know he was back from his self-proclaimed retreat. Wow.” *Smiles.* “He finally got his dream girl.”

Nora: “What? He had a crush on me? I…” *Blushes.*

Shawn: “Really, sis? Still trying to embarrass me at any moment possible.” *Cheeks red as an apple.*

Gail: “It’s my job as your older sister.”

Nora: “Alright, alright, Gail. I don’t know what to say.”

Gail: “You don’t have to say anything. Just promise me you two will take care of each other. Both of you have been through a lot of shit, and it’s about time you guys find love and happiness in your lives.”

Nora: “So you’re not mad?”

Gail: “Why should I be? It feels weird, but you both are still family, now more than ever.” *Laughing.*

Somewhere far away, the Archangel Michael is following a lead. He was told of a location above Russia, on an island called Svalbard, where Uriel and Raphael could possibly be held.

Michael (thinking): Hopefully, this intel is more reliable than the past several thousand leads I followed that ended up being false alarms.

Suddenly, Reeves appears next to him. He updates Michael on the current statuses of Cross, Rayne, Darren, and Louie. After the news, Reeves assists his elder brother in the mission in hopes of finding their captive siblings.

Reeves: "Once we know for certain that the location is correct, what's our next step in freeing them?"

Michael: "Let's just take this one step at a time. We don't want to get our hopes up and end up disappointed."

They arrive on the island. Abandoned houses haven't had residents for quite some time. Snow and ice stretch for miles, as far as the eyes can see. They journey on until they spot a small patch of water that hasn't frozen over.

Reeves: "Michael, what do you make of this?"

Michael: "This might be the place the informant spoke of. The report mentioned an area that seemed out of place, where voices and screams could be heard but no one was seen."

Reeves: "If voices can be heard, then that would mean one of these facilities is being occupied."

Michael: "That's what we're going to find out. I'm not going to stop until we do."

Reeves: "Don't worry, you won't be doing this alone."

Michael and Reeves search all around. None of the houses or buildings seem occupied. Reeves finds what used to be a small clinic blocked by ice and snow that covers half the building. He looks in one of the windows and sees the room is clear of debris. Reeves moves past several windows, then decides to look into another. The floor is covered in blood and skin. Tools of torture are set on a table. He calls Michael over to show him what he found.

Michael: "Did you find something?"

Reeves: "Look at this. One room is clean while the other is covered in body parts and blood. I think this is where our brothers are being held."

Michael: "Good work. Now let's go and look for an entrance."

The angels search for what seems to be several hours. Then Michael falls into a hole covered with snow. Reeves rushes to his aid and realizes that Michael accidentally found an entrance.

Reeves: "Are you injured?"

Michael: "No. You must come down here. This tunnel appears to lead directly to the building."

Reeves jumps down to join his brother. Together, they walk through the passageway, looking for any doors or tunnels that could take them inside. As they continue, several caverns branch off in multiple directions.

Reeves: "What now?"

Michael: "You take the caves on the left, I'll go to the ones on the right. If you find something before I do, call me. I'll do the same."

Reeves: "You got it. Do I get the right to handle any possible threats if they arise?"

Michael: "Do what you must. Remember, our brothers' lives are in the balance."

Reeves: "Believe me, I know the severity of it."

They split up. Michael walks into the first cavern, surveying his surroundings, while Reeves darts through his first cave. He ends up at the final point and contacts Michael telepathically.

Reeves: "These caves were created to make intruders get lost and waste time until the guards show up. Let's meet back up and plan how we should move forward in a stealthy manner."

Michael: "I agree. Good call."

They finish checking their caverns and meet up to discuss a new plan.

Somewhere in the pits of hell, Luna is planning an escape when a figure walks toward her.

Luna: "Who's that?"

Figure: "It's just me, the person that gave you food last time."

Luna: "Oh. Why are you being so nice to me? Everyone else is torturing me. What makes you any different?"

Figure: “I’m not like them. I tried proving that to you last time. Anyway, I’m trying to find a way to get you out of here. You’re not like everyone else, you don’t deserve to be here.”

Luna: “Did Cross or Rayne send you?”

Figure: “I don’t know who those people are. I just want to help you. Is it such a bad thing that I’d like to help free you from this torment?”

Luna: “Kindness is something I haven’t had for quite some time. Who are you?”

Figure: “Let’s just say I’m a friend of a friend. When you’re free, then I’ll reveal who I am.”

Somewhere in the darkest area, the Archangel Raphael tries to plan his freedom.

Raphael (thinking): I must find a way to get out or notify Louie or Michael of my location.

He looks around and sees a shadow figure shackled to the wall.

Raphael (thinking): I wonder who that could be. Whoever it is looks like he was tortured to an inch of his life. Maybe if I try to focus all my energy, I might be able to contact Michael. Well, here I go.

Raphael begins to meditate, focusing his mind, body, and soul. *Father, please grant me strength to contact my brother, or help me become a beacon for a brief moment so he can find me.* Continues to meditate.

Suddenly, the figure wakes up. Raphael looks at it, trying to focus his eyes. It yawns and speaks.

Figure: Yawns. "Hey new guy, you up?"

Raphael stays quiet, his eyes locked on the being. The figure looks back with glowing white eyes.

Figure: "I can see you." *In a teasing tone.* "Why are you staying quiet? We've been stuck here for a while. Stop being an ass."

Raphael: "If we've been here for a while, how is it that I didn't feel your presence until now?"

Figure: "Because I decided to let you know I was here."

Raphael: "Do you know who I am?"

The figure laughs and slips out of the shackles, then walks over to Raphael. The figure looks like he's in his 20s, with long black hair. His body is scarred, and his face is covered with a black mask that has silver eyes glowing white behind them.

Figure: "Hmm… oh, you're that Ark guy, right?"

Raphael (thinking): How does he know that? Why can't I sense him anymore even though he's right in front of me? *Decides to answer.* "Yes. Did someone say my name?"

Figure: "Nope. The reason you can't feel me is because I don't want you to feel my power. It would give away too much, and what's the fun in that?"

Raphael stares at him in shock and disbelief.

Figure: "Oh, you didn't know I can hear your thoughts? Hahaha, that's so much fun. Don't you remember? You spoke to me, then I switched my voice so you'd think I was Uriel." *Laughs.* "Your brother is nowhere near you, never was. By the way, have your prayers been helping you?" *Whispers in his ear.* "Doesn't look like they are. Give up. They can't hear you from here."

Raphael: "Who are you?!"

Figure: "Well, roomie, for now my name isn't important."

Raphael: "How am I supposed to trust you if I don't know your name?"

Figure: "What gave you the idea I want your trust? If you haven't noticed, I can get out whenever I want. They just put me here to keep me occupied until next time I go to training."

Raphael: "So, why are you in here with me?"

Figure: "You seem to be more fun than the others. Plus, they wanted me to try and get some info from you. You're lucky I don't like being told what to do, and those walls you built in your mind are pretty hard to break down, so I'm just going to hang around and have fun with you instead."

Raphael: "Who sent you?"

Figure: "What makes you think I'd tell you? Hahaha. As an Archangel, isn't it kind of your job to know who your enemies are? How the fuck did they catch you anyway? Aren't you supposed to be strong?"

Raphael: "SHUT YOUR MOUTH!!!"

Figure: *Laughing.* "You're so scary. Oh, did I make you mad?"

The figure starts punching Raphael in the face and stomach. Raphael takes every hit with a smile as the figure hits harder and laughs.

Figure: *Laughing.* "I knew you'd be fun!"

Raphael: *Smiling.* "You're going to need to hit harder than that, kid."

The figure grabs Raphael's throat and squeezes. Raphael smiles as he's being choked. The cell door opens and a deep voice calls out.

Deep Voice: "Get your ass over here, boy."

Figure: "What the fuck do you want, old man?!"

Deep Voice: "GET THE FUCK OVER HERE NOW!!!"

As the voice yells, the figure grabs his head and screams in pain, letting go of Raphael and dropping to his knees.

Figure: "GET OUT! GET OUT! STOP!!!"

Deep Voice: "LET'S GO, BOY!"

The figure turns to the door and looks back at Raphael, who has a smirk on his face.

Raphael: *Taunting.* "That's cute, they have their dog on a leash."

Figure: "I'm not done with you. You'll see me real soon, roomie."

The cell door slams shut, and the room is consumed by darkness once again.

In another area of the darkness, Uriel is thrown back into his chamber after being tortured and interrogated. The demons put the shackles back on his hands and feet, then start beating him a few more times before spitting and laughing at him. After they feel satisfied with their actions, the beings turn around, boasting of their feats as they walk away.

Uriel (thinking): They think that I am broken. When they least expect it, that's when I'll attack. Hopefully, I have enough strength to carry this out. I wonder if my siblings are searching for me. What's taking Raphael so long to find me? How much longer must I deal with this? I don't know how long my body can withstand this abuse. Just when my body is about to be fully healed, they begin the torture again. Father, please guide me and protect me from these abominations. Grant me the strength to smite them in your name and please lead my brothers here safely. Please show them the way before it's too late to rescue me.

After Uriel's prayer and meditation, he hears the screams of agony and cries for help from other areas. He nods his head and continues concentrating on healing before they come back for more torture.

Somewhere overseas, Louie, Cross, Rayne, and Darren are about to begin their training before going back to base.

Louie: "I can't wait until we're in the same squad, tearin' shit up and killing those that deserve it."

Cross: "Just like the good old days. Only difference now is that we can all pull our weight, unlike before."

Darren: *Confused and a bit jealous.* "What do you mean, like the good old days? You two never did an actual mission together. Anyway, now I can show you that I'm extremely useful and don't need help."

Rayne: "Same here, but we still need each other. We don't know what we're up against in the future." *Shaking her head.* "By the way, what do you think they were doing when they hung out, duh?"

Louie: "Let's train hard and wait for Reeves' orders once we return to our group."

After hours of training and joking around, they say their goodbyes and part ways, anticipating the next time they'll meet again.

Back at the park, Nora, Gail, and Shawn are walking together. Gail and Shawn begin talking about their childhood.

Gail: "Remember when you tried climbing the tree and miscalculated how strong the branch was, then fell?" *Laughs so hard she begins tearing up.*

Shawn: "Hahaha, very funny. I broke my arm." *In a calm, sarcastic tone.* "You laughed then, and you're still laughing." *Begins to laugh with her.*

Gail: "You do have to admit, the way you landed was funny." *Laughing.* "The final landing was in the form of an Egyptian picture, your arms were positioned perfectly. Broken but still pretty funny."

Shawn: "I guess you're right. It was a cool pose, but I won't willingly do it again." *Laughing.*

Nora: *Feeling left out.* "Hello, I'm still here."

Gail: "Sorry, I haven't seen this knucklehead since he started being a dick."

Nora: "What!?"

Shawn: "No, she means a detective. Wow, how our minds wander when a body part is mentioned." *Chuckles and blushes.*

Nora: *Blushing.* "Sorry."

Gail: "It's alright, I should've worded it differently." *Still laughing.*

Nora: "How long have you been a dick for?" *Chuckles.*

Shawn: *Laughs.* "Over fifteen years. While you and my baby sister were out fighting God knows what, I was in the precinct catching human criminals."

Nora: "Oh, so he does know of our jobs."

Gail: "Of course. He was the one helping out Wolf and Hunt with intel on certain people."

Nora: "So how much do you know, Shawn?"

Shawn: "Enough to know that you two are no one to fuck with. I always told Wolf that he was a lucky man to have such a wonderful woman by his side." *Nora looks down toward her feet, blushes, her eyes watery.* "I apologize, I didn't mean to make you feel,"

Nora (interrupting): "It's fine. I kind of forgot how many people he interacted with."

Gail: *Changing the subject rapidly.* "What are your plans for today?"

Shawn: "Originally, we were going to walk around, catch a horse ride to the ferry, then go to the Statue of Liberty."

Gail: "That's it? Really?"

Nora: "I think it's pretty romantic."

Shawn: "As for your information, Miss Nosey, I have other things in store for today, so…" *Sticks out his tongue toward Gail.*

Gail: “Very mature.”

Nora: *Giggling.* “Are you two always like this?”

Gail and Shawn: “No, only when we’re around each other.” *Everyone laughs.*

Shawn: “Hon, would you mind if my sister comes along with us?”

Nora: “I don’t mind. We’ll make this a family event.”

Shawn: “Awesome. Now to make this day memorable for the two beautiful women in my life.”

Gail: “You are such a kiss-up.”

Nora: “Let’s try to enjoy today, please, for me.”

Gail: “We will. I like bothering my baby brother just to remind him who’s the older sibling. I still need a little time to wrap my head around this whole thing.”

Shawn: “Take all the time you need, sis. Please try to enjoy yourself in the process.”

Gail: “I will.”

Nora: “So, what are we going to do first?”

Gail: “Shawn, let’s find the horses before it gets hotter and we end up behind a gassy, smelly horse and his driver.”

Shawn: “Wow, I picked a gassy horse once.” *Laughs.* “I only picked that one because you liked it.”

Nora: “That’s so sweet. Smelly but nice nonetheless.” *Smiles.*

Reeves: “Did you hear that?”

Michael: "Yes, I did. From the tone in Uriel's voice, he's in distress, and we must find him immediately before it's too late."

Reeves: "Once we find him, we'll be able to find Raphael. Hopefully, he's close to him."

Michael: "We can only pray that it's that easy."

Reeves: "Even if it's not, nothing will stop us from getting to our brothers. I dare them to try to get in our way."

Michael: "Be careful what you wish for. Sometimes our pride can be our downfall, brother. Let's just take it as it comes. Whatever happens, we'll assess the situation and handle it accordingly."

Reeves: "As usual, you're right."

They try to focus their energy to sense where Uriel's signal is coming from. The angels walk forward, the essence of their sibling growing stronger with every step. It leads them to a stone wall. Dark smoke escapes through the cracks, and the smell of sulfur, blood, and burning flesh fills the air.

At the secret military base in Nevada, Louie meets with Vin.

Vin: "I have a message from Reeves. The other three will be joining you in twenty-four hours. Upon arrival, you four will be taken to a disclosed location. Not even the government knows where this place is. Gabriel will greet you. Please treat him with the same kind of respect you would give Michael or Raphael."

Louie: "Nice to see you again. That's great news." *Excited*

Vin: "Also, until that time, I was told to ask if you would kindly assist in the rescue mission of the missing warriors. And remember to be on your best behavior, lives are on the line. Heaven will be watching every move done by all participants very closely."

Louie: "I'll try. Are you serious? Let's go." *Eager to get his mentor, his head military official, and the two damsels in distress.*

In a flash, the two disappear.

Back in New York, Nora, Gail, and Shawn are enjoying their time together.

Nora: "I can't believe you two are brother and sister. Gail, why didn't you say anything about him?"

Gail: "I wasn't sure if I would see him again. We were so young when our father left with him."

Shawn: "I still remember it like it was yesterday." *Looks down.* "I couldn't stop crying. Dad got tired of hearing me cry. Then he said that things in life aren't always perfect, crying isn't going to change anything, and I needed to man up."

Nora: "Did you and your dad get along after that?"

Nora sits closer to Shawn, blushing a little. Gail giggles, it's been a while since she's seen Nora act so shy.

Shawn: "Didn't really have much time with him. He put me into boot camp most of my life. He thought it would make me a respectable man. After that, the next time I saw him was when I turned twenty."

Gail: "Is he still...?"

Shawn: "He's alive. We just don't talk."

Shawn glances at Nora, smiling, then puts his arm around her. He holds her close as her cheeks turn bright red while she smiles.

Gail: "Do you think he remembers me?"

Shawn: "Dad is an ass at times, but you never left his mind or heart. After all, 'you're still the one I was supposed to try to be like,' were his exact words."

Nora: "Why hasn't he called or at least written her a letter?"

Shawn: "Our father has never been known for showing his emotions."

Gail: "Do you think he would like to see me?"

Shawn: "I could give you his number so you can talk to him."

Gail: "Really?! That would be great. So what brings you down this way?"

Shawn: "I have an informant who had a lead, but he got cold feet. Now I'm looking for him. Then I met Nora, and I finally got to reunite with my baby sister. I'm glad he got cold feet."

Nora: *Blushing* "I hope I'm not interrupting your investigations."

Shawn: "There's no other place I'd rather be." *Looks Nora in her eyes and smiles.*

Nora blushes and leans on Shawn's shoulder, smiling. Gail grins and gets ready to take a picture.

Gail: "Say cheese, you two."

Shawn pulls Gail into the picture and snaps it.

Shawn: "That's a good one. May I get you ladies another round?"

Nora and Gail: "Yes, please."

Shawn goes up to the bar and orders their drinks. A man sitting on the stool next to him catches his eye. He grabs the man, drags him outside without the girls realizing, and throws him into the alley next to the building.

Shawn: "Oh Reggie, why do you always show up when I stop looking for you?"

Reggie: "Easy, I wasn't hiding from you or anything. Look, I'll make it up to you. I got some good info for you."

Shawn: "It better be good, Reggie. You know what happens if it's not." *Electricity surrounds his body.*

Reggie: "Come on, Trik. You know me, when have I led you wrong?"

Shawn's eyes light up as bright as the energy around him.

Shawn: "Alright, forget I asked."

Shawn: "Don't waste my time. It's your assistance that's keeping you protected. Remember, you're alive because of me."

Reggie: "I know, I know. I haven't forgotten everything you've done for me. I just need to lay low for a bit. I'll call you in a couple of days to meet up."

Shawn: "Okay, you have two days. Don't make me look for you."

Reggie: "Just expect my call."

They part ways. As Reggie walks away, Shawn looks around, then returns to the bar to get his order. With drinks in hand, he heads back to the girls.

Gail: "What took so long?"

Shawn: "I had to use the restroom."

Nora: "Oh, I told her you probably found your informant and cornered him." *Laughing.*

Shawn: "How'd you... Yeah, that would've been funny." *Thinking.* You have no idea how close you are to what really happened.

Gail: "If that would've occurred, you'd let us know, right?"

Nora: "I know if something happens, he'll notify us, just in case."

Shawn grins, knowing perfectly well that he wouldn't.

Shawn (thinking): *I will not put them in harm's way, no matter how much they think they can help me.*

At Archangel Michael and Archangel Reeves's location, they hit the wall, but it withstands the attack. Suddenly, a bright light appears. When it dims, two figures can be seen.

Louie: "Where do you need me, as support?" *Gets interrupted.*

Michael: "I didn't call for you." *Upset and worried that if something happens to Louie, Raphael would be livid.*

Reeves: "Sorry, brother, I asked for him. Please hear me out. Most of his senses are heightened in the heat of battle. Next, with all those that are missing, he knows what's at stake. As we know, the

Hearts are about family first, but with our backing, this would be an element no one would expect."

Michael: "Fine. Remember, Blak Hart, you follow our word to the letter. Is that understood?"

Louie: *Out of habit.* "Yes, sir!"

Michael: "We must find a weak spot. At this rate, our siblings will die before we make a dent."

Reeves: "Let me search around. I estimate we have five minutes before the guards get alerted because of our attempts."

Reeves searches high and low, but no weak spots are found. He looks further down the hallway and sees ghouls going in and out of a wall that appears to be warping. Reeves turns his head for a split second when a barrage of muscular, gray, flesh-falling-off-the-body ghouls charge at him full speed. He looks at Louie and Michael.

Blak Hart: *Sighs.* "We don't have time for this."

Michael: "Easy, brother. We can't lose our heads. Blak Hart, do you think you can hold them back while we continue to search?"

Louie: "Consider it done."

One of the ghouls jumps and tries to bite Louie, but he punches it, sending it spiraling back into the others. Michael rushes in, draws his sword, and decapitates several of them as Reeves charges in and pins some of the beasts to the wall, triggering them to burst into flames. They keep charging at the angels. The brothers stand back-to-back and unleash a flurry of blades. Louie engulfs his fists in fire, then releases a blast as Michael channels energy into his blades, igniting Reeves' sword and causing a huge explosion, turning the ghouls to ash. They all look at each other.

Reeves: "As quick as it escalated was as quickly as it finished. Very disappointing."

Michael: "That was too easy. Let's keep moving. Seems like they're trying to divert us from the actual area. Keep an eye out for anything that appears off or out of the ordinary."

Back at the base, Gabriel arrives and goes to the hangar, waiting for the arrival of Louie, Cross, Darren, and Rayne.

Cross: "Hey Gabriel, how's the search going?"

Gabriel: "As far as I know, my brothers are following their latest lead. The minute they find something, they'll notify me."

Cross: "Around what time does the rest of the group get here?"

Gabriel: "In a few minutes. Why?"

Cross: "I'm eager to see them again."

For the rest of the time, the two just stand there waiting in awkward silence, neither trying to say something that would offend or anger the other.

Moments later, a helicopter arrives with the group.

Gabriel: "Let's help them with their luggage."

They walk over and help with their belongings. Darren jumps out and startles Gabriel, making him squeal and causing everyone to laugh hysterically.

Darren: "Did I scare you?" *Speaking sarcastically, tears in his eyes.*

Cross: *Chuckling.* "Gabriel, I'm sorry about him." *Pointing at Darren.*

Gabriel: “It’s all right. It could always be worse.”

Cross: “Very true.”

Cross goes over and hugs the group. Gabriel looks at them in amazement.

Cross: “Let’s take you guys to our base so you can settle in. Tomorrow we train, but tonight we’ll party like never before. After Louie gets here.”

Gabriel drives them to a secluded location, an area that can’t be found on any map.

Hours later, once everyone finds their rooms, they begin talking about their individual journeys up to this point.

Rayne: “So, what’s our next move, and where’s Louie? He’s never late. Something’s off.”

Cross: “We have several missions lined up for the next three months. Let’s try to enjoy tonight, because I’m not sure when we’ll have another day off.”

Darren: “Damn, how many missions do we have?”

Cross: “Too many to count, and that’s just for three months. Who knows how many after we’re done with those.”

Darren: “We got this, right?”

Cross: “We were given this job because they know we can handle whatever comes our way. Remember what Louie said: we are destined for incredible things. Why else would we need archangels for, duh?”

Rayne: “I guess so. Still doesn’t make sense why Louie hasn’t arrived yet.”

Darren: *Ignoring what Rayne said.* “Fine, pass me a beer. Let’s get this party started before we get called to do something. If we’re going to die tomorrow, I want to do it with a hangover and a smile.”

CHAPTER 14.
REUNION

The entire night was spent picking up from the time before any of them went into the military.

The very next morning, Gabriel meets with the group.

Gabriel: "Louie won't be joining you at the moment."

Cross: "Why not?"

Gabriel: "He's been sent on a classified mission. He'll join you once it's completed. Until then, here's the assignment. The locals in Colombia are complaining about their crops being destroyed by evil beings called Chupacabra. We're not sure if it actually is or if it's something pretending to be one. Your mission is to find out what it is, then handle the situation with no casualties. Please keep in mind that the villagers are ready to shoot first and ask questions later. Do the job, then head back."

Cross: "The villagers sound more like they are," *gets interrupted by Gabriel.*

Gabriel: "It's not your job to find out what they're doing. Just take care of the matter at hand."

Cross: "Consider it done."

Darren: "What if we come across drug traffickers?"

Rayne: "Like Gabriel said, don't approach or interfere unless we have to defend ourselves."

Cross: "When do we leave?"

Gabriel: "In one hour. Don't worry about packing anything. You're only there for the job; there's no time for relaxation."

Back in New York, Nora, Gail, and Shawn are having dinner at Shawn's favorite restaurant in Manhattan.

Nora: "What exactly have you been doing all these years?"

Shawn: "Where do I start?"

Nora: "From the last time I saw you and Gail together?"

Shawn: "Well, as we all remember, my father took me to live with him. The first month was pretty good. He was trying to buy me with toys and other items." *He flashes back to the moments as he narrates.*

After a while, he started getting paranoid. Since he was the chief of police in Rendville, Ohio, the smallest village in the state, he arrested the most criminals in state history. Other departments from neighboring states called him to capture some of the most violent felons. Due to these reasons, he feared for my life. He began training me on self-defense, offense, and how to handle a gun safely. Once I reached my junior year in high school, he began to push me to head into his profession. After many arguments over this, by the end of my senior year, he had me take the police exam, and then a week after passing it, I was off to boot camp to become a police officer. I survived the eight weeks of training. My father pulled some strings for me to start in his precinct. I started as a beat cop. Four months later, I was assigned to assist in homicide and special victims units. Two years later, I was promoted to sergeant. The following year, I decided that I wanted to become a detective for a more diverse line of work. Once promoted, I was given several assignments that had me working with my dad.

I was able to see a side of him that made me respect him, and I also saw the side that made me lose all respect for him. As an officer, he was someone to fear if he was coming after you. But he was also one of the most corrupt in the force, the main reason why he was so paranoid. Two years after working with him, I despised the job and everything tied to him. I tried everything possible to avoid working with him. Eventually, he was promoted to commissioner of police, which meant I wouldn't have to deal with him anymore.

While on the job, I began to learn about several abilities I possessed. My first one happened when I was on an assignment interrogating a suspect. I did the usual good cop/bad cop routine when I began to notice the individual was lying. He passed the lie detector, but I knew he was lying. Every time I questioned him, I could see what he was doing at the time in question. Once I proved to him that I wasn't falling for his act, he finally confessed.

The second one happened several months later. I was chasing an assailant fleeing on a motorcycle. I ran fast enough to catch him even though I was on foot. I was stunned. My parents never told me about any family members with abilities, so I felt like a superhero. I began using my newfound gifts to help me in the field.

Shawn relives the moment.

About four years ago, the day that would change my perception of my father forever, my father and I were on a case. Our job was to find and stop a drug ring connected to the mob. We arrived at what should've been an abandoned warehouse. The area was well guarded, and my father took out some of them while I snuck in through a vent on the second floor. On that floor were several crates filled to the lid, each with different narcotics, varying in type and potency. I took pictures and noted everything I saw, then went to the first floor to join my father. More pallets held crates containing weapons, enough for several armies, and one skid was shrink-wrapped, concealing briefcases I assumed were filled with money. I slowly walked behind the crates. I saw my father being brought in by the guards. They tied him to a chair and began to beat him. While I tried to get a closer look, I heard a gun cocking back.

I turned around and saw the barrel of a rifle pointed directly at me. He told me to drop my gun. I complied. Two more came up behind me with guns drawn. I put my hands up, and they grabbed me, dragging me to a beam across from my father. While giving him the beating of his life, a couple of them started beating me senseless.

One of the goons: "Well, well, look who we have here. Commissioner Valentine, you're a little short this month. Where's the rest of the money?"

Commissioner Valentine (Shawn's father): "I'll get it to you. You know I'm good for it."

Shawn: *Out of breath and spitting blood* "Really? You're no better than they are." *Coughing.* "How could you?"

Goon leader: *Looks at Shawn.* "Silence. I see we have someone else here. He looks a lot like you. Is this your son?"

Commissioner Valentine: "Let him go. I'll give you the money. Release me, and I'll give you what I owe."

They released one of his arms. Commissioner Valentine reached into his chest pocket and handed him a wad of money.

Goon leader: "You see, if you would've done this earlier this week, we wouldn't have this issue. I could try to look the other way, but you killed some of my guards."

Commissioner Valentine: "If you're not going to let me go, at least let him go."

Goon leader: "No. You need to be taught a lesson. The sins of the father will be paid through the son."

Sounds of screams and cracking bones echoed through the warehouse. They punched, kicked, and stomped on him while he was unconscious. They stopped beating Shawn only when he was bloody and battered, leaving him barely alive.

Goon leader: "Remember this for next time. If it happens again, you'll be coming out of the emergency room to plan a funeral. Do you understand me?"

Commissioner Valentine: "Yeah, yeah. Untie me so I can tend to him."

They released the commissioner, who rushed to Shawn. He picked him up and carried him out far enough to get a signal on his phone to call for an ambulance.

Commissioner Valentine: "Hold on, Shawn. The ambulance is on its way! I'm so sorry that you got wrapped up in this mess."

The paramedics arrived and quickly attended to Shawn. They checked his breathing and his heart.

Paramedic: "It seems something is obstructing his windpipe, but we have to get his heart beating again."

They prepared the paddles and shocked him several times. Finally, his heart started up.

Weeks passed. Shawn's recovery was slow but steady.

Shawn: *thinking* "This is fucking unbelievable. I've been here several weeks. I almost died because of my father, yet he hasn't come by to see me or apologize. What an asshole. Once I get out of here, I'm going to resign and form my own detective agency, cut out the bullshit and handle things my way."

Shawn begins narrating again.

Two weeks later, I found an office space to start my detective agency. My clients were people from the districts I'd helped in the tri-state area. Eventually, my clientele expanded. Before long, I was getting calls from all over the country.

In one of my cases, I was called to assist a gentleman named Jeffrey Finnian. He told me my expertise was needed to help him catch a warlock who stole from him. At first, I was skeptical. As time went on, he helped me unlock other abilities that let me see, hear, sense, and feel beyond what people normally do. He called it the paranormal veil. Unlocking those abilities lifted that veil, allowing me to experience the supernatural realm.

We worked together on that case for three months, and in that time, I learned how to control those abilities. Later that year, I hired him, so I created a night job. During the day, we arrested criminals. At night, we hunted the things that go bump in the night. It sounds impossible, but it's real. The night job is what brought me back here, and I'm glad it did.

Nora: "Jeff is a kind soul. I met him when Will was on a mission with him."

Shawn: "It's a small world, isn't it?"

Gail: "Yes, yes, it is."

They continued their meal and talked about meeting up again. They all agreed. Shawn kissed Nora and then hugged Gail.

Shawn: "I'll see you soon, sis. Nora, my dear, I'll call you tonight." *He blows her a kiss and walks away.*

Somewhere in the underworld, Zima is torturing his prisoner when he realizes something.

Zima: "Where the hell is Hecate?"

One of the demons: "She was seen roaming around the other cells, master."

Zima: "What is she doing over there?"

One of the demons: "I don't know. She's been disappearing a lot and going to that side. Should I get her?"

Zima: "No. I want to see where and what she's doing when she least expects it."

Hours later, Hecate returns to her cell and sees Zima waiting there for her.

Zima: "Well, well, well. Look what the hellhounds dragged in."

Hecate: "What do you want, Zima?"

Zima: "So feisty. Just wondering where you were. Is that so bad?"

Hecate: "Making sure your pregnant guest is fed, my way of befriending her so your plan can go without a hitch."

Zima: "Is that all? Seems like you're getting too friendly with her. Almost like you're hoping that by treating her well, it'll put you in Blak Hart's good graces. Anyhow, by the time he sees you, I'll make sure he hates your guts for betraying him." *Laughs.*

In Colombia, the three warriors have been scouring the fields for several hours in search of the infamous Chupacabra.

Cross: "Be careful where you step. There are traps placed in random locations."

Rayne: "What should we look for?"

Cross: "Bear traps were set for the beast that's been destroying the crops, and outside the fields there are land mines for trespassers. If any of the mines go off, the villagers will come out with guns blazing."

Darren: "So aside from looking for a dangerous, blood-sucking, crop-destroying monster, now we have to deal with bombs, traps, and gun-crazy people. Reminds me of almost every mission I was sent on." *Laughs.*

Rayne: "I'll keep a lookout for any suspicious characters or any unhumanly things."

Cross: "Sounds good. Remember, only fight if we are attacked."

Rayne: "We know."

Cross: "I'm just saying this aloud specifically for me." *Smiles at the group.*

They split up to cover more land. Rayne climbs to a treetop for a better view, Cross slowly walks using his senses to find anything out of the ordinary, and Darren uses his SEAL training to track.

Darren darts toward the end of the field when he's stopped by a bright light. Once it dims, Louie can be seen standing there. He looks down and tells him to turn around.

Darren: "What the hell? I think I found a trail."

Louie: "If you found something, you're supposed to notify them so we can search from this point on. You almost stepped on a land mine. Call the others."

Darren: *He calls Rayne over, then continues talking to Cross.* "How were you able to see the mine if it's almost pitch-black here?"

Louie: "I just did, what the fuck?" *A little startled.*

Rayne: *Just joining them.* "What's going on? And where did you come from?"

Darren: "He stopped me from going forward because he said he saw a land mine. How, if it's dark as fuck out here?"

Blak Hart: "I saw a green mine right in front of him. In fact, I see all of you in a green silhouette. I was asked by Reeves to join them on a recon mission, but Michael sent my ass back without a reason or heads-up. Vin notified me before you guys arrived to meet Gabriel about what this mission was going to be."

Cross: "After all the hunts and all the gifts you have, now you're finally getting the ability to see at night. Welcome to my world." *Laughing.*

Rayne: "Stop being a dick, Cross. Blak Hart, once you learn how to focus your vision, you'll be able to see the same whether it's day or night."

Blak Hart: "Rayne, thank you for the advice. Have you seen anything yet?"

Rayne: "Nothing so far."

Blak Hart: "Cross, has your nose picked anything up?"

Cross: "I've picked up two non-human odors, one going in one direction and the other in the area of a few farms."

Blak Hart: "Are any of the directions in the path where Darren was heading?"

Cross: "Yeah, the one that was destroying the crops went in that direction."

Blak Hart: "Darren, Rayne, follow the path you started, only this time use the flashlight on low beam so you can see where you're going and it won't alert the enemy. Cross and I are going after the other scent."

As Darren and Rayne track the scent into the fields near the farms, Cross and Blak Hart follow the other trail leading to an underground tunnel. The tunnel appears to have been dug by hand, with leaves and vines covering the entrance, blending perfectly with the surroundings.

Cross: "So what really happened?"

Blak Hart: "I didn't lie. Reeves did ask if I wanted to help them with one of the leads they were following. Then Michael turned around, saying he didn't call for me, that if something happened to me, Raph would give him hell for it. Then he transported me into this without properly preparing."

Cross: "Well, he's not wrong. If anything were to happen to you on Michael or an angel's watch, I'd hate to be on the receiving end of Raphael's anger. Also, Vin figured this might happen, so I brought your things in my bag." *Cross digs in his tactical bag and retrieves Louie's things. Louie thanks him and gears up.*

Meanwhile, the other two warriors find a trail of carcasses leading through the village. As they make their way to the center of town, they're shot at by villagers. Darren pulls Rayne to safety and checks their perimeter.

Rayne: "How bad is our situation?"

Darren: "We're outnumbered, surrounded, and everyone is heavily armed."

Rayne: "I'm going to find higher ground so I can take out some of the shooters and open a path for us to move forward."

Darren: "Sounds good. I'll try to distract the others so they won't focus on you."

Darren throws several smoke bombs to cover Rayne's movement, then begins throwing grenades in random areas to confuse the enemy. Every time a bomb goes off, Rayne takes out several people. When the smoke clears, Darren spots a clear path out of the town.

Darren: "Rayne, let's go before we lose our window of opportunity!"

Rayne hops off the rooftop and runs. Darren shoots anyone approaching their direction, giving her enough time to escape. He throws a few more smoke bombs; in the blink of an eye, the two of them are gone.

Once out of danger, they contact Cross and Louie, notifying them of what transpired along with a warning.

Darren: "We were ambushed. It looked like they were waiting for us to follow that thing so they could kill us."

Rayne: "Not only that, they were shooting at us with military-grade weapons. They're not your typical villagers. Stay alert. They're very stealthy, find another way that doesn't involve the town."

Blak Hart: "Affirmative. Thanks for the update and good work. If you two get cornered or find that beast, just call. We'll get there as soon as possible."

Cross: "Did they say anything while they were attacking?"

Rayne: "I overheard one of them saying to lead us away from the animal. Sounds unusual that they'd try to protect it when it's the reason they called for help."

Cross: "We'll try to find out more about this new information. Keep on the trail. Anything we learn, you'll get an update."

At the tunnel, Cross and Blak Hart move deeper inside. It appears that whatever made this was very intelligent and skillful. To pass the time, Cross decides to ask Louie something that's been on his mind.

Cross: "Can I ask you something?"

Louie: "Sure."

Cross: "How exactly were you able to hide your abilities for so long and not go crazy or get caught? Not to sound rude or nosy, but you were on your own for quite a while."

Louie: "At first, I wasn't doing so well hiding them. Anytime I was frightened or couldn't cope with a situation, I used my power. People began calling me a freak of nature, and luckily for me, it was Raphael who found me every time I was about to lose my mind. He might have come to my aid under different identities, but he was the reason I learned how to conceal my gifts and deal with problems, no matter how big or small." *He smiles a little, then looks down, depressed, knowing he still has no word on Raphael's status.*

Cross: *Feels what Louie is going through and puts a hand on his shoulder.* "I didn't mean to depress you. It's just… I had a hard time trying to hide my gifts and not lose my shit. My father was the only one who could help me with it. My mom tried, but it was never the same."

Louie: "Let's just leave it at this: we both had someone who was there. They weren't the one we wanted, but at least they were there when it counted most."

They both nod in agreement as they continue their search. A few steps further, they find the remains of plants and vegetation of various kinds.

Cross: "We must be getting close to whatever this is, but I thought they said it was eating humans."

Blak Hart: "I'm just theorizing, but hear me out. From what Rayne reported, she said the people were trying to protect this thing, and they were armed like military. Who's to say that what we're tracking and what Rayne and Darren are tracking are even related? Not to mention, this sounds like another setup to take us out of future battles. Now I see why I was sent back."

Cross: "That makes sense. We have to find this creature to make sure your hunch is right. As for the setup, yeah, after everything that's happened, I wouldn't be surprised."

They start running through the tunnel. After three minutes, the two warriors are tackled by an unseen force. Cross darts to one side of the cave as Blak Hart sprints to the opposite side.

Blak Hart: "Try to sense it while I look for some kind of echo."

Cross: "I'll let you know what I find. Once we get a lock on it, we'll attack it from each side."

Blak Hart looks around and sees the silhouette of a strange beast. It seems to run toward them, then vanish. He looks up and spots the creature clinging to the tunnel ceiling.

Blak Hart: "Cross, it's above us! Shoot up!"

Cross: "Watch out, it's heading your way!"

It was too fast, and it began attacking immediately. Cross shot at its legs to slow it down while Blak Hart punched its midsection. Cross pinned the beast down and tranquilized it. They pulled the creature out of the tunnel. It looked exactly like the typical description of a Chupacabra. It woke up and began to plead with them.

Chupacabra: "Please don't hurt me."

Cross: "Where did you learn to speak English?"

Blak Hart: "This thing has been killing people, and that's your main concern? Really?"

Chupacabra: "I didn't kill anyone. I only eat vegetables. You're confusing me with the experiment."

Blak Hart: "What experiment?"

The Chupacabra started remembering.

Chupacabra: "The government from where you're from captured me almost a year ago. I learned to speak from the scientists. They didn't try to listen to me. They took blood samples and tried cloning me. The first dozen experiments didn't survive, and the other countries were about to pull the plug on the whole thing. The scientists tried one more time, but it was successful. It was as agile as I am, faster because of their chemicals. The countries wanted to use this for war purposes. They gave it a mission, and when it got hungry, it was permitted to hunt the one thing it needed to survive, meat.

"There was talk of them making more of those things. Whether they did or didn't, I wasn't there long enough to know. When they were going to use me to feed it, I escaped and came here. That was over four months ago. Then, a few weeks ago, it found me. The next town over is worshiping that thing like a god, which it is not. Anyone who opposes it gets fed to it."

Cross: "How do we know you're telling the truth?"

Blak Hart: "Well, it does match up with the report Rayne and Darren sent us. Cross, tell Darren and Rayne to capture it and bring it to our location. If they need help, they can call us. Also, Cross, didn't you notice something? When it was talking, its heartbeat was steady, compared to when it was pleading for its life."

Chupacabra: “If you guys allow me, I’ll help you, with one condition. I’ll destroy it, and you place me somewhere I can’t be found, with enough plant life for me to survive.”

Blak Hart: “Let’s say we believe you. What guarantee do we have that you won’t hurt any livestock or humans?”

Chupacabra: “I only need vegetation. I don’t eat meat.”

Cross called Darren and Rayne to update them on the current situation. They told Cross they had been keeping eyes on the creature for the past hour.

Darren: “From the looks of it, its speed is faster than anything I’ve ever seen. The claws are sharper than any of our swords, and its strength is something to worry about.”

Cross: “Give us your location. We’re going over there with a little present for it. Cross out.”

The Chupacabra led Cross and Blak Hart to the area where the other creature was, using shortcuts away from any homes.

Blak Hart: "Is this how you've stayed under the radar? Pretty clever."

Chupacabra: "This was the only way I could eat the carrots and vegetation without getting caught. I was destroying the other plants they were growing. I was getting sick when I ate them."

Cross: "How much time will this route take compared to the direct approach?"

Chupacabra: "The direct route would take two hours. My way takes an hour longer, but we'll stay undetected. Even with all its abilities, detection is one of its weak areas."

They continued through the rainforest, passing poisonous snakes and insects. In the midst of the journey, Blak Hart and Cross found themselves admiring the scenery, almost forgetting why they were there.

Blak Hart: Talking to Cross. "Who would've figured we'd be getting a tour from the infamous Chupacabra?" Chuckles and looks at the beast. "By the way, what's your name?"

Cross: Smirks. "I know, right?"

Chupacabra: "Why do you want to know my name? You're just going to kill me after you get what you want, like the typical military."

Blak Hart: "The reason I want to know your name is because you're intelligent, and instead of calling you Chupacabra, I can properly greet you by your name."

Cross: "How many soldiers do you know who'd bother learning the name of the one they're about to kill? Normally, that's too much effort."

Chupacabra: "You have a very good point. My name is Paca. That's the name one of the scientists gave me. Once the others found out, they fed him to the experiment." He lowers his head.

Blak Hart: "I apologize if I brought back bad memories. That wasn't my intention."

In the village, Rayne and Darren were watching everything unfold when army trucks arrived. One of the villagers approached one of the vehicles.

Darren: "I'm going to get a closer look, hopefully I can hear what they're talking about."

Rayne: “Instead of going there, why don’t you use one of my brother’s drone creations that looks like a bug? It’ll record everything.”

Darren: “Damn, I forgot about that. Thanks for reminding me.”

Rayne took out the bug drone and handed it to Darren, who set up the remote. He sent the drone to perch on top of the truck. The trucks left the area, and Darren called the drone back.

Upon receiving it, they reviewed the footage and quickly sent it to Cross. The footage showed villagers talking to soldiers, each truck carrying troops from various countries. They were discussing releasing other test subjects in different parts of the world. One of the men, referred to as Mr. Deville, appeared to be one of the people behind the operation. He was pleased with how the experiment was going but complained about the poor effort to retrieve “Experiment C.”

This experiment had been the first successful one. They were preparing to start on Experiments A and B, hoping for the same results.

At the angelic brothers’ location, Michael and Reeves were quickly approaching an iron door. It looked rusted but sturdy enough to keep intruders out.

Archangel Michael: “We must channel our energy into our attacks to break this door down.”

Archangel Reeves: “Ready when you are. You do know Louie would’ve been perfect for this mission. You didn’t have to send him back without warning.”

Archangel Michael: "I know he would've been a great asset, but the others needed him more than we do. Brother, you forget that we have backup if we need it, the kids don't. Louie is the equivalent of several backups. I'll apologize next time I see him."

They charged their energy and combined it into their weapons. Together they attacked the door. It exploded without much effort. When the dust cleared, creatures with bone and muscle exposed rushed at them, screeching as they attacked. The brothers were outnumbered, but while they fought, an unseen force seemed to assist them. When the battle ended, the mysterious helper was gone.

Archangel Reeves: "Whose energy was that? I wanted to thank them, or at least find out who it was."

Archangel Michael: "The energy felt familiar. I was too busy fighting to pinpoint who it belonged to."

Archangel Reeves: "Could it be one of our brothers?"

Archangel Michael: "No. They wouldn't hide if they helped us. This mysterious helper is… interesting."

Back at Cross's household, Nora and Gail were drinking coffee when they received a call.

Nora: "Hello?"

Jeff: "Hi, Nora. It's Jeff. Just wanted to see if you and Gail could come over, we found something you might want to see."

Nora: "Sure, give us an hour."

Jeff: "See you then."

An hour later, there was a knock on Jeff's door.

Jeff: "Hello, who is it?"

Nora and Gail: "It's us, silly."

Jeff: "Sorry, can't be too sure these days."

Nora: "What did you find?"

Jeff: "Kate was looking around in the attic and found several old letters sent by Hunt and Wolf. We received them a couple of days after their deaths. Some of them have interesting information about your godson."

Nora: "What is it? Does it have to do with any prophecy?"

Kate walked into the room, hugged Nora and Gail, then filled them in on her findings.

Kate: "Like Jeff said, these letters were put aside when they both passed on. We must have forgotten about them when we moved. Anyway, Hunt's letter asked us to put a protection spell on you two, the kids, and Aggy. He wasn't specific why. Usually, when he asks us, it's because he doesn't want to bother you or alarm you."

Gail: "What do the other letters say?"

Kate: "Wolf's letters were about his research on the Heart family bloodline. He wrote that Hunt was barely around his father, his father disappeared when he was twelve. Wolf's father searched for decades but had no luck. He thought the disappearance meant his death.

"The last two letters mentioned Wolf continuing his father's and Hunt's work in finding Sparta. He was close to figuring it out. Then something incredible happened. On one of their missions, they found a bald man who looked a lot like Hunt. The man attacked them in the same way Hunt would, but before they could retaliate, he vanished without a trace."

Nora: "Could it be possible they found Louie's grandfather?"

Jeff: "That's not even the most important part. Sparta's father vanished when he was twelve, the same as his grandfather disappearing on his father's twelfth birthday."

Gail: "That's a weird pattern. Could it be some kind of ritual? If it is, I've never heard of it."

Kate: "In medieval times, if a curse was placed on a family, the man of the house would train his son to take his place in protecting the family. At twelve, the age a boy was considered a man, the father would go to the witch or warlock who cast the curse. By giving his life, the curse would be temporarily lifted for that generation, or until the first son had a son of his own."

Nora: "In this day and age… if that's the case, then if Hunt didn't agree with the ritual, that's probably what killed him."

Jeff: "That could be possible. But these are all theories for now. Now do you see why it's important?"

Nora: "How can we help Louie avoid this curse?"

Kate: "Honestly, if it is a curse, I believe there's almost nothing we can do without finding the originator. I'm not saying it's impossible,"

She was interrupted by Jeff.

Jeff: "We'd have to look for spells and curses made and bound before the medieval era. Even then, we might have trouble finding the right one."

Gail: “To find who cursed the Heart bloodline, it must’ve been someone they trusted. But who?”

Nora: “We might have better luck if we knew where Hunt and Wolf found the bald man.”

Jeff: “Remember, there’s no guarantee it was Sparta. Wolf never wrote down where they saw him.”

Kate: “If you want, you search on your end, and we’ll search on ours. If either of us finds something, we’ll meet up again.”

Everyone agreed. Nora and Gail hugged Kate, then left. In the car, Nora spoke softly.

Nora: “I need to find the spell before it’s too late.” *Sobbing.* “I can’t lose him again.”

Gail: “Don’t worry, Nora. I’ll do everything I can to help you prevent the worst from happening.” *Hugs Nora.*

Back at Jeff’s house, the couple talked about what had just been said.

Kate: “Do you really think we can help Nora save Louie from the curse?”

Jeff: “I doubt it.”

Kate: “Then why did you give them hope?”

Jeff: “There’s always a way, but I don’t think we’ll find what’s needed in time.”

Kate: “The problem is, you made it sound like we could help them trace the origin.”

Jeff: "That's the easy part. The hard part is finding what spell was used, which materials kept it bound to the bloodline, and how it became so powerful."

Kate: "From hearing what happened to Hunt, I fear what could happen to Louie if this curse is as old as I think. Who knows what might happen since he's one of the chosen, with a curse as powerful as this…"

CHAPTER 15.

THE SEARCH FOR RAPHAEL AND URIEL

In Washington, D.C., President Garrison is talking to General Devin Deville.

The President: "What's the reason for this sudden meeting, Dev?"

Devin: "I wanted to know if you've found someone to fill the spot for the Vice-Presidency?" *smiles*

The President: "Not yet, I was holding off, hoping that you would change your mind. I guess you have."

Devin: "I also have important news which I figured you would like to know."

The President: "What's it about?"

Devin: "It's about what's going on in Colombia..."

Suddenly another general comes rushing in, interrupting the conversation.

General: "Sorry for the intrusion, Mr. President, Raul Ortega the ambassador of Colombia is on line one. He says it's urgent."

The President: "Transfer the call."

General: "Passing the call now."

The President: "Hello."

The Ambassador Ortega: "Mr. President, your soldiers are attacking without cause. If this persists, that would be enough to cause a war."

Devin: *whispering in the President's ear* "If you permit me, I could talk to them as your Vice President."

The President: *whispers back* "Good idea." *goes back on the phone* "Mr. Ambassador, my Vice President is handling the situation. If you like, I can have him talk to you."

Ambassador Ortega: "It doesn't matter." *already irritated*

Devin: "Hello, Ambassador Ortega, I understand the matter at hand is very serious. This was a treasonous act by a few members of the military, I will personally make sure that these individuals get court-martialed and imprisoned for their actions. I'll be sending my best soldiers to arrest these traitors."

Ambassador Ortega: "Thank you for your understanding and diligence, Mr. Vice President Devin." *smile comes across Devin's face*

At Archangel Michael and Reeves's location; after many hours of searching, several paths leading in various directions are found.

Archangel Michael: "I think I'll call one of Heaven's ultimate warriors to assist us."

Archangel Reeves: "It's not who I think it is, is it?"

Archangel Michael: "He's the only one who can help us with any situation and has the ability to adjust to any fighting style. He is one of the best at tracking or finding; lucky for us, he has returned from the mission that was given to him from the prince years ago."

Archangel Reeves: "Do you think he's rested enough to join us?"

Archangel Michael: "He's the one who contacted me a while ago, he said in case of anything he would be honored to fight by our side."

Archangel Reeves: "Correction, I am the one who is honored to be working side by side with my two of the three legendary brothers known for being the king's greatest warriors. Out to save the third of the king's warrior." *star struck and extremely excited but trying to hide his contentment* "I am at a loss for words."

Archangel Michael: "Very flattering but, I think the best way to honor us is to make sure we save our siblings and survive this whole ordeal."

The Archangel Michael contacts the Archangel only known as J.A.L. (Only three beings know his full name, one is the king, the prince and the other is the Archangel Michael.) He notifies him of the situation and their location, J.A.L. responds stating he'll be there as soon as possible. Then, Archangel Michael begins speaking in the ancient language that has only been spoken by the elder angels in heaven. He laughs then disconnects the call.

Archangel Reeves: "Well, what did he say?"

Archangel Michael: "He'll be here very soon."

Archangel Reeves: "Is that all? I saw you laughing, for you that's rare."

Archangel Michael: "It's something that he reminded me about, it's between the two of us. I'm sorry, I can not relay what it was about."

Reeves looks down feeling left out, but he understands that it was a private conversation. He wished he was just part of something that his older brothers were able to joke about and share.

Back in Colombia, Blak Hart and Cross arrive to their destination. Their guide, the famous Chupacabra is pointing out multiple ways to get into the compound with the least detection possible.

Cross: "What do you think our plan of action should be?"

Blak Hart: "Our guide has given us various options; the question is, which way can we take to meet up with Darren and Rayne, so we can find the other Chupacabra, sorry the experiment with the least resistance."

Cross begins to notify the others of their arrival while, Blak Hart discusses the most effective way with the new ally.

Blak Hart: "The only way to get in is on the western sector of the facility, it runs right beneath Darren and Rayne. Once they're with us, we can proceed in capturing the cause of this whole mess."

Paca leads them to the location where the rest of their team are waiting. Once they are reunited, the five continue down the tunnel.

Paca: "My foe is a few feet north of our location, I sense there are four guards protecting him."

Blak Hart: "Rayne, take out the guards closest to our location. Cross, cover her in case of anything. Darren, I will head toward our objective. I need you to cover me; as for you," *points at the Chupacabra* "you're going with me, so we can knock it out before it can make a sound."

Rayne follows the plan along with Cross, Darren does his part. When it's time for Blak Hart and his ally, the plan goes south when the Paca goes into a blind rage toward the other Chupacabra.

At the torture chambers, somewhere in purgatory, Hecate is lurking around the cells.

Hecate: "Uriel, am I correct?"

Uriel: "Why? Are you here to continue my torment, witch?"

Hecate: "Easy there, I am here to help you out."

Uriel: "Why?"

Hecate: "Don't ask questions, just accept the help. I only have a limited time before Zima gets back."

Uriel: "What's your angle? Why help me?"

Hecate: "Well... If you must know, being around Louie for so long has rubbed off on me in more ways than one. Along with being the mother of his unborn child, he has made me realize that it is more ratifying helping and saving than destroying. Please hurry, I've been making sure Luna is being treated well and your brother Raphael has been holding his own. He won't be able to do it for much longer, once I loosen the shackles, wait until I leave before you make your escape. Raphael is on the other wing, you'll know which room he's in. It's the only one that still has the light of hope emanating from it."

Uriel: "What do you want in return?"

Hecate: "To let Louie know that I am bearing his child and I would like for someone to rescue my baby from this before it's too late. I might not survive but our baby might have a chance."

Uriel: "Who should I say sent him this message?"

Hecate: "Catherine."

At the Cross residence, Nora and Gail have been researching all the books left by Hunt and Wolf.

Gail: "We've been at this for several days and still nothing, have you called Jeff or Kate?"

Nora: "I have, the only thing I was told is... this is magic from the beginning. From the time when magic was at its purest, Jeff found a book that has similar traits, but it still needs the one who cursed the individual. That person must have died centuries ago."

Gail: "Not necessarily, witches that powerful can find a way to slow down the aging process to the point of almost being immortal."

Nora: "Kate said that if we find that person, then there's a chance to save Louie."

Gail: "What we need to find out is, how much time do we have before..." *puts her head down.*

Nora: "I'll call her now."

Nora darts to her phone and speed dials Kate, sadly, no response. She looks at Gail.

Nora: "Please try to call Jeff, see if he'll pick up. Kate's goes directly to her voicemail."

Gail: "Ok, hopefully everything is fine over there." *worried*

Gail calls Jeff, same thing happens. She tries again, similar results.

Gail: "I don't like this, they always answer or the phone rings. It never goes directly to their voicemail."

Nora: "I hope Shawn is close to their area."

Gail: "Definitely, call him. Who knows, he might have a job over there."

Nora calls him, on the third ring he answers.

Shawn: "Hey sexy." *using his flirtiest tone.*

Nora: "Hi babe, are you working in Sayville today?" *blushing.*

Shawn: "Why? Are you in the area?"

Nora: "No, I wish..." *Gail takes the phone from Nora.*

Gail: "Hey bro, just answer the question... Please... We need your help with a possible situation. This is not the time to be flirty."

Shawn: "Sorry sis, I'm currently in the next town over, what's going on?"

Gail: "We've been trying to get in contact with Jeff and Kate, they're not responding. It goes directly to voicemail."

Shawn: "I'm leaving right now to go there, I'll call you two once I get there." *thinking, if they're not answering it has to be serious.*

Gail: "Thanks, be safe. I love you bro."

Shawn: "I will, I love you too sis."

Shawn runs to his car and speeds to Jeff's house. Ten minutes later, he arrives to the property. He rushes out of the vehicle, then darts toward the door. As he begins to call Nora, Shawn realizes that the door was broken.

Shawn: *whispering* "Nora, I'm at the location. I haven't gone in but at first glance I noticed the door was broken into. It's not looking good."

Nora: "Gail and I are on our way, do what you can to help them, please... and whatever you do, be careful."

Shawn: "I will, my angel."

Shawn slowly pushes the door open, he sees broken glass shattered all over the floor. The furniture is flipped as if it were thrown.

Shawn: *thinking* "There definitely was a struggle here. I hope they're fine."

He proceeds to creep through the house, the living room, kitchen, dining room and patio look as if a tornado pass by. As he continues to walk around he hears a faint yell and arguing coming from the basement. Right before he walks to the basement door, Shawn spots droplets of blood that start from the back yard leading past the cellar door. He goes to open the door when the smell of sulfur breezes through the cracks of the door. He grips the door knob and twists; another scream is heard. This time he knows who that scream came from.

Shawn: *thinking* "That sounds like Kate, she's in agony. The other voice I don't recognize but doesn't give me a vibe either. I still haven't heard Jeff's voice, I pray he's ok."

He opens the door instantly, then rushes down the stairs. Once he reaches the bottom, Shawn looks to the right and sees Kate on the ground; bloody, tortured, and unconscious. Standing over her is a shadow figure.

Back in Colombia, Blak Hart tells everyone else to start moving out and taking out the rest of the guards. Rayne goes east, Darren darts west, Cross runs north, while Blak Hart heads southward keeping the two beasts within sight. As he shoots the guards trying to help the Chupacabra that was on their property, Cross calls him.

Cross: “Blak, heads up; there's something going to your location extremely fast!”

Blak Hart: “What the hell is it?”

Cross: “I don't know, it's moving too fast for me to get a good look. All I can sense is that it is probably a B level entity.”

Blak Hart: “Thanks for the heads up, it's in my view now.”

Blak Hart charges toward the unknown being. When he finally arrives to it, the being was choking several guards with its massively long chain. Louie gets a good look at it; it's a spirit, a man almost seven feet tall. He has an old 1800's black hat, tattered black shirt and pants, brown boots, long black hair, no eyes, the skin that was exposed through the shirt was grey with black veins, and has chains wrapped around his body. He uses the chains for attacks.

Blak Hart continues toward him. The being notices him and throws the chain in his direction. The chain flies toward Louie like a flying snake with precise aim and direction. As it gets close to him, the warrior blocks the chain causing it to wrap around his arm. The spirit begins to control it and pulls, dragging Blak Hart closer. For an instant they both lock eyes and the entity begins speaking to him in an unknown language. Louie pulls back causing the spirit to lose balance. Suddenly, gunshots go in his direction, distracting him. When he turns back to look at the being, the chains that were around his arm are gone along with the specter.

Blak Hart: "What was he trying to tell me? I've heard demons talking in the dead Latin language, but this is one I've never heard of."

He turns and returns fire, then rushes back to aid his guide. Cross joins him along with the rest of the group.

Cross: "Did you see what that thing was?"

Blak Hart: "Yeah, it was a spirit trying to warn me about something. Of what exactly, I'm not sure. It was spoken in a language that I never heard of before."

Rayne: "What being?"

Darren: "I didn't sense anything. Probably wasn't that powerful."

Blak Hart and Cross (at the same time): "Wrong! That was a level B entity."

Blak Hart: "It was strong enough to control heavy chains, tie my arm and pull. If it wasn't for me pulling, it would have done what it intended."

Cross: "Who's to say it didn't? Maybe he just came to let you know something and leave."

Blak Hart: "Maybe."

The group's guide breaks the neck of the Chupacabra and brings it over to the four.

Paca: "I'll give you some samples of my blood and compare it to this poor excuse of my clone. This will prove everything we spoke about. I held my end of the deal, now what about you?"

Cross: "We always keep our word. When our extraction gets here, you will come with us. Then our first stop will be the location we spoke of."

Paca: "That thing you were fighting seems very familiar. I saw something similar in Puerto Rico. It would come out every night after midnight to terrorize the villagers. Anyone found around that time would be killed by his chains."

Blak Hart: "So, what's it doing over here?"

Paca: "It is known to terrorize several places, Puerto Rico, because that's where he died, the Dominican Republic, Mexico, and all of South America."

Darren: "That's impossible, to be everywhere every night."

Cross: "Could there be a possibility of several specters doing the same thing?"

Paca: "It's possible. The origin of the specter that drags the heavy chains that shackled him in life through the very town he lived in started in Puerto Rico. If there are more of them, each one will have a distinct difference that shows where they're from. The one that confronted Blak Hart seems to be the one from Puerto Rico. It might have attacked out of habit, but he made sure that after speaking to him, the specter disappeared."

At Jeff's house, the shadow figure turns around looking at Shawn. Shawn realizes that the figure looks like Jeff. The difference between Jeff and this person is the face has burn marks, he is a bit shorter, and has a bulkier frame. Just as the Jeff look-alike walks toward Shawn, the outside door slams open. Gail and Nora arrive calling Shawn's name.

Gail: *worried tone* "Shawn! Shawn! Where are you?"

Shawn: "Stay up there! I'm here with the perp."

Behind the girls, Jeff comes running in.

Jeff: *frantic* "Kate! Kate, where are you?!"

Nora dashes to the basement followed by Gail and Jeff. Loud laughter can be heard coming from the basement. Once they get downstairs, the look-alike speaks.

Jeff look-alike: "Nice to see you again, big bro. I'm keeping your wife company."

Jeff runs toward Kate, cradling her and holding her, as his brother laughs.

Jeff's brother: "Jeff, even as your wife bleeds out, you still can't harm me."

Nora: "Maybe he can't, but we can."

Gail and Nora begin chanting. Gail sends a few electrical blasts in his direction. The man laughs.

Jeff's brother: "You can't harm me, witch. None of you can. The only one that has a possible chance is too weak to do anything." *maniacal laughter* "And what possible thing can a chanting shapeshifter do," *snickers* "along with a supernatural detective? This is too funny."

Jeff looks up angrily and lets out a loud yell. He places Kate gently on the ground, then stands up.

Jeff's brother: *sarcastic tone* "Oh no, I'm in trouble now. Really?" *mocking* "What do you think you're gonna do, bro?"

Jeff's shirt rips off his body. He pulls the cross hanging from his chain. He grips the cross by the top; the bottom of the cross starts to stretch. His brother's eyes widen slightly.

Jeff's brother: “Is this the only way you think you're going to finish me off?”

Jeff: “No. You've had a hold on me for many years. This stops now!”

The cross changes into a cross-shaped sword. Along with the sword changing, so does Jeff. Three sets of beautiful white wings manifest from his back, then he emits an angelic glow.

Jeff's brother: “I can do that too.”

From the brother's back comes a set of dirty black wings with dusty-looking feathers. Jeff lunges at his brother, swinging the sword, slicing the shirt and revealing a cross burned onto the chest. Then he tackles the brother. The two fight, breaking the wall of the basement and exposing the backyard.

Jeff's brother: “You couldn't finish the exorcism last time, remember? I still have the mark of what you left.”

Jeff: “Gail, Nora, and Shawn, keep chanting. Don't stop!”

The brother punches Jeff in the jaw, then Jeff kicks him in the gut. Right after, he charges at him, driving the sword into his brother's stomach.

Jeff's brother: “This isn't over, not by a long shot.”

In a blink of an eye, Jeff's brother disappears in a puff of thick grey smoke.

Nora: “What the hell was all that about?”

Jeff: “My twin brother, he is something else.” *attending to Kate's wounds, then he begins to heal her with his angelic power.*

Gail: "The wings are new." *trying to lighten the mood.*

Jeff: "This was supposed to happen several years ago, but my brother did something to prevent my angel powers from developing."

Nora: "I guess it had to take something this severe to activate it."

Shawn: "No, it was the power of pure love and his faith to get him through a dangerous time for him to evolve into all this coolness."

Jeff looks at Shawn and laughs as Kate is waking up.

Jeff: "I don't want to sound rude..."

Nora: *interrupts* "No need to explain. She will be groggy, shaken up, and confused. She's going to need you now more than ever. Alright gang, let's go home. We'll see you during the week."

Shawn, Nora, and Gail leave the property happily knowing that Kate and Jeff are okay, but the thought of Jeff's brother still lingers in their minds.

Nora, Shawn, & Gail (thinking simultaneously): "If he could harm his sister-in-law with no remorse, imagine what he'll want to do to us for getting involved. No matter, let him try. He'll regret it." *They all smirk, then look at each other, wondering what the other is smirking about.*

At an undisclosed location, Vice President Deville is golfing with the President and his delegates when, out from behind the golf cart, comes a well-dressed man walking toward Deville.

Well-dressed man: "Mr. Vice President, may I have a word?"

Vice President Deville: *talking to the other golfers* "Excuse me, gents, I'll be right back."

He walks with the man several feet away, then begins.

VP Deville: "What do you want?" *angry tone.*

Well-dressed man: "Sorry to disturb you, Mr. Vice President, but the four in Colombia have destroyed Project EX13. They have found a way to get the original of that project to join them."

VP Deville: "Where are they now?"

Well-dressed man: "They are heading back to the States. As for the test subject, it disappeared."

VP Deville: "Good job. Get your men to pick up the remains of EX13, then bring it to the lab."

Well-dressed man: "Yes, sir! Don't you remember me, sir?"

VP Deville: *ignoring the question and answering it with his response* "Until I need you again or if you get wind of something is the only time you come out of your slumber. Is that understood, Amon?"

The Vice President rejoins his delegates and continues to play golf. Later that evening, he contacts one of his associates from his office.

VP Deville: "I need you to place a bounty on the following individuals." *looks at the military files of Cross, Darren, Rayne, and Blak Hart* "A reward will be given to those who bring me evidence of their demise."

The person on the other end of the phone: "Will do, Mr. Deville."

VP Deville: "That's Mr. Vice President to you. Now find the people who will do the job; it must be clean with no witnesses. Is that understood?" *disconnects the call thinking* "These kids will not mess up what I worked so hard to arrange. Let's see how they measure up to the world's most infamous hitmen." *maniacal laughter*

At an undisclosed location, the four warriors finally return to their base. Upon arrival, Archangel Gabriel is waiting for them.

Archangel Gabriel: "Welcome back, good work in Colombia. You have raised a few eyebrows; it's good for our purpose letting them know that there are people out there who will stop them. But, there are those that are very upset. Please keep your eyes peeled, we don't know who is going to plan something."

At that moment, Louie receives a call from Sargent Gray.

Louie: "It's been a while, how are you guys?"

Lt. Gray: "This isn't a social call, I wish it was."

Louie: "Why? What's going on?" *everyone begins looking at him.*

Lt. Gray: "Whatever you all did in Colombia pissed off someone with high connections."

Louie: "Good, now they know if they fuck up again, we'll be there to stop them."

Lt. Gray: "That's not the point, sir. That person or group sent out a hit list on all four of you. They don't want to capture you, they want evidence of your deaths. I'm giving you guys a heads-up because this just got posted for all mercenaries in the world. The reward was said to be in the millions if all four were killed, then and only then will the reward be given. Here's the kicker, there can only be one recipient."

Louie: "Wow, I feel honored yet intrigued."

Lt. Gray: "Please sir, don't take this lightly. Just like our squad, there are plenty of mercs out there with abilities, human and supernatural."

Louie: "Affirmative, once again thank you." *Lt. Gray hangs up*

Cross: "Who was that?"

Louie: "It was Gray letting us know that due to our actions in our last mission, Gabriel is right. We ruffled some feathers and now every mercenary in the world and underworld wants our heads on a silver platter for a big payday."

Rayne: "So what is our next move?"

Darren: "Attack right, I mean they can't possibly be as strong as us."

Archangel Gabriel: "Wrong, there are many that haven't used their abilities out in the open, but with these turn of events; they will stop at nothing to kill you."

Louie: "You're suggesting we hide our tails and run, not happening."

Archangel Gabriel: "I'm not saying that..." *Louie interrupts*

Louie: "Can you please do us a favor before you continue to suggest things? Get Gail and Nora to a safe location until we can figure out what to do."

Archangel Gabriel: "Sure. Before I go, just lay low at the other location that isn't on any maps. Only five of us have the full maps, Michael, Raphael, J.A.L., Reeves, and me."

Louie: "That sounds like a plan for now until we can figure out a way to handle this appropriately." *pulls away*

At Uriel's current location, he stretches then looks toward the area where Hecate was, but she was gone.

Uriel: *thinking* "I better hurry before she changes her mind. What could Louie have done to make a demon-witch want to change her ways? It doesn't matter, I hope Raphael is still fighting." *He gets out of the shackles then heads toward the door.*

On the other side of the torture facility, Hecate has already loosened the shackles for Luna. Now Luna is planning when to run to freedom.

Luna: "Catherine told me to wait until I hear a loud boom, then run to the south of my location."

At a hidden location, Cross's phone begins ringing.

Cross: "Hello?"

Person on the other line: "It's Lieutenant Lockhart, sir. I have a location on Luna."

Cross: "Good to know, Lieutenant. Please send me the coordinates of Luna's location. Get your squad and meet me there. I'll be there at 2000 hours."

Louie: "What's this about leaving?"

Cross: "I got a tip of Luna's location. I don't care about the consequences. For her I'd," *gets interrupted by Louie.*

Louie: "You don't need to say anything else. I have to stay back here to make sure no one suspects. Do you need any of my squad to back you up?"

Cross: "I already have my squad meeting me there."

Louie: "Please keep me posted. If you're in a tight jam, just call me and I'll be there."

Cross: "What are you going to do about everyone else?"

Louie: "Let me worry about the things here. You go and get Luna. Please keep an eye out for Catherine." *Cross smiles at him as a sign of thanks, then leaves.*

In a flash, Cross rushes to a contact of his.

Cross: "I know this is short notice, but I need those weapons I put the order for three weeks ago."

His contact: "Damn, no 'how are you?' 'How's your day going?' Just fuckin' with you, they've been ready two weeks ago. I sent you a message to notify you."

Cross: "Sorry, Angel. I've been extremely busy. Currently, I'm in a rush and my mind is going a million miles a minute."

Angel: "Is this the big mission you've been preparing for?"

Cross: "Nope, but it's a very important one."

They shake hands and part ways. Now that Cross is fully equipped, he can focus on the mission at hand, rescuing the love of his life, Luna.

Back at Michael's location, the ground is shaking.

Archangel Reeves: "Something powerful is coming."

Archangel Michael: "Not something, someone."

A loud explosion can be heard. All that was seen was a blast of light. Within the rubble and smoke, a muscular figure with huge wings holding a sword that can only be described as immense emerges from it.

Archangel Reeves: *excited and honored* "It's him! It's J.A.L.!"

Archangel Michael: "Yes, yes, it is." *Smiling*

J.A.L.: *smiles back as he walks forward, passing the carcasses of demon bodies from his arrival* "Guess they weren't expecting me." *laughs* "Let's finish what's left of these demons and save our brothers. Then we can enjoy a nice barbeque celebration and maybe a lasagna or two."

The three angelic brothers get into their battle stance as they prepare for the next wave of entities to arrive. The dust clears; various demons charge at them.

J.A.L.: "Michael, take the left side. Reeves, take the right. And I'll take everyone else in between."

The angels look at each other and nod. Michael jumps with sword in hand, then cuts off the head of the first entity. Reeves blasts several of the enemies with angelic energy while J.A.L. tackles the whole middle section of fiends, finishing off with a swing from his blade that splits them all.

For the next several hours, every clash from their blades to any of the foes' weapons causes the Earth to quake. Blood, body parts, and shrieks fill the air.

Not too far from their location, Uriel is looking around for the glowing door. Suddenly, a group of centaurs are heading in his direction. He finds an open cell to hide in until they pass. As the last one runs by, the angel grabs it by the neck and breaks it in one motion. Uriel takes his enemy's weapon. Now with a medieval sword at hand, he rushes down the corridor, still searching for Raphael. He reaches the end of the hall, finding the glowing cell that is guarded by two demons surrounded by thick black smoke.

At Luna's location, rocks are falling from the ceiling and cracking the walls. Luna was stunned for a few minutes due to the loud bang; the constant tremors were making it difficult to escape. She finally gets to the door. Hordes of demons are running to the north.

Luna: *thinking* "This is my chance. Whatever that noise was must have these things going crazy. Catherine knew this was going to happen. I just wonder when did she give birth? If this is Louie's child, he has to know. Anyhow, I gotta get out of here if I want my child to be born in a safe area." *She looks to the side and sees a figure waking up in her cell.*

Figure: "Back here again? When will they learn that this place can't hold me?" *It looks at Luna.* "Hi. Before you go, you're going to need this." *Hands her an amulet, it looks like a round crystal with a bright blinding light from within the gem.* "When you're out of this, we'll see each other again. Take care, Luna."

Luna: "Who are you?"

Figure: "That's not important now. What's important is for you to get out of here before your due date. Now go, you don't have much time."

Luna tries to get a good look at him, but the darkness hides his appearance. Then she glances at the gem one more time. She looks up and sees him running out of the cell. She waited a few minutes, thinking that the demons would catch the guy and bring him back. Once there was no sign of anyone coming back to the cell, she checked outside of the door. Seeing that the coast was clear, she ran as fast as she could. She turned the corner but was blocked by wendigos and duwendes.

At Cross's current location, he finally gets to the facility where Luna is being held, fighting through multiple enemies while opening doors, hoping he finds her in time. His squad takes out anyone trying to sneak from behind while he pushes forward.

Several hallways later, Cross comes up to a door that was wide open. A familiar aroma emanates from it.

Cross: *sniffing the air* "Luna was here." *Cross contacts Louie.* "I'm getting close to her. I'll give you more info later. Cross out."

He slowly enters the cell looking for a clue when several demons jump down and ambush him. He punches one through a wall; more rush in. The warrior is surrounded, then the clashing of metal and bone stops when a loud yell is heard at the end of the hall.

Yelling voice: "GET AWAY FROM HIM!!!!!"

Cross glances where the voice came from. Tears roll down his cheeks as one of the demons stabs his shoulder with its long blade-like nails, pinning him to the wall.

Cross: "Shit! LUNA!!"

Footsteps are heard running down the hall. Cross breaks free. He tries to run toward the footsteps, but the demon bites down on his leg, tearing flesh. Then a white glowing light blinds the demons. Cross looks up and sees Luna standing in front of him, her hair glowing white. She kicks the demon off, then gets him to his feet. He grabs Luna; she gazes at him. Her eyes glow a light green color; she tears up. Cross looks deep into her eyes. She kisses him passionately, then wraps her arms around him. A huge pulse of aura freezes all the demons in their place.

Cross: "Let's get out of here, I'm not going to lose you again!" **Luna:** "I knew you would find me. Took long enough, but I knew you wouldn't stop."

They give each other one last kiss before making the escape. In front of them and behind them is Cross's squad, making a path and protecting from behind, ensuring that this rescue is a complete success.

On the other side of the facility, Uriel tries to speak to Raphael telepathically.

Uriel: *quietly speaking* "Brother, I'm here. I need a distraction to get you out. Are you strong enough to assist me?"

Multiple seconds go by when suddenly he gets a response.

Raphael: "Give me a few to figure out what distraction I'll do so we can get out. How many do I have to deal with?"

Uriel: "For now, all I see is two, but they are surrounded by thick dark smoke."

Raphael: "Wait for my cue."

Uriel: "Ready."

Raphael concentrates, then creates a blinding ball of light from his hands that blasts the door and the guards, pushing them back. Uriel sees creatures heading in their direction; he begins to attack them.

Raphael: *thinking* "Damn, if I would've known I had stored enough strength to do this, I could have been out a long time ago."

Raphael begins fighting alongside his brother. Uriel slashes as Raphael continues to blast his enemies without hesitation. As they run to the end of the hall, Raphael looks to his right and doesn't see Uriel. He turns around and sees Uriel on the ground with a blade through his chest.

Raphael: “NO!”

He runs to his brother as demons’ bodies fail to stop him.

Uriel: “Touch my temple.”

Raphael taps Uriel's head, flashes of things that he has done, missions that Raphael knew about and certain ones that he didn't. He looks at Uriel with anger and a certain message that needed to be passed on.

Uriel: *spitting up blood* “Brother... please... forgive... me... I'm... sorry.”

At Michael's location, the three angels are battling their way to free Raphael and Uriel. J.A.L. blows up walls to make a direct path to the cells where the brothers are located. Four walls later, they arrive to see Raphael holding Uriel in a bright bubble, then a bright light leaves Uriel.

Michael, Reeves, J.A.L.: “What happened to him?” **Raphael:** *looks up with tears in his eyes but also filled with anger* “He died telling me we had three traitors. First let's mourn our brother, then we'll handle the traitor.”

Michael: “Who are they?”

CHAPTER 16.
WHO ARE THE TRAITORS?

Somewhere in hell, Zima gets a call from Amon who's on the other side. He is notifying him about the prisoners escaping.

Zima: "Don't worry about that, I already have a backup plan. Just deal with whatever job you're supposed to be doing and remember your place. You're only alive because I choose it to be."

Zima appears in the area and sees everything destroyed. He checks the cells where Luna and the angels were, the only thing he finds are the shackles that held them.

Zima: *thinking out loud* "So, she found a way to escape. The only person attending her was Hecate." *laughs manically, then he calls his slaves* "Find Hecate and bring her to me alive. Don't be afraid to rough her up a bit. She'll regret the day she ever met Blak Hart." *An evil smirk appears on his face.*

At the secluded area, the Angels along with Darren, Rayne, Cross, Luna, and Louie are preparing for the ascension ritual. This is done for the warriors or Angels who have died in battle.

Luna holds Cross tightly as the ritual begins. She looks at Louie with tears filling her eyes.

Luna: "Louie, I'm sorry that I wasn't able to bring her with us. She told me to leave because she had things to take care of."

Louie: "Was she okay, did she say anything else?"

Luna: "We'll talk more after the ritual."

The ceremony commences. J.A.L., Raphael, Reeves, Gabriel, Vin, and Michael begin with a prayer while everyone else bows their heads out of respect.

At the White House, Vice President Deville is having a meeting with several mercenaries and delegates from multiple countries while the President is overseas meeting with the British Prime Minister.

V.P. Deville: "Gentlemen on my left, each of you have a folder placed to your right. Please open them. Now, everyone has a task. The jobs are all designed for the special traits you have mastered. If all is done to the letter as stated in the files, a handsome deposit will appear in your accounts. Please do the job discreetly, without casualties and no witnesses. The cleaner it is, the easier my job will be. To the gentlemen on my right, we have much to discuss."

At the Cross household, Nora calls Gail to notify her of the news.

Nora: "Cross called me last night to say that he found Luna." *excited*

Gail: "That's great news! Were they able to find Louie's girl in the process?"

Nora: "They told me that she was seen but wanted to create a distraction for them to escape. I'm going to be a grandmother."

Gail: "Congratulations! Do they know what it's going to be?"

Nora: "I think they want to keep it a secret. After all, they've been away from each other for so long. What I don't understand is, she's been gone for quite some time, how is she still pregnant?"

Gail: “Most likely, time moves much slower for humans in the location where she was at. For demons, time for them is the same as if they were out here. I've only heard of places like this in books. My guess is she was somewhere close to the underworld.”

Nora: “You mean close to hell!”

Gail: “In the books, it talks about several locations on this plain that were doorways. Some unlucky souls accidentally found them and were never seen again, all that was found were their notes. Personally, I never believed it. I guess once the doorway is discovered, it changes location so it's never in the same place for too long. For us, it's been several years that she was gone. For her, it probably felt like an eternity.”

Nora: “Poor Luna, I can't begin to imagine what she went through.”

At Archangel Gabriel's secret location, the angels are talking to the group about the recent actions.

Archangel Michael: “Why were you there, Cross?”

Cross: “To save Luna. I got intel of her location, so I left.”

Archangel Michael: “It could've been a trap. Why did you think the intel wasn't a plot to capture you?”

Cross: “My informant is very reliable, and I had my old squad with me for this rescue mission.”

Louie: “If I may interject…” *Archangel Michael cuts him off*

Archangel Michael: “No, you may not. How can you let him go?”

Louie: "Simple. He said he was going to save Luna, and I said fine. I would've done the same thing. You should understand where he's coming from. You would and did do anything for the ones you love, so why shouldn't Cross? I have faith in his skills, maybe you should too. Come on…" *Archangel Michael tries to interrupt but Louie doesn't let him* "Stop trying to cut me off. After all that we've done and everything you guys taught us, the least you could do is trust us. Don't send us back or shield us from the things we're destined to battle. You sent us out for the past several years battling God knows what, and you were fine with that. Now he goes after someone he loves and you have a problem with that? Really?"

Archangel Raphael: "Brother, Cross did have a valid reason for doing what he did. It was no different than what you did for me."

J.A.L.: "It's over and done with. He made it back safe with Luna. Let them have their reunion. Then…" *looking at Luna* "we have many questions to ask you, young lady."

Luna: "I understand."

J.A.L.: "Brothers, let's go and plan out the next things that are about to unfold. Raphael will come back to take over for Gabriel after our meeting." *Archangel Michael tries to change his brother's mind but gets ignored* "Let's go, Gabriel, we have much to discuss." *A blinding light is seen, then all the angels are gone.*

Rayne, Darren, Cross, and Luna all sit at the dining area discussing what happened while Luna was gone. In the living room, Louie glances at how happy everyone is that Luna has returned. He pulls out his wallet, then takes out a picture of Catherine with him at an amusement park in New Jersey, dated weeks before his deployment.

Louie: *thinking* "Wish they would've got her out too. I know she was making sure that Luna would escape no matter what... But damn, I just hope we'll be together again. I'm happy for Cross, but maybe if I went with them, I probably could've convinced her to come with us." *He sighs and his eyes water. Right at that moment, Luna calls him over. He wipes his eyes, then heads over to the group.*

At Nora's house, her phone rings. Gail picks up.

Gail: "Hello?"

Caller: "Hi Gail, it's Jeff. Is Nora close by?"

Gail: "Hi, and yes, she is. I'll get her for you."

Jeff: "No need, I'm on my way to her house now."

Gail: "What's wrong?"

Jeff: *urgent tone* "I'll explain everything once I get there."

Ten minutes later, Jeff arrives at the house with Kate. Nora greets them at the door and invites them in.

Nora: "What happened?"

Kate: "We found something that I think you both should know. It may be difficult to take at first, but we'll try to break down for you the importance of it."

Gail: *eager and worried tone* "What is it? You both are keeping us in suspense."

Jeff: "We found a few books that talk about Louie's curse. While searching, Kate found several books that have two people that were considered great warriors in their time."

Kate: “I think you should see the pictures of the great warriors. You might be surprised to find out that the appearance of these men pops up in different eras all around the world. They work for the same causes. Also, the people they associated with are painted in some pages.”

Nora: “What's the news about the curse?”

Gail: “What's in the books that we would be surprised about?”

Jeff: “As for the curse, there is only one way to end it.” *Nora interrupts*

Nora: “Well, what is it? What can I do to help him get rid of it?”

Jeff: “If you let me finish, I'll tell you. You can't do anything, it has to be him. He's the only one that can break it. The question is, can he do it?”

Gail: “Do what? He's been through a lot. What could possibly be so difficult that he might have trouble with?”

Kate: “What Jeff is trying to say so delicately is... he must kill a loved one, the death of his father or grandfather, solely by his hand.”

Nora: “There must be another way! His grandfather disappeared over thirty years ago; his father has been dead for over twelve years. And even if he were alive... I don't think he would've gone through with it.”

Jeff: “Sorry, this is the only way.”

Kate: "On a lighter side of things, these other books were brought to us by a good friend from the Vatican. Nora might know him, Father Knight."

Nora: "He had a website years ago for anyone who wanted spiritual advice to email him at royalknight777@holy.advice.com. How has he been?"

Jeff: "Dealing with a few things here and there. Aside from that, doing well. These books from the Vatican are very helpful for giving us information about your kids in more ways than you'll ever imagine. Look at a few pictures and a few books so you can see what I mean."

Gail and Nora look at a few pictures. The first one is of a crusader warrior charging, the person resembles Blak Hart. The rest of the pictures have two warriors in battle.

Blak Hart in the Crusaders. Cross in Crusaders

What interests them most is the uncanny resemblance to Cross and Louie. In different eras, these two fighters are seen. There are scrolls that describe them and their journeys as far back as prehistoric times. The warriors have different names every time, but the cause is usually the same. The girls keep looking and find some pictures with the two warriors and their two alliance members who look just like Darren and Rayne.

Nora: "How far back are there pictures of the warriors?"

Kate: "As far as humans have been around."

Gail: "Are there any more of Darren or Rayne?"

Jeff: "Only the Louie and Cross look-alikes are the ones so far. I'll check to see if Darren and Rayne go as far back as them."

The four have dinner, share stories, and when they look at the time, it's 2 in the morning.

Gail: "Wow, I didn't realize it got so late."

Kate: "I know, we should do this again."

Nora: "Definitely."

They say their goodbyes. Nora walks them to their cars. She watches as they pull off and begins to wonder.

Nora: *thinking* "Could that be them in a previous life? Was I in those lives with them? So many questions and no person that can give me a straight answer without having to pull any teeth."

At the secluded base, the angels and the four warriors are listening to the chatter on the radio. They search through various frequencies until they find one with familiar voices.

Archangel Michael: "This station is talking about four fugitives that are wanted for crimes against the government."

Archangel Raphael: "It's hilarious how these guys are being poorly hunted while the real criminals are deep in the system." *Chuckles*

Archangel Reeves: "What about the message that Uriel sent us before his demise? What is going to be done about that?"

Archangel Raphael: "Currently, nothing. That will get dealt with soon enough. We must handle one thing at a time."

Archangel Michael: "Since you're eager to find out if what he said is valid, take Gabriel with you in search of the truth."

Archangel Gabriel: "Finally, we get to work together on something other than humans. Traitors, beware! We're on the job."

Archangel Reeves: *thinking and sighing* "Something else I must train this kid on."

Suddenly, Darren's phone rings. He talks very low, then walks over to Cross.

Darren: "Cross, Rayne, I gotta go to see a contact about the current info about my mom and if he's heard anything about us."
Cross: "Please be careful, you already heard the angels bitching about me leaving."

Darren: "I'll be back before they realize anything." *kisses Rayne*

He sneaks out the back taking a jet ski to head to his destination. Fifteen minutes later, he arrives at the meeting spot. He parks the jet ski then walks to a dark area behind a container yard. Several lights flicker as he gets closer to his location. Several feet away, a fancy black vehicle with tinted windows waits for him. Darren walks to the car, the back window rolls down enough for the smoke to seep through the crack toward Darren's face. A voice can be heard telling him to enter, so he does.

Hours later, Darren gets out of the vehicle. He walks to the jet ski and rushes to the base, he looks around then darts to his room.

Back at the container yard, the gentleman in the vehicle with peppered color hair, minor wrinkles, and Victorian era clothes smoking a cigar talks to his driver.

Man in the car: "This kid thinks I'm looking out for him. Little does he know, he'll be my bitch soon. Until then, I'll wait patiently. I have more than enough time on my hands." *laughing manically as his eyes begin to glow red*

In the fiery pits under Mount Dutch, a shadow of a woman can be seen running away from a slender dark figure. She trips over a rock then gets partially sucked into the ground.

Shadow woman: "Give me back my baby!"

Dark figure: "I don't have him. He's better off without you." *evil laughter* "You wouldn't be a suitable parent for him. Look at you getting maternal, how cute. He's mine now." *sarcastic tone.*

Shadow woman: "Tell me where he is, NOW!"

Dark figure: "What do you think threatening me will do for you? Do you really think you can beat me? I'm not Blak Hart, I will beat you to a bloody pulp with a smile on my face and your blood on my fists, you stupid witch."

Shadow woman: "You can look similar to him, but you'll never be half the man he is."

She begins to concentrate, then levitates off the ground. She charges at him, he dodges all the attacks then hits her off a cliff causing her body to land on a sharp rock that rips through her stomach as she screams in agony.

Dark figure: "Next time, think before you act, Hecate." *he yells toward her*

At the Cross household, Nora continues to look through the books. She finds out that no matter what lifetime, the two warriors constantly had traumatic childhoods and in some lives their adulthood wasn't any better.

Nora: "Poor kids, this life doesn't look any easier. On a good note, at least they always have each other's back."

Gail shortly arrives, she notices that Nora has been crying. As Nora begins to cook, she tries to find out what was wrong.

Nora: "Cross and Louie have had horrible past lives, and" *getting watery eyed* "this life is ending up just as bad."

Gail: "How bad could it have been?"

Nora shows her the pages on the books that talk about their lives and the result of all that they did. She also read what lasting impression the two left to those who were affected by them. There was a message in an unknown language.

Gail: "Nora, we have to find out what this note says."

Nora: "Doesn't look like any language I've ever seen."

Gail: "Me either. Maybe the angelic brothers can read it."

As they research for the language used, Gail tries to give Nora positive reinforcement.

Gail: "Even though the boys had hard lives, they left their time as legends and heroes. Just like in this lifetime, they are doing work to benefit the safety of others even if it's not being publicized."

Nora: *frustration in her voice* "The only people recording the activity are the ones at the Vatican."

Gail: "Since when do you know of anyone of the Hearts or Cross bloodline liking the spotlight?"

Nora: "You do have a point, but I just wish they would get appreciated for all their hard work." *she puts her head down*

At VP Deville's secret location, he receives a phone call.

Caller: "Mr. Vice President, I suggest you look at the screen. There is something you must see."

Deville looks, he sees the town that the President was visiting destroyed.

The news anchor: "The search continues for survivors. Once again, the area that was known as Lesotho, South Africa, where the President of the United States was staying at, has been destroyed. No one knows at this time what country or group did this or why it was done. The only thing we were told was that several missiles were seen before the explosion. We'll keep you posted with the latest information."

Deville turns off the TV

Deville: "How did this happen?"

Caller: "The town was bombed by one of our allies. I don't know why or the reason for the attack."

Deville: "Meeting at the White House, now! We are not going to tolerate this behavior."

Caller: "Do I call for a return strike or for a full-fledged war?"

Deville: "I must first make a speech to the public, then our meeting will explain our next step."

Caller: "As you wish, Mr. President."

Deville: "It's not official yet, but thank you." *a smirk crosses his face*

Back at the warriors' hidden location, Darren continues to think of what his contact was saying.

Darren: *thinking* "Maybe he's right. Everyone listens to Cross and Louie but dismisses anything I say."

Luna runs into the area where Darren is.

Luna: "Is everything ok? We're all joking around and stuffing our faces. Why don't you join us?"

Darren: "I'm fine, I'll be there shortly. I have a lot on my mind."

Luna: "Well, when you're ready, you know where we're at."

Darren: "Thank you for checking up."

Luna walks over to Cross.

Luna: "I think you should talk to your friend, he seems a bit off."

Cross: "He's always a bit off."

Luna: *concerned tone* "I have a weird feeling, it's a bit more than normal."

Cross: "I'll go in a few; Louie and Raphael are making stupid challenges to see who can complete them all." *laughing*

Luna grips his arms tightly.

Cross: "What's wrong, love?"

Luna hugs onto him as he holds her.

Luna: "Rayne told me that Darren's been acting extra weird. He's been disappearing without letting anyone know. Haven't you noticed anything different about him?"

Cross: "We've been busy so I can't truly tell you, but he does seem to be acting differently."

Luna: "Back in the cells, I thought I saw him there passing by wearing some strange armor and he was talking to someone that sounded like Louie."

Cross: "Do you remember the conversation?"

She buries her face in Cross's chest and nods no.

Luna: "It's hard to remember, too many things went on, but I'm pretty sure I saw Darren. He just looked a little different."

Cross: "I'll keep an eye on him."

After Cross watches for a bit longer, he walks over to Darren. The closer he gets, Cross hears Darren talking to someone on the phone.

Darren: "I'll try to head toward you in a couple of days, I must figure out a way to sneak out." *he listens to the person on the other end* "Don't worry about Blak Hart or Cross finding out, I can handle them. As for Rayne, I have her eating out of the palm of my hand, so I'll be fine." *waits for the caller to respond, then continues* "The angels are too busy dealing with other matters at this time, just let me worry about this end. Make sure you have what I asked for, if you don't, you know what'll happen."

Cross slowly creeps behind Darren, then after he's done with the call, Cross startles him.

Cross: *in a loud voice* "Hey, what are you doing?" *chuckling as Darren tries to compose himself after screaming like a girl in terror*

Darren: "Dude, not funny. What's your problem?"

Cross: "Were you talking to your mom?"

Darren: "No, why does that matter who I was talking to?"

Cross: "It doesn't. You've been acting very weird lately. What's wrong?"

Darren: *getting very defensive* "Nothing. Strange how? I've never been normal."

Cross: *noticing his body language and his eyes glowing as a defensive impulse* "More than your usual weirdness, come on. You know you can tell me anything. What's with the secret meetings and private phone conversations? My sister is in the next room."

Darren: "It's just things that I'm working on and it's missions that might help us out in the long run."

Cross: "Should I get Louie over here, so we can help plan the mission and get it done quickly?"

Darren rudely interrupts

Darren: "NO, I don't need any help! Especially from him!"

Cross: "A day away from here would do us good. A bit snippy for nothing being wrong, yeah right." *sarcastic tone with a bit of anger in his voice* "Also, what do you have against Louie?"

Darren: "Sorry, I just have plenty on my plate. I would appreciate it if you don't tell the gang. I'd rather handle this alone. Nothing against you, Louie always takes control of the situation."

Cross: "Fine, but if anything goes wrong, make sure you call me."

Darren: "Will do, thank you." *thinking* "I am strong enough to do missions without any of you. You couldn't protect Luna and Louie couldn't save his girl."

Cross walks to Luna.

Cross: "You were right, he's up to something."

Luna: "Like what?"

Cross: "I don't know, but whatever it is, it's going to get him into some serious trouble, and probably bring it to us as well."

Luna: "Are you going to let Louie know?"

Cross: "I'm not sure if I should tell him. I can't jump the gun without knowing all the facts."

Luna hugs him tightly while she slightly trembles.

Luna: “Please be careful. After everything that’s happened, I’m so on edge.” *getting watery eyed* “I don’t know what I would do without…” *before she could finish, Cross kisses her passionately, looking deep into her eyes, then he smiles*

Cross: “Don’t worry, my love. I won’t let anything happen to you or me. I won’t lose you again.”

Luna: “There’s something I have to tell you.”

Cross: “What is it, babe?”

Luna: “It’s about Louie’s girl.”

Somewhere in hell, Zima is talking to a dark figure.

Zima: “How’s the babysitting job going?”

Dark figure: “It sucks, but at least Blak Hart doesn’t have the baby. Hecate isn’t going to come looking for the baby, is she?”

Zima: “No, I already took care of that problem some time ago. Just raise the kid to my specifications, then when he’s of age, I’ll handle the rest.”

Dark figure: “Why do I have to take care of the brat? This kid has been blasting me with fireballs.”

Zima: “Because you look a lot like Louie.”

Dark figure: “Is that the only reason?”

Zima: “No, also because I don’t want to take care of babies and that I don’t have time for that shit anyway,” *sarcastically* “and you have great motherly instincts.” *annoyed* “Do as you’re told and stop bitching!”

Zima walks away from the dark figure. Several feet away, another figure gets stopped by him.

Zima: "How's your training coming along?"

The figure: "To mimic the way the person speaks is difficult, the rest is easy. The fighting style the person has is very interesting because it's various combined."

Zima: "Keep training, you'll be working very soon. Go to Kain, he'll train you in other ways. He will show you how to beat the one you're going to take over, with ease." *the figure smiles*

Back at the angels' secret location, Rayne is talking to someone on the phone.

Rayne: "How is the area you're living at?"

Person on the other line: "Decent, I still appreciate everything you've done for me."

Cross passes by.

Cross: "Who are you talking to?"

Rayne: "Lil, you don't know her." *talks back to the phone* "I'll talk to you later, bye."

Cross: "Where did you meet her?"

Rayne: "On one of my missions, her town was destroyed along with her family. I rescued her and found her a better place to live."

Cross: "That was nice of you, how did she survive?"

Rayne: "I have no clue, but she was able to get far enough from the battle when I found her."

Cross: "Do you know anything else from her past?"

Rayne: "She is young, how much of a past can she have?"

Cross: "By now you should know, don't judge a book by its cover."

Rayne: "She's harmless, you'll see."

Cross: "Did you at least research who her parents were?"

Rayne: *annoyed* "I didn't feel the need to do that. Don't be so overprotective, I'm not a little kid anymore. I can handle myself."

Cross: "I'm sorry, I'm just being an older brother, and it's 101 of research and rescue."

On the other side of the facility, Louie walks over to Raphael.

Louie: "Please tell me you have a mission for me, recon, surveillance, swatting flies, something please. Being confined to one place makes me edgy."

Raphael: "I'm looking."

Louie: "Any info on Catherine's whereabouts? Can I just go check on Nora and Gail?"

Raphael: "We sent several of our brothers over there to protect them. Gail's brother has been with them on and off. He's not you, but he can handle himself. As soon as I find something, I'll call you over. As for Catherine…"

Cross joins them in conversation, suddenly interrupting.

Cross: "Any word on how my mom and Gail are?"

Raphael: "Like I told Louie, they're fine. My brothers are protecting them."

Cross: "Oh, thank God. Raphael, could you excuse us for a moment?" *pulls Louie to the side*

Louie: "Yeah, what's up?"

Cross: "I have some news about what Luna saw when she was captured."

Louie: "What did she see? It must have been some crazy shit."

Cross: "There are some things that I think you need a heads-up on. First is about Catherine."

Louie: "She did tell me that she saw her. What happened?"

Cross: "She spoke to her, and on several occasions Catherine helped her, even on the escape itself. So, on that part it was a good thing that she was there. Second, she was pregnant with your child. Before the escape, she wasn't anymore…" *interrupted by Louie*

Louie: "What happened to the baby? Was Catherine ok?"

Cross: "She was going to cause a diversion so Luna could get out. As for the baby…" *looks down* "She never told her."

Louie: "I have to get her…"

Cross: "Slow down. The other part to that is, her real name is Hecate. She is a half-demon witch."

Louie: "Ok, what else?"

Cross: "She was sent to earn your trust, then destroy you emotionally and physically. But she fell in love with you. Once she began loving you, her plans changed. She did everything possible to keep her boss and his henchmen away from you. By doing so, they brought her back to that area to keep an eye on her."

Louie: "I fell in love with a demon, but I never felt threatened. She never did anything to hurt me." *bows his head* "How do I find her? Once I find her, she can tell me what happened to our child."

Cross: "That place was destroyed by the angels, and from what Michael was saying, they change their location whenever it gets compromised. I can talk to my contacts to see if they can come up with something."

Louie: "Cool, then I can talk to my group so they can help. Is that all that Luna said?"

Cross: "No, she thinks she might have seen Darren down there a couple of times."

Louie: "Was he part of her being there? Did he attack Catherine?"

Cross: "No to both. She stated it looked like him, and the person he was talking to sounded like you."

Louie: "If I would've been there, Luna would've been rescued a while ago and Catherine would still be with me." *feeling insulted*

Cross: "She's not saying that it was you. She said she heard someone that sounded like you. As for Darren, that is still up for debate."

Louie: "Have you confronted Darren about it?"

Cross: "No, I don't want you to either. I want to catch him in something shady before jumping the gun. We need more info about the one that sounds like you." *thinking* "I hope he listens to me this time."

Louie: "Good idea. I'll talk to Raphael to see if there is anyone out there that can possibly sound and look like Darren, or like me." *looks at Cross* "I give you my word, whatever I decide to do, you'll be the first person to know."

Louie walks back to Raphael to converse about the new information.

At the nation's capital, Acting President Deville is watching the news with the rest of his staff. On the screen shows two individuals setting off bombs and missiles that were the cause of the late President's demise.

President Deville: "Is there any way to zoom in on the video feed taken?"

Top CIA official Richard Durand: "Yes, there is. We can get a better look at the perps from the pictures and the videos."

President Deville: "Well, what are you waiting for? I want to see the son of a bitch that killed our previous President, a great humanitarian and my best friend. I will not rest until we find and bring them to justice."

President's advisor Gospodin Vasiliev: "Please relax, Mr. President. You need to be composed and look confident when you make the speech in about two and a half hours."

President Deville: "Mr. Vasiliev, you are right. Thank you for reminding me of what's next. Mr. Durand, I want to know the names of the terrorists before I make the speech. Our people must know who they are so they can help us in apprehending these criminals."

Agent Durand messes with the video feed until he can finally visualize the two clearly. Immediately after, she searches the database to see if any facial matches are in the system.

Three hours later, Agent Durand calls the President over.

President Deville: "Did you find out who these people look like and their names?"

Agent Durand: "Yes, I did…"

Back in New York, Nora and Gail are trying to decipher the markings that were pictured in the book. As they search through books and the internet, a knock is at the door. The women stop what they're doing and head to the door.

Person at the door: "Let me in, I have news about the things you're researching."

Gail: "It's my knuckleheaded brother."

Nora: "Shawn!" *excited voice*

Nora opens the door and tackles him to the ground.

Gail: "Damn, I guess she misses you." *laughing and giggling*

Shawn: *shocked that he just got leveled by Nora* "Hey babe, I missed you too but can I get up and hug my sister?"

Nora: *blushing and slightly embarrassed* "Uh, yeah sure. Sorry, it's just that I missed you so much."

Shawn: "I know, hun." *still trying to get up*

Shawn gets to his feet just to get tackled by his sister.

Shawn: "Damn, really sis?"

Gail: "Just letting you know I missed you too. Duh." *Smiling*

Nora: "What information did you find out?"

Shawn slowly gets up, making sure that he doesn't get attacked again.

Shawn: "The language that is in the book hasn't been used since the time before Moses. I found several alphabets that look similar to the one you showed me, but I can't be too certain."

Gail: "Only one way to find out, let's try them out. Nora uses one, you use the other one, while I use the next. Then we try all of them to see which one makes the most sense."

The three start off with different versions. Once they finished those, they started on the next ones. Hours go by, still nothing made sense. Four books left; three of them were languages used by priests at the time, and the last one was of a long-ago demon language before it was added into the Ancient Latin language.

In Washington, D.C., the acting President gives his speech.

President Deville: "Citizens of this great country, we have found the culprits who caused the death of our great President. On the screen you will see footage of the killers. The video has very graphic nature. If you cannot handle what is about to be presented, please look away."

Agent Durand looks at Deville.

Agent Durand: *in a low tone* "Mr. President, you haven't viewed it yet. Are you sure I should publicly show this?"

President Deville: "Yes!" *smiling deviously*

At Nora's house, Shawn takes a break from searching and decides to watch television.

Shawn: *turns on the TV, the news is on* "Girls, I think you should come over here and see this!"

Nora: "What is it?"

Gail: "What is going on?"

Shawn: "The President is giving his speech. Along with that, he has footage of who caused the previous President's death."

Nora: "Why should we care?"

Shawn: "Cause the two that caused it…" *in deep shock*

Nora and Gail: "Who could they possibly be?"

Shawn: "I recorded it."

They look at the video, then replay it over and over again.

Nora and Gail: *with tears in their eyes* "It… it can't be…" **Nora:** "This must be some kind of joke. They would never do something like this. It's impossible… It's… It's…"

CHAPTER 17.
THE UNKNOWN

At the secret location, the angels are watching the news. Raphael calls Louie over.

Archangel Raphael: "Louie, come over here. There is something you must see!"

Louie: "What is it?"

Raphael: "Don't ask questions and come here!" *demanding tone*

Louie rushes over, then looks at the screen. A flood of emotions courses through his mind.

Louie: "Who the fuck are they?"

Archangel Michael: "First of all, watch your language. Second, I think it's your parents."

Louie: "That's impossible, they're dead. We all know that."

Cross enters the room, confused.

Cross: "What's going on?"

Louie: "The current President is calling for a manhunt for two people who were involved in the murder of the previous president. That's not the worst part, the two look like my parents."

Cross: "Damn, how are you taking this news?"

Louie: "I don't know what to think. This week has been a mind fuck. First, I find out that my girl is a very powerful witch/demon who was having my child, then Darren is doing some questionable things. And to top everything, two people that seem like my dead parents are causing havoc. All I know is that I have to find out if it's really them and why they're doing this. If it's not, then they have to answer for their actions."

Michael: *jumping into the conversation* "Don't be hotheaded, what if this is a trap? Someone might have planned this to get you out and frame you as a co-conspirator in the crime."

Cross: *puts his hand on Louie's shoulder, then speaks to him low enough for only him to hear* "If there's anything I can do just say the word."

Louie: "I know and I appreciate it. If I think of something, I'll let you know."

Raphael: *pulls Louie to the side then speaks to him in a low tone* "I'll help you make a plan and together we'll get to the bottom of this. Don't worry about my brothers, I'll deal with them when the time comes. Together, you and I will find the truth. I felt like I failed them and you. Now we can make it right or protect their honor."

Louie: "I respect and appreciate that you would put your neck out against your brothers for my parents, if that's them, but I think this is something I should do alone. In case things go out of control, you won't get blamed."

Raphael: "They weren't my blood, but they were close enough. I need to do this too, please don't do this alone."

Louie looks at him, then realizes that they both need answers. The only way they can is by working together to make sure it gets done properly with the least possibility of damage or errors.

At Nora's house, Gail and Nora are still in shock at what they just witnessed on the video.

Gail: "There must be some explanation for their actions, I just wonder what it could be?"

Nora: "Shawn, what do you think?"

Shawn: "I can't truly give you my opinion, I only knew them through work. I never had a personal relationship with them."

Nora: "Gail, I think we're going to make a trip over there to see if we can uncover something."

Shawn: "I think you should leave that to the angels or the authorities to deal with. You two should stay here while I try a couple of my people to see if they can find something out for us."

Nora and Gail: *lying* "Sure, we'll wait here until you get back."

Shawn: "Ok, then it's settled. While I'm gone, see if Louie knows what's going on and what is being done on his end."

Nora: "That's not a bad idea, I can only imagine what he's going through."

Gail: "He must be going out of his mind, I don't want to be in his shoes right now."

Nora calls Cross.

Cross: "Hi mom, I guess you heard about the Hearts."

Nora: "How is he dealing with this whole scenario?"

Cross: "Honestly, I can't tell. I sense his anger, pain, frustration, and confusion. The weird part is, he's not showing any of those emotions. I know he's up to something, I just don't know what."

Nora: "How about Raphael?"

Cross: "I think if anyone can understand Louie, it's him. I know Louie told me that when he thinks of something he would let me know, but I think Raphael is the only one that can."

Nora: "I don't think Michael would let them."

Cross: "You're right, but if there's one thing I learned being around Louie and Raphael when they set their mind to something; there is no one on this planet that can stop them individually. I can only imagine both of them in that state of mind, holy shit!"

Nora: "Well, keep in mind if he needs you; he will call for you. How are Rayne and Darren?"

Cross: "They both seem off. There is no way to explain it, seeing is the only way."

Nora: "How's Luna? Do I have a grandchild yet?"

Cross: *chuckling* "She's crabby but fine and no, she hasn't popped yet." *laughs*

Luna walks into the living room.

Luna: "Is that your mom?"

Cross: "Yes, it is, would you like to talk to her? Also, can you fill her in on Louie and Catherine's situation?"

Luna: "Sure."

Cross hands her the phone, Luna begins to talk to Nora and walks away with it.

Cross: *thinking* "At least that hasn't changed."

Somewhere in hell, Zima is with Hecate/Catherine.

Zima: "You want to be so human, by now Blak Hart knows who you truly are. You have nothing to live for, no kid, no mate. You're just a waste of space."

Hecate/Catherine: "I have a kid, you took him from me! Do what you want to me but let my baby go!"

Zima: "Correction, I have a kid, you just have the memories." *laughs* "You are in no condition to try to defy me, so why try?" *tortures her a bit by reopening her wounds* "Your so-called lord got his plan in effect, little does he know; my plans will shadow his and probably put a dent into the grand design he has set into motion."

Catherine: *in pain* "Your plans will fold the minute Blak Hart finds out." *thinking* "I hope Luna told Louie already."

Zima: "It's cute how you think he'll be your knight in shining armor." *laughs* "By the time he finds out anything, it'll be right before his death." *Catherine cries* "His parents couldn't stop me, he definitely can't. He's not strong enough. The chosen four will fall, one by one. When it's all said and done, they will be either dead or on my side fighting the angels that swore to protect them." *maniacal laughter*

In New York, Shawn is going to each of his contacts asking for any information they can involving the current status of the Hearts. Several of his contacts tell him of a bounty on their heads averaging anywhere from half a million dollars to several million dollars depending on how valuable the information is or the apprehension alive or dead. The only thing they can all agree on is the whereabouts of the duo is unknown. It's like they vanished the same way they appeared.

Back in the angels' secret area, Louie is on the phone.

Louie: "Hey, it's me. I wanted to know if you have time to make special arsenal for the Archangel Raphael and for me. I need weapons that can pierce through illusions and resist demonic corruption. Raphael needs something that can sever soul ties."

Person on the other line: "Sure, by when do you need them by?"

Louie: "We're going to your location, so we can give you the specifics."

Person on the other line: "No problem Lou, see you then."

Louie hangs up then walks to Cross.

Louie: "Do you need any weapons specially crafted?"

Cross: "How are you getting to Angel?"

Louie: "Raphael will help me find a way, when we go to check on the two criminals, keep your phone on. Just in case we need backup."

Cross: "No doubt, as for weapons; I'm good for now. Have you seen Darren or Rayne around today?"

Louie: "I haven't seen either one. I think we have to deal with them at a later time."

Cross: "When do you think we should have that talk with them?"

Louie: "After the whole ordeal with the lookalikes gets taken care of."

The Archangel Michael interrupts them.

Michael: "My brothers and I are going to find out all the information we can on the current issue."

Louie: "What are we going to be doing while you guys are out?"

Michael: "Look on social media and all news to find out any new reports of sightings or unusual occurrences."

Cross: "OK, take your time. We'll see you later."

As the angels leave, Raphael pops in.

Raphael: "Louie, are you ready?"

Cross: "I'll keep you posted if they start asking for you."

Louie: "Thank you, if things get a bit hairy; I'll call you. You see, I am behaving, well sort of." *smiles then in a flash of light Louie and Raphael are gone*

On land, Darren is with his contact. They are walking around in the cemetery when suddenly plasma demons begin attacking Darren, slicing him in both of his arms. He takes out his blessed swords, and Darren swings at them. It goes right through them; the demons weren't affected.

Darren: *thinking* "He left, it's like he knew they were coming. These demons aren't getting fazed by any of my attacks."

He starts to concentrate, focusing all his energy on his hands. He blasts the demons with a strong intensity of electricity, electrifying half as the rest retreat.

Darren: "I wonder where he went, damn. These cuts itch, and I feel a little bit dizzy."

He tries to retreat to the base but ends up falling. The poison that was passed on to him through his wounds causes him to lose consciousness. Before he fully passes out, he sees a dark shadowy male figure walking over to him.

On the other side of the world, Rayne is visiting Lil. They walk together to the park.

Lil: "You would never let anything bad happen to me, right?"

Rayne: "You have nothing to worry about. They would have to go through me first before anything were to happen to you."

Lil: "Oh ok, I just wanted to know. Why are people looking at us funny?"

Rayne: "Maybe cause they're jealous of us." *Lil laughs*

Lil: "I can hear them talking garbage about us, makes me want to kill every single one of them."

Rayne: "What have I told you? They can talk all they want, now if they lay a hand on you then protect yourself at all costs. Has anyone hurt you?"

Lil: "You won't get mad if I tell you right?"

Rayne: "Why would I get mad at you? I would punish those that hurt you. Tell me, who hurt you?"

Lil: "The neighbors to the left of me and the house in front, they pushed me, hit me, and made me feel like I was never good enough for anyone."

Rayne: "Let's go over there now, we'll teach them how to properly treat you. If they have any problems, I'll be right there to deal with them."

Lil: "Why not just kill them so they don't hurt anyone else?"

Rayne: "Let's just see what happens, I'll deal with it as it happens."

At Darren's location, Darren wakes up. He realizes that his body is bandaged up and that he is lying on a hospital bed.

Darren: "Where the fuck am I? Seems like a doctor's office."

He gets up, puts on his shirt and jacket, then walks out of the office. As he gets out, the place appears to have been abandoned for quite some time. He creeps down the hallway, and sounds of people whispering can be heard in the last room to the right. Darren rushes over to the room and kicks open the door. When the door opens, no one is there. He scans the room, but he still hears the voices.

Darren: "Where are the whispers coming from?"

Suddenly, one of the voices begins to speak to Darren directly.

The voices: "You can't escape us, Darren. We have always been with you, no matter what you do, you will be like your father."

Darren: "I am nothing like my father. Who are you and what do you want with me?"

The voices: "We are the part of you that your mother tried to keep from waking up. We are now awake, whether you like it or not, you are more like your father than you want to admit." *Darren tries to talk, but the voices aren't letting him* "You are doing things you know Louie and Cross wouldn't approve of, you are giving information about your group to a person whom you barely know. Why would you do that if you know that this data can hurt or possibly destroy them? HMMM? No words, this is the kind of act your father did years ago that caused the death of the Hearts and Wolf. The story repeats itself." *begins laughing*

Darren: "That's impossible. I'm nothing like him!"

The voices: "Keep telling yourself that, to make yourself feel better. You may not approve of your destiny, but you are heading in the same path he did."

Darren: "That will never happen, I AM NOT NOR WILL I EVER BE!" *getting defensive and angry*

The voices: "Sure, keep lying to yourself. Instead of arguing with us, get out of here before the person who brought you here returns."

Darren: "Who was that person?"

The voices: "We don't know. Do you really want to find out? We think not, just find a way out, and rejoin the group before questions begin to arise."

Darren searches for the exit, all the doors are locked. He throws a chair at the window breaking it, then jumps out of it landing safely on the grass. He runs to the edge of the property and turns around.

Darren: "What the fuck? This place looks like Alcatraz, how the hell did I get here?"

At Louie's location, Raphael and Louie arrive at the weaponsmith's area. An old friend of Louie, Angel, greets them.

Angel: "It's been a long time, how's everyone?"

Louie: "Like a dysfunctional family," *laughs* "aside from that, fine for the moment. How about you? Oh, by the way, Angel, this is the Archangel Raphael."

Raphael: "Pleased to meet you."

Angel: "Likewise. So, how can I help you guys today?"

Louie: "I brought you the schematics for the things we need."

Angel: "How soon do you need it?"

Louie: "Yesterday."

Angel: "Damn, these plans aren't easy, but I can try to have it done in eight hours."

Raphael: "How about if we lend you a hand, how fast will it get done?"

Angel: "Well, normally I prefer to do it solo. Since a bit of your essence is needed and this has a lot of details, with all of us working, it might be done in half of that time."

Louie: "Let's get to it then."

The three proceed to work on the weapons, Louie forms them, Angel cuts and designs them, and Raphael adds a bit of his angelic essence to them. While all of them are working, Raphael receives a call from Michael.

Michael: "Raphael, where are you? You need to have a plan to find out if the two are the Hearts."

Raphael: "Don't worry about that, I'm already on my way to finding out. I already have a plan in the works."

Michael: "We're supposed to be doing it, why are you going alone with Louie? There are too many emotions attached to this."

Raphael: "Brother, who better to get the job done than the bloodline of a Heart and myself? I'll keep you posted, brother, please have faith in me and that I know what I'm doing, after all, I had the best teachers heaven provided."

Michael: "You're letting your personal feelings get in the way, you're not thinking clearly."

Raphael: "On the contrary, we have thought of all the possibilities and outcomes. We have formulated a plan for every possibility that can happen. Like I said, please have faith in me. This is one of the first times I can truly say that Louie and I are treating this as any other mission with the mindset that lives are on the line if we fail this mission."

Michael: "You're not fully healed, brother please rethink this." *pleading to Raphael in a worried tone*

Raphael: "I have thought this over and over, my end result is finding out the truth. Before any other action is taken, I will call you to give an update for advice."

Michael: "Fine, keep me posted." *skeptical, annoyed, and worried*

Raphael: "Will do and thank you for believing in me."

At Laguardia Airport, Gail and Nora are waiting for the call to board the plane. After going through the metal detectors and passing the suitcase through the X-ray, an announcement can be heard.

Voice on the intercom: "Now boarding for the 2 pm flight 732 for Bloemfontein, South Africa."

Gail: *Gail looks at Nora* "Is this the closest to Lesotho, South Africa that you could find?"

Nora: "It's the closest and cheapest, not to mention if anything happens we have a place to lay low."

Gail: "You do remember that we're going there just to investigate, right?"

Nora: "Yeah, but what if we do find something? What then, do we go back home and do nothing or get the ball rolling? Remember our mission to save the guys in Paraguay?"

Gail: "Yeah, but just like back then, if, and only if we find something, we call Louie then do whatever we have to."

Nora: "Ha, you see what I was saying."

Gail: "Yeah yeah, shut up." *laughing* "Does Louie, Cross, or Rayne know what we're up to?"

Nora: "Nope, I was hoping to update them after we finish what we're doing."

They get on line to board the plane. As they walk onto it, they look for seat numbers 1207 and 1208 next to the window.

Nora: "Get ready, this is a 16 to 18-hour flight. If the flight attendant offers liquor, take it."

Gail: "We shouldn't, we should research and prepare for whatever pops up. While I search online for equipment, you can start looking for religious places in the area that might have ideas or maybe heard of something."

In Manhattan, Shawn calls Nora and Gail. No one is answering their phone.

Shawn: *thinking* "Why aren't they picking up? They better not have... nah, they're probably researching. They would've let me know. I just hope they aren't going overseas."

He gets in his car and then proceeds to speed to Nora's house. Hours later, he arrives at Nora's house, since no one is there; he gets to Gail's house. He ended up with the same result.

Shawn: *thinking* "I must find them, who knows what can happen to them in Africa. I'm scared they might not like whatever they find, I have to hurry."

He picks up the phone and begins calling airports for the next flight to Africa, the only airline that has a flight available to go is a delivery company. He pays them and gets on the plane.

At the base, Darren finally gets there. He tried to sneak in, but the entrance wasn't locked. He continues to walk in, only to realize that the only people in there are Cross and Luna.

Darren: "Where did everyone go?"

Cross: "Why does that matter to you? You never gave a fuck if anyone needed your help or let anyone know where you would be in case we needed you."

Darren: "What's the difference between what I'm doing and what you did?"

Cross: "The difference is, that everyone knew where I was. Whenever you go, no one knows where you disappear for days on end."

Darren: *feeling offended* "Why does anyone need to know my whereabouts anyway? No one needs me, the strongest person here and I get treated like the weakling I was years ago!"

Cross: *irritated voice* "In case of an emergency, we need to know where to find you. And as for being treated like a weakling, you would be nowhere near my sister if you were and these missions wouldn't have had you with us."

Darren: *getting defensive* "What about Louie? Does anyone know where he's at right now?"

Cross: "He's with the Archangel Raphael, the Archangel Michael knows where they are and so do I." *feeling defeated, Darren powerwalks away*

Overseas, Gail and Nora get to Africa at 8 am. They meet up with a Hertz agent to get a car, then head to their hotel. Moments later, they get to a bed and breakfast in the town of Bloemfontein. They are greeted and shown to the room.

Receptionist: "Breakfast is ready from 6 am to 9 am, lunch is served from noon to 2:30 pm, and dinner is from 6 pm to 7:30 pm. If you want something else to eat, our chef will accommodate you."

Gail and Nora: "Thank you."

Receptionist: "Only breakfast is complimentary with your stay. If you need any assistance of any kind, please feel free to call us."

Gail and Nora both thank the receptionist and then settle in. While Gail gets her clothes ready for their mission, Nora begins to plan out their routes to and from the designated area; just in case things get tough.

Gail: "How many possibilities are you preparing for?"

Nora: "You know as well as I do, it is always good to have multiple plans set in place for any situation."

Gail: "We can survey the area while acting like tourists, that way we can make an assessment of what plan of action we're going to take."

They begin to quietly chant to make a protective barrier over themselves. Once it is done, they drive around town checking out the sights and routes. After several hours, they drive to the border of Lesotho. Gail parks the car, and the ladies proceed to walk to the disaster site. As they walk toward the destruction, a child runs to them yelling.

Child: "Help me! Help me! I'm trapped in here!"

Nora: "Gail, I'll grab him. You free the kid."

Gail looks at the kid and then at Nora, she nods. Nora grabs the child while Gail talks.

Gail: "Demon, why are you possessing this poor child?"

Demon: "I wasn't given the option, help us, please."

Gail: "Why should we help you?"

Demon: "Because this child doesn't deserve this and neither do I. Free us from each other then at a later date I'll explain what happened."

Nora: "Demon, why should we believe anything you say? The kid we'll save, as for you, that's another story."

Demon: "Please give me a chance to show you that I can be very helpful. I can sense that you are a witch, but as for the one holding us, she confuses me."

Gail: "You do understand until we know what you are, we can't trust you."

Demon: "I understand."

Gail begins to do an exorcism with a bit of a change. Instead of sending it back to where it came from, she's having it go into a wine bottle. In a blink of an eye, a thick black mass leaves the kid's body through the mouth and into the bottle.

Nora: "Damn it, we should've asked if it knew anything."

Gail: "We will later on at the hotel, right now check the boy."

They check his vitals and take him to the closest hospital.

Nora: "He's barely alive, I don't know if the possession has him weak or the explosion."

Gail: "The explosion killed anyone in close proximity, how could he have survived? Unless he was possessed before the blast."

Nora: "Let's hurry up and take the child to the hospital so they can attend to any wounds he might have, hopefully, they can find out if he has any family. We go back to the hotel, I'm eager to find out if that demon knows anything useful to us."

Gail: "The hospital might call the authorities to identify him while they do the checkup. As for the demon, I'm also eager, but we have to be smart about this. We can't let our emotions get the better of us and we end up being reckless in asking the questions. Patience, Nora. We never got the job done by doing things half-assed."

They tell the doctor that he was found close to the rubble. Moments later, the doctor came out to let them know his condition, but they had already left.

Nora agrees, they begin to walk back to the car when they are approached by two officers.

Officer: "Where did you find the boy?"

Nora: "At the site where the rubble was."

Officer: "What were you doing there?"

Gail: "We were sightseeing and got lost when suddenly the boy was calling out for help. We couldn't just leave him there by himself, so we brought him here so they can attend to him if he's injured and hopefully you guys can find out if he has any family."

Officer: "That area is off-limits, you two must come with us."

Nora: "We're not from here, how are we supposed to know?"

Officer: "There were signs posted."

Gail: "We don't understand the writing on there."

Officer: "That's no excuse, come with us. Don't make us use force."

Back at the airport in New York, Shawn's flight got delayed because of engine failure. He is escorted off the plane and then is told that he will be waiting for several hours until the engine is repaired. He begins to call his associates to find another way to get there.

Associate: "If you had called two days ago, I would have given you my personal plane. But, I rented it out for the rest of the week. As for anyone else, their planes or helicopters are already paid in advance for the next two months. How soon do you need it by?"

Shawn: "Yesterday, I needed to be overseas as soon as possible."

Associate: "There might be another way, we can talk to our contacts in the military. When they drop off their soldiers, they can let you off there too. The only problem is, you will have to parachute out to your location."

Shawn: "That's not an issue."

Associate: "Give me half an hour, then call me back. By then, I will have their response."

Shawn: "Sounds good."

Back at the base, Rayne arrives to see Cross over Luna, who's lying on the floor screaming in pain.

Rayne: "What happened to her?"

Cross: "She's having the Braxton Hicks Contractions."

Rayne: "Damn, I remember what that's like. How can I help?"

Cross: "Just get me ice chips so she can have them until the pain subsides."

She rushes back with a cup filled with ice chips and stays with them until Luna is strong enough to stand. They assisted her to the room, Cross massaged Luna, then covered her. She falls asleep shortly after.

Several blocks from the rubble, Archangel Raphael and Louie are surveying the area. While Louie is looking for any supernatural residue, Raphael is searching for echoes left by the criminals and victims. They walked the blocks around the site, and after they were done, they went directly to the scene of the devastation. The residue of the horrific event just kept replaying in the echoes: the sounds of the victims' screams and the laughs of several dark entities.

Louie: "Those entities don't look like my parents, but what are they?"

Raphael: "They look like class-A demons, pay attention to the two beings in the corner."

Louie watches as the two at the edge of the room place something on the walls and then sprint out seconds before the explosion. He keeps observing the two as it replays over and over, suddenly he realizes that the male looks at him and points to the other side of the room.

Louie: "Did you see that?"

Raphael: "See what?"

Louie: "The male at the corner of the room looked at me and then pointed to the other side with a smile."

Raphael: "I'll try to focus on him while you look at the other side to see what he's showing you."

Louie: "I don't think it was showing me a person, if it was my father; he might have left me something there."

Raphael: "If it's not, it could be a trap of some sort."

Louie: "True, but it doesn't hurt to try."

Raphael: "That's the problem, it can hurt or kill if we're not careful."

Louie: "You're telling me that you're not the least bit curious?"

Raphael: "I didn't say that, we must be very cautious on every move we do at this location."

Louie checks where the male pointed to, under the rubble he found rolled-up papers. He picks them up and then skims through them. They seem to be letters that his parents wrote him over the years.

Louie: "Raphael, look at this." *showing him the letters* "What do you think? Could it actually be them?" *suddenly remembers one of the coded writings that his father taught him and reads it* "If you're reading this, it means we failed. But you still can win."

Raphael: "How do you know the language? Anyhow, right now, we must get out of here before we get spotted by the military."

On their way out, the military police sees them.

The military police: "Stop or we'll open fire!" *looks at the other officers* "Once they stop, shoot at will."

The two look back and then begin running, and the police warn them again. Since Louie and Raphael didn't do as they were told, the MP began shooting at the two. The warriors talk as they escape.

Louie: "Why the hell? We didn't do anything to get shot at."

Raphael: "They are supposed to arrest, not shoot, they are hiding something. How are they involved with the incident?"

Louie: "I'm going to throw a flashbang, that will give us time to get out without having to use force. I guess this is a very elaborate cover-up. I wonder who's involved in it and how far does it go?"

Raphael: "We can come back another time, for now our main objective is to get to base and verify if those letters are legit or forged by someone else."

At the military jail, the women were taken to the basement. The strong smell of mildew and rotting flesh was in the air. The police handcuffed Gail and Nora to the railings next to the holding cells.

Gail: *yelling* "Are you guys going to tell us why you are detaining us?" *no response*

Nora: "I don't think they'll tell us, they're just going to charge us on some bullshit crime and keep us here until they choose to or your brother sends someone for us."

Gail: "Or we can try something else, try to contact the Archangel Michael while I try to get Raphael."

The two ladies begin praying, Nora asks for Archangel Michael as Gail pleads for the Archangel Raphael to send Louie to break them out.

Abruptly, the military police barge in.

MP: "You will be executed tomorrow at eleven for terrorist attempts against our nation."

Gail: "That's bullshit, and you know it. What are you covering up?"

MP: "If you like we can execute you now." *waits for a wise remark from them* "I didn't think so."

Nora: "When are you going to feed us?"

MP: "Tomorrow before you two die." *laughs*

As the MP leaves the room, the two continue to pray.

Gail: *thinking* "Hopefully, Raphael hears me and has Louie get here before it's too late."

Raphael hears the prayers.

Raphael: "Louie, we have to turn back. Gail and Nora need us."

Louie: "Let's go."

Back at the hideout, Darren goes to apologize to Cross when he sees them helping Luna.

Darren: "How can I help? Shouldn't we take her to the hospital?"

Rayne: "We're fugitives, remember."

Darren: "She's not, can't we have Vin take her there?"

Cross: "Well, you kind of have a good idea. We could call him to come to us instead."

Luna: "That's the best news I've heard all day, what are you waiting for? Do it!"

Cross calls Vin, he answers.

Vin: "Hello?"

Cross: "Hi, we need you." *hyperventilating* "Luna is about to give birth."

Vin: "I'll be right over. Whatever you do, do not leave the hideout. It's not safe for any of you to try to go to the hospital."

Cross: "We weren't planning on leaving."

At Raphael's location, the archangel and Blak Hart rush to Nora and Gail's, in hopes of getting there in time. As they get close to the area, they are approached by Michael.

Raphael: "What happened, brother?"

Michael: "I heard the call from Nora, but I can't go assist you two."

Blak Hart: "Why not?"

Michael: "Luna is about to give birth, and Vin is on his way. I'm going to be there as protection. Go save Nora and Gail, Godspeed." *he disappears*

The two warriors reach the prison, which seems well-guarded.

Blak Hart: "I'll get the guards on the left; you get the right."

Raphael: "Remember not to kill them, they are just doing their job."

Blak Hart: "If any of them are possessed, I'll do a quick exorcism."

They stick to the plan; Louie takes out the ones on the left, rendering them unconscious. The archangel handles the right, he takes one of them and begins to interrogate him.

Raphael: "Where are the women that were brought in?"

The guard: "Why should I tell you?"

Raphael: "If you value your life, I suggest that you tell me where they are."

The guard: "If anyone is brought in they would be in processing."

Raphael: "Where exactly would that be?"

The guard: "In the basement."

He knocks out the guard and proceeds to join Blak Hart.

Blak Hart: "Do you know which way to go?"

Raphael: "Just follow me."

Back at the hideout, Vin gets prepared to deliver the baby. Luna gets hooked up to a heart monitor for her and the baby. Cross holds Luna's hand, Michael paces back and forth throughout the base, checking all the perimeters, while Darren and Rayne wait in another room.

Vin: "Ok, breathe then when I tell you, push as hard as you can." *Luna does as she is told and screams*

Cross: "You're doing great, Hun." *she squeezes his hand and crushes some bones in his hand*

Vin: "I see the head, one more push." *she pushes as Vin gets the baby*

Vin: "Good job, the baby looks healthy. It's a beautiful baby..." *the heart monitor flatlines*

Cross: "What's going on? Is she?"

Vin: "I need you to back up. Just let me do my job so I can save her."

He charges the paddles and then shocks her. No response, he tries again, but her heart still doesn't start beating. The machine's sound is still heard. Flatline tone...

Chapter 18.
Could it Get Any Worse?

The heart monitor still has the sound of a flatline, and Vin shocks her one more time. Her heart starts beating.

Vin: "That was close." *wiping the sweat off his brow*

Cross: "Is she going to be ok?"

Vin: "Right now it's hard to say, her body went through a tremendous ordeal. Let her rest, and I'll stay here until I have run all the tests I can to find out how that time without breathing or blood flow has affected her."

Cross: "Is there a good chance that she could come out of this with no problems?"

Vin: "There's always a possibility, but there could also be complications. It could go either way; I'll be able to give more information after I run the test. Until then, go and hold your child."

Cross kisses Luna on her forehead and then is given his kid from Vin. He walks to the next room to introduce his kin to Rayne, Darren, and Michael.

Cross: "Everyone come meet the new addition to our family."

Back at the prison, Raphael and Blak Hart take out anyone in their way. They finally get to the basement, and what was seen enraged them. Louie couldn't take it any longer.

Blak Hart: "I see you like beating and torturing women, why don't you try that on us?" *he charges at the guards as Raphael frees Nora and Gail*

Once everyone is out of the prison, Raphael turns to them.

Raphael: "We must get to the base quickly; Luna was in labor."

Nora: "Did she give birth and is she alright?"

Raphael: "I don't know yet."

Gail: "What are you waiting for? Let's go."

They finally get there; Raphael can feel a mixture of emotions that is currently in the environment.

Louie & Raphael: *at the same time* "What's going on?"

Michael walks over to his brother and Louie as Nora runs to Cross and Rayne, while Gail heads to Darren.

Michael: "There were complications with the pregnancy, the baby was born healthy. In the process, Luna's heart failed, and when she was brought back... She is in critical condition."

Louie: "How is Cross taking it?"

Michael: "Why don't you ask him yourself?"

Raphael: "Do you know what caused this to happen?"

Michael: "My theory is that the baby is like the father and her body wasn't strong enough to handle the power."

Louie walks over to Cross.

Louie: *puts his arm on Cross's shoulder* "Congratulations, what's the baby's name?"

Cross: "Due to all the events that were going on, we haven't given a name yet."

Louie: "Didn't you two have names picked out already?"

Cross: "We did, if it was a girl, Solay; if it was a boy..." *looks down then glances at the room where Luna is*

Louie: "Well, what's the name?" *eager for the unveil*

Cross: *seeing how proud Louie is to be an uncle, he smiles* "Lucas."

Louie: "I have a nephew." *hugs Cross and congratulates him again* "Can I hold him?"

Cross: "You have to get in line, Rayne is being greedy and not letting my mom hold him." *they both laugh*

Louie: "That's Rayne. What's Luna's condition so far?"

Cross: "I needed that laugh, no change yet. I can't lose her, bro."

Louie: "I know, remember... miracles happen every day. Not to mention, we work with angels, duh." *they both start laughing*

Cross: "We've been around them so often that I completely forgot about that."

Days pass, and Luna's condition remains the same. Vin runs various tests, none of them giving any useful information on what's keeping her in a coma. Suddenly, Michael storms in.

Michael: "Emergency meeting right now!"

Everyone is confused and rushes to the meeting room.

Raphael: "Once everybody is settled, then we will begin." Cross finds Rayne and Darren, they're the only ones missing.

Cross: "On it."

At Shawn's location, he waits to meet with Jeffrey Finnian at the restaurant. Jeff was already there before him.

Jeff: "Damn, took you long enough." *joking*

Shawn: "Any news on the possible doppelgangers of the Hearts?"

Jeff: "No, they vanished. The only chatter is that the spot that was destroyed opened portals in four different deserts where hordes of demons have been reported coming from. By the way, how's Nora and Gail?"

Shawn: "She called me once Raphael and Louie rescued them, at this moment they're in a safe location. Who else knows about this and why aren't the archangels on this?"

Jeff: "They are trying to find the doppelgangers."

Shawn: "Why?"

Jeff: "They believe that those are the Hearts."

Shawn: "Are you telling me that there is a possibility that it's actually them?"

Jeff: "Big possibility, they're trying to make sure that if it is them, they want to find out what their new powers are and if they can be turned back to our side."

At the White House, President Deville makes a phone call.

President Deville: "Is everything ready to start?"

The person on the other line: "Just waiting for your orders, sir."

President Deville: "Make sure your squad is prepared to attack."

The person on the other line: "We have gotten word that many different beings have been running loose in the locations you specified, would you like me to handle it?"

President Deville: "No, no," *smirks as he talks* "just let them cause a little bit of havoc for a while. Eventually, the leaders of those areas will call upon my help to solve their problems."

The person on the other line: "What do we do about the angels and Blak Hart snooping around there?"

President Deville: "Nothing, they're not going to find anything. By the time they have any clues, the plan will already have been in motion."

At the base, Rayne receives a phone call.

Rayne: "Hello?"

Caller: "Hi Rayne, it's me, Lil. Why haven't you called me?"

Rayne: "Sorry, Lil. I have been busy with work and I am an aunt. I haven't forgotten about you, just making sure that when I did call, you would have my undivided attention."

Lil: "Oh," *giggles* "I thought you were mad at me."

Rayne: "Why would I be, like I said; I've been busy. You haven't done anything for me to be upset with you."

Lil: "When's the next time we're gonna hang out?"

Rayne: "I'll sneak out of here in a few, then we can chill."

Lil: "Won't you get in trouble?"

Rayne: "If the guys can do it, so can I. I'll call you later." *Cross can be heard in the background*

Cross: "Rayne, who were you talking to?"

Rayne: "A friend, damn, you are so nosey." *defensive*

Cross: "The angels called an emergency meeting that we must attend."

They walk together to find Darren, before walking into his room, they see him talking to himself.

Darren: "Leave me alone!"

Voices: "You must kill them all, they aren't as powerful as you. Remember, you were able to shock Blak Hart on the battlefield before being this powerful; imagine now." *He laughs*

Darren: "He's a friend and that was an accident."

Voices: "That was an accident that made you feel strong; at that moment you knew that if he ever annoyed you, he could be dealt with."

Darren: "Yeah, it did."

Rayne and Cross walk in.

Rayne and Cross: "Who were you talking to? You're alone."

Darren: "No one, what do you think I'm going crazy?" *defensive*

Cross: "Yeah, just a bit."

Rayne: "You were yelling, was a demon trying to attack you?"

Darren: "No, why are you two here?"

Cross: "Michael has called an emergency meeting, and everyone must attend."

They join everyone else in the conference room.

Michael: "We have reports of four portals that have opened in the following locations: the Namib Desert, Negev Desert, Thar Desert, and Rub'al Khali. We need to destroy anything coming out of them and close it, who knows how many demons have come out by now. Raphael has the assignments."

Raphael: "Thank you for the intro," *Louie chuckles* "each place gets checked; once it has been cleared of demons, Gail will close the portals. This will be done at all the locations. Michael will transport Gail to a safe area next to the location to close the portal. So, get ready Gail, you'll be going to all the locations when it's safe. Cross and Rayne will go to the Negev Desert, Darren and Gabriel will go to the Namib Desert, Louie and I will go to the Thar Desert, and then we'll all go together to the last location. We'll have Reeves there waiting for us, we leave here at 0400, so set up your gear now, because later on we won't have the privilege."

Nora: "I have the best job of them all, I'll be here with Lucas and Luna." *in a bragging tone while smiling, Cross looks at her and smiles*

Rayne watches everyone go to the cafeteria and then makes her way to the water cycle. Fifteen minutes later, she meets up with Lil.

Lil: "Hi Rayne."

Rayne: "Hey Lil, so what do you have planned?"

Lil: "I learned about meditating, would you try it with me?"

Rayne: "Sure, I meditate every now and then. I might be able to show you how to do it properly."

They close their eyes and begin. While Rayne is deep in meditation, Lil opens her eyes and moves closer to her ear. She starts whispering and chanting, Rayne's body glows for a few seconds then Lil moves back in her spot. She waits for Rayne to get out of meditating then opens her eyes.

Rayne: "You see, if it's done right, you'll feel extremely relaxed."

Lil: *smiling* "Yeah, I see."

Rayne: "I gotta go, I have a busy day tomorrow early in the morning."

Lil: "What are you going to do?"

Rayne: "Very important stuff that I can't speak of. Maybe I'll tell you after it's all done."

Lil: "I won't tell anyone, you can trust me."

Rayne: "I know but I don't want you getting hurt, for your safety I'll tell you after."

Rayne heads back to base hoping no one realized that she was gone. At four A.M., the gang gets ready to set their sights on what's about to happen and what they will encounter.

Suddenly, across from the four, two figures emerge from the smoke on the west side.

Blak Hart: "What the fuck?"

Cross: "What is it?"

Blak Hart: "It's my parents! It can't be." *eyes get watery and confusion sets in* "They're dead, it's impossible."

Rayne: "Are you sure? Maybe it's demons trying to imitate them, so you can lower your guard."

Darren: "Either way, get ready. I know they may look like your parents, just keep them in mind."

Cross: "He's right, let's go and find out."

Blak Hart: "No, I must find out alone. We have other things going on around us. I'll check if it's not them... I will take care of it."

Michael contacts Gail.

Michael: "While all this is going on, I need you to begin chanting to close the portal. The ground is shaking, and I sense it's below us, try to close it before any bloodshed causes it to grow."

Gail: "Consider it done, please make sure you all make it back to base safe."

On the north side is Mezun.

Cross: "Oh shit, it's Mezun. Are you ready, Darren?"

Darren: "Cross, I'll get my grandfather. Help Rayne."

Mezun: "Boy, you're going to need all the help you can get to try and hurt me. But at least I see you have your father's rage and reckless reaction. You could be by my side to rule."

Darren: "I'm nothing like him or you."

Mezun: "That's funny, you're more like us than you know, kid. Last chance boy deny me and it'll be your downfall."

Darren: "Well, let's get this over with. I have better things to do with my time than listen to your shit."

Mezun: "So be it, such a waste of untapped potential. See you're just like your father, a disappointment and a weakling."

The two dart toward each other with such speed that all that can be seen are blurs. Blood and dust fly everywhere; to anyone watching, the appearance of a mini tornado squirting blood and black ooze all over.

On the east of them is a smoke demon. Rayne heads toward it. As she runs toward it, she realizes that it appears to look like a toddler. Cross runs over to assist her when the same dark smoke appears in front of him, blocking his path.

The entity: "You think you can defeat me? Wait till you see what your friend is about to turn into." *begins to show his true appearance*

Cross: "You are not Darren." *the entity appears to look like Darren with a few differences, his face has black veins all over, and the eye color is completely black, which looks like moving sludge in the eye socket for the color*

Darren doppelganger: "You're right, I am just a small part of him, this is what he will become."

Cross: "Let's see what you think you can do to me."

While Cross battles the doppelganger, Rayne is having multiple battles. One in her mind, and several attacks from various demons.

Voice in Rayne's head: "Let's go and leave them to fight."

Rayne: "I can't leave them, they're my family."

The smoke demon in the form of a toddler speaks.

Smoke demon: *voice is that of a child, then mid-conversation turns deep and demonic* "So was I, you left me to die so why not them? I guess I wasn't that important to you mom." *Rayne begins to cry*

Rayne: "You're not real, you're not real."

Voice in Rayne's head: "You see, I told you we should have left." *arrogant tone*

Smoke demon: "I would've been with you but you let me die. Now I will return the favor." *as it charges toward Rayne, it begins to grow into a seven-foot demon*

Rayne: "Lil, this isn't the time for this shit. Just shut up and let me figure out what I'm gonna do."

She charges at the demon, remembering a few chants Nora taught her, and she begins to fight it with every emotion she has. But it continues taunting her.

Voice in Rayne's head: "Ignore it, just use your whip and pray."

At Blak Hart's battle, he runs over to them; they begin to attack him. Louie charges over and counterattacks. With tears in his eyes, not using his full ability, he fights these beings that resemble his parents, even to their minor scars and mannerisms. The battle continues, but Blak Hart secretly hopes that this is his parents.

Unexpectedly, another puff of smoke manifests, and out of it comes Catherine. She just stands there and watches as Blak Hart battles on. He turns around.

Blak Hart: *thinking* "Where did she come from, why is she here?"

Being off guard, the beings cut his body with their swords. In anger, he draws his sword and begins to plead.

Blak Hart: "Please don't make me kill you, what is wrong with you two? I'm Louie, your son." *tears roll off his cheeks*

Behind Louie, the archangels are assisting the other three in their fight. The angels will prevent anything else from distracting them from their mission. Michael fights off enemies from the east, while Raphael takes care of the ones attacking from the west; Reeves battles all that come from the south, and Gabriel guards from the attacks coming from the north.

As the angels are destroying the hoards, Blak Hart waits for the beings' reply. They both looked at him and laughed.

Blak Hart: "Since I see you can't possibly be my parents, prepare to die." *begins to pray* "Heavenly Father, please forgive me for what I'm about to do. I believe these two are my parents, but they haven't given me any signs that they are. In your name, I'll send these two from where they came from."

They sprint in his direction, attacking him with swords and a bo staff. He tries to remind them of certain things that happened in his life and how it was because of what they taught him that he was able to survive.

Suddenly, as Blak Hart looks at them, their eyes turn back from glowing red to brownish green. The one that looks like his father speaks.

Albert: "I don't have too much time, we just want you to know that we're proud of you and that you must kill us, son."

Blak Hart: "No, I won't do it." *images of his mother singing lullabies, his father's warm hand patting him on the back, then the fire*

Aggy: *whispers to him* "You have no choice, son." *continues to whisper*

Just as he's trying to convince them, Catherine calls out to him, temporarily distracting him. Aggy goes behind Albert, Albert holds onto Blak Hart's sword and pulls himself and Aggy, causing the blade to pierce them as Catherine laughs. He looks at his parents, and with one powerful thrust, they force him to ram his sword through them. Suddenly, he hears...

Aggy: "Louie..." *spitting out blood*

Albert: "Well done. You are... the only..." *whispering*

Aggy: "One that... has to do it... We are proud of who... you've become... always remember... we love you." *whispering*

Louie: *crying and bleeding out* "No, what have I done?" *while holding his parents, he falls to his knees*

Albert: *whispers in Louie's ear* "Always remember, Hearts can only..." *gets interrupted by Catherine unknown to Catherine, Albert told Louie what he had to say before dying, as his parents were passing, they slowly started dissolving and their soul-like smoke went into the blade without Blak Hart realizing*

Catherine: "This is no fun." *she chants and changes her form into a black mist resembling a male demon, the one who resembles Blak Hart, only difference is he is skinnier*

On the other side, Darren kept exchanging powerful hits and slashes until Mezun threw several ooze balls at Darren. He dodged most of them and quickly glanced over to Rayne to make sure how she was doing, and one of the balls hit Darren's face, temporarily blinding him. Panicking at the situation, he swung his blessed blades wildly, luckily killing his grandfather by cutting his head clean off.

Not too far off, Rayne defeats her foes combining her skills with her whips and chants to weaken her enemies right before her bladed whips slice and cut off limbs and chunks of flesh with ease until the bodies of her enemies turn into a puff of smoke. Suddenly, the ground rumbles and a loud explosion is heard, followed by the ground under the Heart family crumbling. Louie lets his parents go as he falls into the fiery pit that has just opened. As he's falling, he looks up, then sees the entity looking at him, and disappears. He continues to free-fall but doesn't try to save himself. Feeling defeated and not understanding why his parents wanted him to kill them, guilt consumed him, and he decided he wasn't worthy of being the chosen one that the angels thought he was. Just continuing to fall further and further into the abyss.

The others rush over to help Blak Hart out, but it's too late. He has fallen deep into the pit, too far to be seen. Suddenly, the ground closes as quickly as it opened as if this was done with strong magic. Cross stands over the area where the crater was along with Rayne and Darren. He tries to dig in an attempt to make the crater open again. Sadly, he couldn't. Once the angels finished off the remaining demons, they sprinted over. Raphael falls to his knees and screams in anger.

Raphael: "NOT AGAIN! I have failed them all!" *tears rushing down his face* "Maybe he's still alive. Michael, transport Gail over here so she can open the crater."

Michael: "I very much doubt it, no one can survive a fall like that, especially with all those rocks and boulders landing on top of them. As for the crater, it's not a portal."

Raphael: "She could still try!" *yelling*

Darren: "Fuck this. The so-called leader was defeated by his enemies." *chuckling* "Figured that would happen, everyone acting like he was so incredible, the chosen hero to save the world." *laughs* "What a joke. We surpassed him a long time ago, don't mourn for him, that piece of shit got what he deserves." *flipping off the direction Blak Hart was at* "We don't need him and I don't need any of you!"

Darren turns around and walks away from the group. *Looks at his mother*

Darren: "Gail, don't look for me. Consider me dead, no one is even worth shining my boots."

Cross: "Where are you going? This isn't the time to pull this bullshit." *he continues to look in the direction where the crater, hoping to see Louie trying to climb out or a sign that he can still be saved.*

Gail: "Darren, come back. What are you doing?" *looks at Raphael* "I tried to reopen the crater, I'm sorry, there's nothing I can do." *crying for Louie and for what her son just said*

Darren ignores them and continues to speed out of sight. *silently smiling*

Rayne: "No, it can't be... He can't..." *crying*

Hours later, below the feet of the tired warriors began to tremble, but nothing else happened. The fighters left heartbroken and disappointed. Raphael along with Cross kept digging with their bare hands until their hands were shredded and bloody. Michael pulled his brother, while Rayne grabbed her brother.

Michael: "At this rate, you guys will die from exhaustion before getting anywhere near his body."

Raphael and Cross both look at Michael with eyes filled with tears and still have a slight gleam of hope that somewhere, Louie is alive and fighting to be back.

In hell directly under the location, the demons rush to the rubble to scavenge over anything that might have fallen. As they dig, they smell flesh and blood, then excavate quicker. Suddenly, in the cracks of the debris, they see a pair of red eyes that begin slowly turning into fire. The fiery-eyed being grabs the demons; several others go in after it to eat whatever pulled the fiends. Shrieks of all kinds can be heard, blood gushes out deep from the boulders. Then the others, waiting for anything to come out, hear a loud burp. Shortly after, the last demons were killed, glowing red eyes and a creepy smile can be seen, along with an evil laughter that was heard

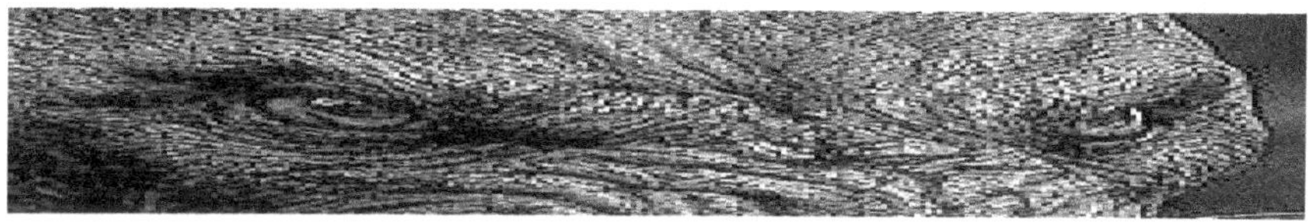

from the rubble. The other lower-class demons ran in horror, not wanting to know what kind of monster could kill several dozen higher-level demons in mere seconds.

A week later, the Archangel Michael called over everyone that was touched in some way by Louie. A funeral service was to be made. People from Blak Hart's Black Ops squad were to attend, the Cross family, Gail, Archangel Gabriel, Archangel Vincent, Archangel Reeves, and Angel. In two days, this was everyone's way to say their last goodbyes.

Nora: "Michael, what about Raphael? Louie was like a son to him."

Michael: "He's refusing to go, he's in denial saying that he was right about Albert and Aggy. Now he says he has the same feeling about Louie. We all saw it right in front of our eyes. I wish he was right, but after what I saw... It not only would take a miracle, but to survive an injury like that is definitely impossible."

The smell of rotting flesh fills the air; no one truly knows the torture my job is, I finally see why I was given this position. No matter how difficult it may be, I will see it through. Only my brother knows how this feels, few were chosen, but only I am one of the destined to do it. Walk an hour in my boots to see how my life will make you.

Michael Cross

I am supposed to be a beacon of light in a world filled with evil, hatred, deceit. It's being an example when I was surrounded by everything negative, then an earthbound angel showed me love. He has his own cross to bear, yet he still has enough strength to give me hope, bringing out the light inside me that was hidden for so long.

Luna Light

www.ingramcontent.com/pod-product-compliance
Lightning Source LLC
Chambersburg PA
CBHW031837200326
41597CB00012B/179

* 9 7 8 1 9 6 8 8 9 4 3 8 2 *